Editorial Policy

for the publication of proceedings of conferences
and other multi-authorvolumes

Lecture Notes aim to report new developments - quickly, informally and at a high level . The following describes criteria and procedures for multi-author volumes. For convenience we refer throughout to „proceedings" irrespective of whether the papers were presented at a meeting.
The editors of a volume are strongly advised to inform contrlbutors about these points at an early stage.

§ 1. One (or more) expert participant(s) should act as the scientific editor(s) of the volume. They select the papers which are suitable (cf. §§ 2 - 5) for inclusion in the proceedings, and have them individually refereed (as for a journal). It should not be assumed that the published proceedings must reflect conference events in their entirety. The series editors will normally not interfere with the editing of a particular proceedings volume - except in fairly obvious cases, or on technical matters, such as described in §§ 2 - 5. The names of the scientific editors appear on the cover and title-page of the volume.

§ 2. The proceedings should be reasonably homogeneous i.e. concerned with a limited and welldefined area. Papers that are essentially unrelated to this central topic should be excluded. One or two longer survey articles on recent developments in the field are often very useful additions. A detailed introduction on the subject of the congress is desirable.

§ 3 . The final set of manuscripts should have at least 100 pages and preferably not exceed a total of 400 pages. Keeping the size below this bound should be achieved by stricter selection of articles and NOT by imposing an upper limit on the length of the individual papers .

§ 4. The contributions should be of a high scientific standard and of current interest. Research articles should present new material and not duplicate other papers already published or due to be published. They should contain sufficient background and motivation and they should present proofs, or at least outlines of such, in sufficient detail to enable an expert to complete them. Thus summaries and mere announcements of papers appearing elsewhere cannot be included, although more detailed versions of, for instance, a highly technical contribution may well be published elsewhere later.
Contributions in numerical mathematics may be acceptable without formal theorems/proofs provided they present new algorithms solving problems (previously unsolved or less well solved) or develop innovative qualitative methods, not yet amenable to a more formal treatment.
Surveys, if included, should cover a sufficiently broad topic, and should normally not just review the author's own recent research. In the case of surveys, exceptionally, proofs of results may not be necessary.

§ 5. „Mathematical Reviews" and „Zentralblatt für Mathematik" recommend that papers in proceedings volumes carry an explicit statement that they are in final form and that no similar paper has been or is being submitted elsewhere, if these papers are to be considered for a review. Normally, papers that satisfy the criteria of the Lecture Notes in Biomathematics series also satisfy this requirement, but we strongly recommend that each such paper carries the statement explicitly.

§ 6. Proceedings should appear soon after the related meeting. The publisher should therefore receive the complete manuscript (preferably in duplicate) including the Introduction and Table of Contents within nine months of the date of the meeting at the latest.

§ 7. Proposals for proceedings volumes should be sent to the Editor of the series or to Springer-Verlag Heidelberg. They should give sufficient information on the conference, and on the proposed proceedings. In particular, they should include a list of the expected contributions with their prospective length. Abstracts or early versions (drafts) of the contributions are helpful.

Further remarks and relevant addresses at the back of this book.

Lecture Notes in Biomathematics

96

Managing Editor:
S. A. Levin

Editorial Board:
Ch. DeLisi, M. Feldman, J. B. Keller, M. Kimura
R. May, J. D. Murray, G. F. Oster, A. S. Perelson
L. A. Segel

S. A. Levin   T. M. Powell   J. H. Steele (Eds.)

Patch Dynamics

Springer-Verlag
Berlin Heidelberg New York
London Paris Tokyo
Hong Kong Barcelona
Budapest

Editors

Simon A. Levin
Princeton University
Department of Ecology & Evolutionary Biology
Eno Hall
Princeton NJ 08544-1003, USA

Thomas M. Powell
Division of Environmental Sciences
University of California
Davis, CA 95616, USA

John W. Steele
Woods Hole
Oceanographic Institution
Woods Hole, MA 02543, USA

Mathematics Subject Classification (1980): 92-XX

ISBN 978-3-540-56525-3   ISBN 978-3-642-50155-5 (eBook)
DOI 10.1007/978-3-642-50155-5

Typesetting: Camera-ready by authors/editors
46/3140 - 5 4 3 2 1 0 - Printed on acid-free paper

FOREWORD

A century from now, humanity will live in a managed—or mismanaged—global garden.

We are debating the need to preserve tropical forests. Farming of the sea is providing an increasing part of our fish supply. We are beginning to control atmospheric emissions. In a hundred years these separate aspects will need to be integrated into a single management system. We shall use novel farming practices and genetic engineering of bacteria to manipulate the methane production of rice fields world-wide. The continental shelf, especially off Asia, will be developed to provide food, energy, and, probably, living space. The capture of any remaining wild marine animals will be regulated like deer hunting.

To make such intensive management possible will require massive improvements in data collection and analysis, and especially in our concepts.

A century hence we will live on a wired earth. Like the weather stations that form a network over the land's surface, the oceans of the next century will have a three-dimensional lattice of sensing stations. The crust of the earth will also receive the same comprehensive monitoring now devoted to weather. Thus earth, air, and sea will be continuously sensed and their interactions modeled in order to anticipate major events such as El Niño, hurricanes, earthquakes, volcanoes, and climatic fluctations.

As the peoples of Asia, Latin America, and Africa approach the levels of wealth of Europe and North America, environmental fatalism and modest demands for food will be replaced by impatience with the accidents of nature and intolerance of mismanagement of the environment—particularly the living resources that are the focus of our material and altruistic concerns. The need for careful global management will become irresistible. Our control of physical perturbations and chemical inputs to the environment will be judged by the consequences to living organisms as individual species and as interacting systems. Above all, our human ability to affect life in all sectors, aquatic or terrestrial, brings these aspects together.

The problem is: How can we provide the factual and theoretical foundation needed to begin to move from our present, fragmented knowledge and our limited abilities to a managed, wired—and beautiful—global garden a century from now?

Joel E. Cohen
New York, New York

ACKNOWLEDGMENTS

This volume is based on lectures and work carried out at a summer school on Patch Dynamics held at Cornell University from June 23 to July 19, 1991. Support for the summer school was provided by the following organizations:

> National Science Foundation, Division of Ocean Science and
>   Environmental Biology
> Department of Energy, Environmental Sciences Division
> National Aeronautics and Space Administration, Earth Sciences and
>   Applications Division

through NSF grant OCE-9024396; and by

> Ecosystems Research Center at Cornell University, which is funded by Cooperative Agreement CR-812685-03 from the U.S. Environmental Protection Agency; the U.S. Army Research Office through the Mathematical Sciences Institute at Cornell University; the Agricultural Ecosystem Program at Cornell University, which is funded by Cooperative Agreement 91-34244-5917 from the U.S. Department of Agriculture; the Center for Applied Mathematics at Cornell University; the New York Sea Grant Institute; and the National Oceanic and Atmospheric Administration, National Marine Fisheries Services.

We are especially grateful to Colleen Martin, who bore the brunt of the arrangements at Cornell. The task of turning diverse formats and writing styles into a coherent book was carried out most effectively by Mary Schumacher. Lastly, the organizers wish to thank the participants, who, during the month at Cornell, provided us with a stimulating and enjoyable environment. It was much more fun—and more successful—than we had any right to expect.

Simon A. Levin
Thomas M. Powell
John H. Steele

CONTENTS

6. Description and Analysis of Spatial Patterns (García-Moliner et al.)

Part III.   Concepts and Models

7. Ecological Interactions in Patchy Environments: From Patch-Occupancy Models to Cellular Automata (Caswell and Etter)

8. Spatial Aggregation Arising from Convective Processes (Leibovich)

12. Stochastic Models of Growth and Competition (Durrett)

13. Mechanisms of Patch Formation (Deutschman et al.)

Part IV.   Ecological and Evolutionary Consequences

14. The Ocean Carbon Cycle and Climate Change: An Analysis of Interconnected Scales (Denman)

18. Ecological and Evolutionary Consequences of Patchiness: A Marine-Terrestrial Perspective (Marquet et al.)

PART I

COMPARING TERRESTRIAL AND MARINE ECOLOGICAL SYSTEMS

John H. Steele, Steven R. Carpenter, Joel E. Cohen, Paul K. Dayton, and
Robert E. Ricklefs

SUMMARY

We have entered a period where the study of the earth as a total system is within the reach
of our technical and scientific capabilities. Further, an understanding of the interactions of
earth, sea, and air is a practical social necessity. These interactions encompass physical,
chemical, and biological factors. The biological or ecological components are critical not only
as parts of these processes, but as a major and direct impact on man of the consequences of
global changes in the system. Yet the possible nature and direction of ecological change is
the most difficult aspect to predict and to relate to the other, physical and chemical, processes.

So far the terrestrial and marine sectors[1] have been considered separately. There can
be good reasons for this lack of integration. The practical logistics (ships versus jeeps) are
one reason for this separation. The organization of research institutes and of the federal
funding exacerbates the dichotomy. But the critical question is whether the science itself
requires this division. A workshop in Santa Fe in 1989 was held to address this question
specifically and to propose measures to bring the components together. The need for such a
meeting was evident from the discussions. The participants agreed that they all acquired new
and useful ideas from the exchange of information and concepts. More significantly, these
discussions revealed many topics that required and would benefit from more detailed and
extensive consideration.

The scientific interests and excitement of generalizing across sectors was the dominant
theme. For example, is the correct comparison between the longest-lived components—trees
and fish—rather than at the same trophic level? We were also aware of the societal
importance of understanding the very different consequences of human disturbance. Thus,
assessments of waste disposal options in each sector of the environment and at local, regional,
or global scales demand comparative study. Especially, we were conscious that any real
convergence in ideas and integrations of theories would be a long-term process involving the

---

[1]It is recognized that freshwater coastal estuarine environments are of intrinsic importance and particularly
significant in these comparisons. In the following text, "terrestrial and marine" is often used as a shorthand for the
complete range of systems.

removal of institutional and funding barriers. Therè was no doubt, however, that the perceived need to view our world as a single system requires ecological theory and practice to achieve a strong common basis.

At this preliminary meeting we sketched some major topics for comparative studies (food web structure, patchiness, biodiversity, etc.) and methods for promoting convergent evolution (workshops, summer schools, paired collaboration, production of texts, etc.). The summer school at Cornell in 1991 was the direct outcome of these discussions. It is intended to be the first in a series that will cover the topics listed in this introductory section, which draws on the report of the 1989 meeting and is intended as background to the subsequent material.

PRESENT STATUS

General concepts such as Global Geoscience presuppose some ability to integrate ideas and research in the aquatic, terrestrial, and atmospheric sciences. Thus, the physics of the atmosphere, the ocean, and even the interior of the earth come together under the auspices of geophysical fluid dynamics, even though the research programs and facilities are quite separate and distinct. Programs are under way to study the fluxes of carbon, nitrogen, and other elements through the atmosphere, ocean, and land interfaces. These fluxes involve interactions that encompass physical, chemical, and biological factors. In particular, various flux rates are determined by ecological conditions. But the ecological components of these global studies are critical not only as part of these processes, but also because they are seen as having direct impacts on our own economic or aesthetic values.

Changes in plant and animal distribution and abundance are seen as the consequence of our large-scale interventions, and these perceived changes provide the basis for societal concerns and actions. Yet the underlying processes that cause ecological changes are the most difficult to identify and to relate to physical and chemical changes on land, in the atmosphere, and in the ocean.

Considering the urgency of the global problems, there is distressingly poor communication among ecologists. Even scientists studying the same habitat from different perspectives—ecosystem or population biology—ask different questions in different languages. For example, a population biologist might study crabs or birds and have no interest in the nitrogen cycling that is fundamental to the local existence of the animals. A worse division separates "pure" and "applied" ecologists. The former carefully avoid situations influenced by man, although agricultural and fisheries biologists ask similar questions of their systems. As a result, they have different professional societies and journals.

But nowhere are differences greater than those existing between terrestrial ecologists and biological oceanographers. They belong to different professional societies. There is less than 10 percent overlap between the memberships of the Ecological Society of America and the American Society of Limnology and Oceanography. Certainly their systems are different, and so are their questions and methods of study. For example, most marine ecologists have no grasp of the ecological diversity of insect species, or of the ubiquitous coevolutionary relations of terrestrial systems. Few terrestrial ecologists have any appreciation of the intricate and dynamic relations between physical and biological factors in oceanic systems.

Recently there has been evidence of better communication among ecologists working in the same general habitats. However, the terrestrial and marine fields seem to be growing further apart. The organization of research and its funding exacerbates this dichotomy. Are there also conceptual reasons for this separation?

Although the atmosphere and ocean are governed by the same dynamics, the processes operate at fundamentally different space and time scales. Probably the most important consequence is that marine adaptations have evolved in situations where the populations are closely dependent on physical features. Pelagic marine populations are faced with ever-changing physical habitats and are motile and usually capable of rapid reproductive responses. This contrasts with terrestrial adaptations, which often respond to much longer time scales and deal with atmospheric variability as short-term noise.

How can such differences be bridged? It is critical that the scientific community become aware of the different perspectives and the various strengths and weaknesses of the several disciplines. The following examples illustrate how the strengths in one discipline could be imported into another.

1) Terrestrial ecologists have long been very effective in developing evolutionary paradigms. Marine systems have equally fascinating but very different evolutionary patterns that could be exploited profitably with theories and methods developed for terrestrial systems.

2) Marine ecologists have developed sophisticated methods to study and analyze physical and biological coupling across space, time, and size scales. These approaches might contribute to a better understanding of atmospheric/biotic relations of dispersal and behavior at boundaries.

3) Terrestrial and freshwater ecologists have a considerable body of knowledge and theory about foraging behavior and biology. This has led to increased understanding of the role of specialists and generalists in food web dynamics. Marine research could apply some of these theories to the foraging of higher-order predators.

4) Marine studies of patch dynamics as a mix of physical processes and biological behavior are well developed. Many of these concepts would be appropriate to problems in the terrestrial realm on large time or space scales—especially climate-related phenomena.

5) Terrestrial workers have a long history of controlled (or intrusive) field experiments. Manipulations of intertidal situations have been undertaken for 50 years but only recently have been applied to benthic populations. Experimental control of pelagic systems is very difficult but can be used to test carefully posed hypotheses.

6) Freshwater processes are intensively studied. These aquatic ecosystems are capable of controlled (and uncontrolled) manipulation. Although questions of mobility and of scale appear to separate them from marine and terrestrial systems, freshwater studies should provide opportunities for conceptual and technical links.

Finally, we need to be reminded that there are many common issues and questions. Cross-system comparisons include boundary layer communities in different fluids; maintenance of pattern at different temporal and spatial scales; and the role of disturbance, ecotones, succession etc.

MAJOR THEMES

Why Are Marine and Terrestrial Ecology Different?

Marine and terrestrial researchers function in different institutional and granting situations. But their divergent approaches appear to arise from perceived differences in the physical environments and in the manner these affect organisms and biological interactions. Biological oceanographers, in particular, view the physical characteristics of the marine environment as primarily responsible for pattern in biological communities, relegating the intrinsic pattern-generating capacities of biological systems to a minor role. Terrestrial ecologists, while recognizing the dependence on the physical background, emphasize that dynamical properties of populations and communities generate pattern within ecological systems independently of the physical environment.

Biological interactions in the ocean, such as predation, are viewed as important; but the major determinants of spatial and temporal variation in biological populations and processes are usually considered to be imposed by corresponding patterns in the physical system, especially variations in temperature, salinity, light, and nutrients. In particular, the spatial and temporal scales depend on the pertinent scales of variation in the physics. Most of the energy in the marine environment is stored in physical forms—temperature gradients and water movement. Thus, fluid and thermal properties of water dominate these biological systems.

Terrestrial ecologists stress the storage of energy in biomass and organic detritus and so decouple biological and physical components to some degree. The influence of the atmosphere on temporal patterns is moderated by the storage of biomass. Furthermore, spatial variation is under primary control of topography and soil, whose temporal variation (without human influence) is of very long scale compared to both atmospheric and marine processes of similar spatial scale. It is usually assumed that the dynamics are mainly demographic interactions between populations. For example, time lags in the response of populations to environmental changes can initiate population cycles, but their periods and amplitudes depend on biological characteristics. Finally, terrestrial systems are considered to be strongly organized by evolutionary interactions. Host specialization, mutualism, mimicry complexes, and other evolved arrangements among species are thought to be far more prevalent in terrestrial than in marine systems, where consumers are seen as more generalized (algal-coral symbioses notwithstanding). Evolutionary ecology is predominantly a terrestrial discipline.

While biological components of marine and terrestrial systems are subject to the same general processes, the expressions of these processes, especially as a function of space and time scales, differ greatly due to the physical nature of each environment. This fact has reinforced the separation of ecosystem studies but also offers the potential for evaluating and testing general theories of ecosystem processes that could predict these major differences between ecological sectors.

Dimensions for Comparisons

No single "axis" can bring together the contrasts among marine, terrestrial, and freshwater ecosystems. For example, the contrasts between systems dominated by sessile and mobile organisms are at least as marked as those between terrestrial and aquatic regimes. The two-dimensional structure of sessile systems is determined mainly by topography, while mobile systems are subject to the three spatial dimensions of hydrodynamics.

"Pelagic" organisms in air or water are influenced by the temporal scales in each medium. At all spatial scales, temporal change is slower in the ocean than in the atmosphere. In particular, the major eddy systems responsible for much of the variability in each environment have very different scales. Atmospheric eddies (high- and low-pressure systems) are about 1,000 km in diameter and move a distance equal to their diameter in two or three days. Ocean eddies are much smaller (ca 100 km) and can move this distance in about 30 days. Consequently, the weather fluctuations of the two environments differ by an order of magnitude in both temporal and spatial scale.

Parallel distinctions exist for major biotic processes. In mobile systems, patterns are set by passive advection and active migration, and the use of these alternative mechanisms depends on the relation between biological and physical scales in each environment.

In the sessile components of systems or of life cycles, spatial pattern depends heavily on biogeographic ranges and on *in situ* competitive, predatory, and mutualistic interactions. Succession sets the tempo of community variation. Thus, the mobile-sessile axis in the context of environmental scales can integrate seemingly disparate features of different environments. This axis must include "boundary-layer" communities whose patterns are determined by both topography and hydrodynamics.

Studies in freshwater ecology provide remarkably clear examples of the perspectives that derive from pelagic and benthic ecology. Recently two parallel workshops (supported by NSF) were convened to assess progress in lake and stream ecology. In the lake workshop report, predator-prey interactions and temporal variability were the major issues, with only one chapter dealing mainly with spatial patterns. At the stream workshop, disturbance, spatial heterogeneity, and biogeography were dominant topics and only two chapters dealt with interspecific interactions.

Another "axis" received significant attention at the Santa Fe workshop—the scales of body size, turnover time, and trophic status. In aquatic systems, the size of organisms and population turnover time increase up the food chain while unit growth rate (Rmax) decreases. In terrestrial systems, body size and turnover time often decline up the food chain while Rmax increases. Compare phytoplankton and trees. These opposite trends have important implications for stability and temporal variability. They are especially relevant to the degree and manner of coupling or decoupling between physical and biological processes. Thus, in aquatic systems, nutrient enrichment will have an immediate effect but the temporal pattern of subsequent community response can depend on predator turnover time. In contrast, the quasi-cycles of spruce budworm outbreaks appear to be set by the rate of recovery of the forest canopy between outbreaks. Thus, cycling rates are often governed by large biota having slow turnover times but with very different trophic status (forests or fishes) in different systems. The general implications of opposite trends in turnover time with trophic position are worthy of future study in non-linear food chain models.

These "dimensions"—(1) space/time scales of physical processes, (2) mobile/sessile life styles, and (3) size/growth rate/trophic position—provide systematic methods to define the differences between the marine terrestrial and freshwater sectors. The participants consider that they form a basis not only for qualitative comparisons of observations but also for more detailed future conceptual integration.

Common Issues

There are many topics in aquatic and terrestrial research where some common definition of the concepts, or comparisons of data sets, would be useful. Thus one way to illustrate the need for more interaction is to list briefly common issues faced in the study of ocean, terrestrial, and freshwater ecosystems. This list is not intended to be comprehensive.

> Cross-system parameters: What variables should be used to make possible comparisons among different ecosystems?

> Biodiversity: How many different species or phyla are there on land and in the sea? Is biodiversity best described by Linnean taxonomy, or would other functional concepts, such as body size, be equally or more useful?

> Disturbance: What roles do anthropogenic and natural disturbance play in changing the diversity in different ecosystems?

> Dispersal: What is the nature and importance of the movement of organisms across ecosystem boundaries?

> Coevolution: What is the importance of coevolution in different environments?

> Food webs: At what level of detail are the trophic structure of marine and terrestrial food webs similar—or different?

> Patchiness: What are the mechanisms underlying spatial patterns and what are their predictable or stochastic consequences?

> Energetic and material balances: How are the dynamics of energy and material flow related to ecosystem structure?

> System aggregation: What are the trade-offs in describing ecosystems at various levels of aggregation?

> Remote sensing: How can we assimilate the dense data sets from satellites? How do we combine them with *in situ* observations?

> Long-term data: What human and natural records are available from aquatic and terrestrial systems, and how do we compare them?

> Boundary layers: What are the special fluid dynamic conditions that characterize communities living at the interfaces and utilizing the solid and fluid media?

Scale dynamics: A very general question. How should dynamics on widely different scales be linked in theory or in numerical models?

PRESENT PROGRAMS

The previous sections have illustrated the wide range of common issues and also the difference in scales at which aquatic and terrestrial systems respond. If we are to study the interactions across time scales, then long-term data sets are necessary. At geological time scales, pollen analyses on land show the trends in forest and grassland distribution since the last ice age. In the sea, oxygen isotope analyses of calcareous shells in deep ocean cores demonstrate the temperature changes since the last ice age (and earlier). It is assumed that at the very long periods, we are observing the response of a globally coupled system.

At historical time scales, the long-term data sets are nearly all associated with and affected by human activity—forestry or viticulture on land, fisheries in the sea. Can we compare tree-ring data and fisheries statistics? Are the longer-lived components the main determinants of ecosystem structure? We require long-term studies at the community or ecosystem level. Some of these exist. There are the Hubbard Brook Forest Program (30 years), for example, and the Californian Current Surveys (25 years), which provide both space and time coverage. Can these be compared in terms of ecological processes, scales of variability, response to environmental change?

For terrestrial and freshwater systems, the Long-Term Ecological Research (LTER) network, supported by NSF, is an emerging source of data and ideas for cross-system comparisons. Studies across LTER sites are under way, focusing on the identification of parameters and processes that can be used for quantitative comparisons. At the Santa Fe workshop there was substantial interest in expanding such cross-system studies to include sites that are not part of the present LTER network. It was considered that a major advance would be the inclusion of marine systems in this expansion. There have been comparative reviews, particularly of fishery systems, but a more systematic progress is required. One program on global marine ecosystems (GLOBEC) is being developed with the aim of defining the physical/ecological relations that affect population dynamics for a wide range of scales and a diversity of species. Thus assessment of previous marine data sets and of pending programs in the context of the terrestrial studies would close the information gap among marine, terrestrial, and freshwater systems.

OPTIONS FOR ACTION

What topics require active collaboration by researchers in terrestrial and aquatic ecology? Based on discussion in the 1989 workshop, the following set was selected. It is not exhaustive but represents the range of subjects where significant benefits to science would result from effective interaction of active researchers.

Long-Term Data Sets

A primary requirement is for the different research communities to appreciate the nature of the data available in other sectors, the way in which observations are made, methods of analysis, the underlying hypotheses or conceptual models, and the future plans. This must be the basis for cross-system comparisons of global ideas or specific theories. The LTER network provides timely examples and growing experience with the types of comparison that are needed. It is essential to broaden these efforts by combining them with relevant and appropriate marine studies. Sustained comparisons of marine, terrestrial, and freshwater ecosystems are a major recommendation of the workshop.

Body Size, Trophic Structure, and Community Dynamics

Numerous observers of aquatic food chains have pointed out the steady increase in body size from phytoplankton through herbivorous zooplankton to carnivores. Other observers, at least since Elton in 1927, have remarked that many terrestrial food chains, or portions of these, proceed from very long-lived primary producers such as trees or shrubs to short-lived organisms such as insects and their parasites. Coupled with these patterns of increasing or decreasing body size are many other physiological or ecological variables such as rate of growth and length of life. These divergent patterns are often cited as the basis for the very different dynamics of each system.

At the same time, patterns in the topological structure of food webs have been discovered in recent decades that seem to transcend these distinctions between aquatic and terrestrial ecosystems. For example, the fractions of top, intermediate, and basal species appear to be independent of total species numbers. These fractions do not seem to differ significantly between the two kinds of systems.

How is topological structure invariant for systems with very different dynamics and scale relations? Do food webs with increasing body size respond to perturbations differently from those with decreasing body size? These questions are of considerable theoretical interest. They are also of practical importance, in view of our concerns about anthropogenic perturbations at global and local scales.

Methods of Analysis of Community Structure

General comparisons are very dependent on the methods for collecting data on community structure and on techniques of analysis. The geographical extent of a community and the position of its boundaries are difficult to define because the species inhabiting a particular place extend or contract their ambit at a wide range of scales from the diurnal to seasonal, to successional, to evolutionary periods. The underlying processes are very different in each environment, including passive dispersal patterns determined by physical dynamics, active migration, and alteration of the environment as well as adaptation to it. The common usage of terms such as *population*, *community*, and *ecosystem* for descriptions in the different sectors can conceal significant differences implicit in underlying concepts.

Interdisciplinary studies could usefully focus on techniques for measuring scale relations and defining the dimensions of populations and the coupling and exchange between communities. These couplings have practical consequences in terms of the definition of fish stocks, the design of nature reserves, and the identification of "damage" from pollution and other disturbances to natural systems. They are also important to our understanding of the role of evolutionary dynamics and speciation in marine and terrestrial systems.

The products of these studies would deal with comparisons of analytical techniques, examples of analyzed systems, scales for definition of community structure, and the consequences for community development and evolutionary processes.

Experimental Manipulation of Ecosystems

Large-scale experiments have been remarkably successful in resolving controversy and achieving insights that would take far longer through observational or laboratory scale experimental studies. Whole lake manipulations are a good example. Evolving statistical and modeling techniques can provide a rigorous foundation for detecting change in large unreplicated experiments. Freshwater and terrestrial habitats provide virtually all the examples of such controlled large-scale experimentation. In the open sea, such direct experiments are not practical. The consequences of extreme over-fishing can be viewed as very large exclusion experiments and can provide valuable insights into community responses. But over-fishing obviously does not allow rigorous definition of cause-effect relations, particularly in the context of natural variability.

Partial manipulation in fjords has been carried out. Mesocosms (enclosed volumes up to 3000 m^3) have been used, but the value of this approach and the interpretation of results have been controversial.

It would be valuable to have comparisons of the opportunities for, and the limitations on, manipulations at various scales, methods of analysis, and interpretation of results from these different "experimental" approaches. The potential for future work would be considered. For example, whole estuary experiments may be both feasible and critically important for predicting impact on near-shore regions—both land and sea.

Disturbance

The general role of disturbance is of very great interest. The term is difficult to define exactly. Disturbances include coarse-grained, infrequent events such as hurricanes, landslides and fires, as well as finer-scale events such as tree falls, ant mounds, and badger diggings. Predation in a very broad sense can be an important disturbance by changing the size and age frequency of the prey or by altering the spatial mosaic. The effects of disturbance have become an important component in the study of terrestrial, freshwater, and benthic systems. While these effects on the patch dynamics of two-dimensional systems are dramatic and ubiquitous, there may not be a comparable effect on ocean planktonic systems. Extreme alterations by man in density of fish stocks have no detectable link to observed fluctuations at lower trophic levels. Are these differences a matter of definition of "disturbance," of the data sets, or of different ways in which each system responds to irregular forcing? This topic—the modes of response to disturbance—would be a valuable focus of comparative and collaborative workshops.

Origin and Maintenance of Diversity

It has been suggested that diversity at the species level is generally greater on land but at the phylum level is larger in the sea. Such divergent patterns, if confirmed, require examination of the process responsible for their origin and maintenance. Major issues include the degree to which local diversity is determined within the context of the local physical environment, as contrasted with rates of species production resulting from migration of populations between regions. Another important issue is the relationship between local and regional species diversities that are coupled by the turnover of species between habitats (beta diversity). If marine communities are delimited primarily by physical processes and terrestrial (and benthic) communities exhibit greater influence of species movement and habitat selection, one might expect to find different patterns of beta diversity and perhaps differences in the influences of various processes and local and regional diversity.

Such comparisons would likely reveal gaps in our understanding of diversity and elucidate general patterns and the processes responsible for them. The inclusion of paleontologists would contribute an important historical perspective.

Patch Dynamics

In all environments, it is recognized that spatial and temporal variability—patches and population outbursts—are not merely noise but essential features of the food web dynamics ensuring adequate feeding rates and reproduction. However, methods of observation and analysis differ significantly between environments. In the sea, continuous spatial records are obtained from ships, and spectral analysis is used to define the biological patterns and compare them with physical observations. Moored recording systems provide comparable temporal data. Satellite data now extend the scales and display the complex interactions of physical and biological dynamics. For obvious logistic reasons, such methods cannot be used on land, and in turn different methods of analysis and description are used. As with other aspects, the primary focus in the open sea is on the physical forcing, whereas on land the ecological interactions are considered most important. Freshwater and benthic communities provide significant examples with alternative and sometimes conflicting explanations.

Aggregations of organisms imply that, locally, the system is far from a general equilibrium state. The behavioral mechanisms by which aggregations are formed and the consequences for the dynamics of the populations are important topics. At present, terrestrial and marine studies of these phenomena are conducted independently. The theoretical descriptions are quite separate. This is a major topic where useful comparisons can be made.

Boundary-Layer Communities

Exploring ecological processes may be most meaningful if contrasts are made among communities that reside within similar physical settings. In a moving fluid (water or air), the "boundary layer" is that region adjacent to the boundary (e.g., seafloor or forest floor) where there is a gradient in velocity perpendicular to the boundary due to the drag of the surface on the flow. All boundary layers are similar in structure but differ in their thickness, the shape of the velocity profile (the shear), and the mixing characteristics, all of which are functions of the flow velocity, fluid viscosity, and, in some flows, the roughness of the boundary. Communities residing within a boundary layer may be defined at several spatial scales. In the ocean, for example, a relatively thick boundary layer forms over the seafloor, due to steady, large-scale ocean circulations; thinner boundary layers form over local features, such as a rock ledge in an otherwise sandy bottom; and even thinner boundary layers form over organisms (e.g., kelp blades and mussel beds) that come into direct contact with the flow. Similar scale changes occur for desert, grassland, or forest systems.

Organisms residing within boundary layers in air or in water have many common problems. For example, erect plants and animals must be able to withstand fluid drag without being damaged, attached organisms may have spores or larvae that disperse in the fluid and must somehow make it back down to the surface again, and organisms that feed on suspended material must live in fluid regions with a high suspended food flux. The specific adaptations of organisms on land or on the seabed will differ because of the much lower fluid viscosity of air versus water. Fluid velocities tend to be much larger and mixing processes much faster in air than in water. Contrasting the ecology of boundary-layer communities living in different fluids should provide meaningful insights into the coupling between physical and biological processes in the evolution of population and community characteristics.

Scaling Up and Scaling Down

The problem of scale interactions is now a central theme in ecology. The advent of satellite observations has enlarged the range of spatial scales over which ecologists can describe their systems. On land this has increased the scales at which patterns are observed. In the oceans the reverse is true. We now see complex patterns at 1 - 100 km scales, where previously we assumed relative uniformity. Thus, one of the dichotomies separating land and sea studies is removed. One problem in both regimes is to assimilate the small-scale heterogeneities into descriptions of larger systems. The patchiness in the observations and the non-linearities in the processes do not permit simple averaging. Are there emergent properties? Can the fine structure of ecological processes be parameterized into the larger biochemical relations required by regional, or even global, studies of flux dynamics? What are the corresponding time scale changes?

Once again, the general questions are similar even though the detailed methodologies differ. If we are to have a comparative discipline permitting us to appreciate the effects of change at different scales from short-term episodic events to decadal climate trends, then we need to understand the range of responses available in the biosphere and especially the ways in which these responses occur at quite different scales from those of the forcing processes.

MECHANISMS FOR ACTION

Fostering new perspectives that integrate marine and terrestrial points of view will require a breaking down of traditional intellectual and institutional barriers. To some degree this may be accomplished by enlightened scientists and innovative funding. But major shifts in any discipline are more likely when students are encouraged to pursue new directions. We require the establishment of specific mechanisms involving faculty and students from both marine and terrestrial backgrounds.

The specific topics and options discussed in previous sections deal with very diverse aspects of ecology where there are overlaps or, more frequently, gaps in our understanding of common features in different environments. The topics cover the need for systematic data comparisons and availability of different analytical methods, as well as theoretical or conceptual issues. Various mechanisms for achieving a more integrated view will be required.

First, the conduct of field research is best carried out by the groups or institutes specializing in each sector. Thus, we do not recommend new field programs. This does not mean that such research groups or individuals will not benefit from interaction with colleagues in the other sectors. Quite the opposite. We have noted that such interactions are notoriously absent, restricting the sources of ideas for analysis and for generalization.

Secondly, these deficits are longstanding, being based on the separate organization and funding of research in each sector. Integration will not be achieved by a single large conference or symposium. Such large meetings tend to exacerbate rather than remove the separation of interests. So the need is to bring together relatively small groups over a relatively long period of time, allowing sustained interaction.

Thirdly, progress in increasing the dialogue should involve those near the start of their careers as well as the more senior researchers. The latter may be the generalists, but they are also often set in their separate ways.

Lastly, the federal agencies should be brought in, not only because their funding is the basis for action, but also because their present structures are significant factors in maintaining the separate directions. The need for restructuring is recognized in the emerging patterns of inter-agency support for global change research. An involvement of program managers would be very helpful in ordering specific project developments to take account of cross-system integration.

PART II

METHODS AND DESCRIPTIONS: AN OVERVIEW

Thomas M. Powell

Two broad themes characterize the six chapters in this section. The first is the role played by formal techniques that are common, even necessary, items in the bag of tricks carried by any investigator of spatial patches. Early studies of patches (e.g., Cassie 1959) used a strongly statistical approach. The tradition of a heavy dose of quantitative formalism in "patch studies" persists to this day. And all six authors (or groups of authors) address, or apply, some formal analytical approaches in their contributions. The second is an implied comparison between marine and terrestrial environments. That is, how do patches differ in marine and terrestrial habitats (and what techniques do different researchers use in the two environments)? All the authors discuss detailed examples of spatial patches in several terrestrial and marine habitats. They note the critical ecological processes that are at work, or phenomena that are (or could be) affected by the patchy environment.

The section also has some biases that some may consider unfortunate. The contributions that address formal analytical techniques are brief; they were not designed to be substitutes for the many works on formalism that are referred to in the chapters of this volume. In particular, there is little discussion of time series analysis (and its straightforward generalization to one-dimensional spatial transects). The time series approach has dominated studies of patchily distributed organisms, especially in aquatic environments (e.g., Mackas et al. 1985), for nearly twenty years. Excellent modern introductions to these techniques are widely available (Chatfield 1984, Diggle 1990, Shumway 1988). And applications are widespread in the ecological literature (e.g., see Shugart 1975, Jassby and Powell 1990). We note that, despite our neglect, the time series techniques remain a staple for one's understanding of spatial pattern. Marquet et al.'s concluding chapter to this volume underscores that point. Further, though the six pieces describe widely diverse habitats and organisms (including, but not limited to, soils, marine phytoplankton, the larvae of benthic invertebrates, gopher distribution in a grassland, krill, and evergreen forests), that coverage is not encyclopedic. Each reader may find no mention of his or her favorite habitat, population, etc. But the ecological literature on patches is vast (and continues to grow rapidly). For example, there is little overlap between our volume and recent works on disturbance and patch dynamics by Pickett and White (1985), patchy environments by Shorrocks and Swingland (1990), and ecological heterogeneity by Kolasa and Pickett (1991). A comprehensive review of *all* studies on spatial pattern is not our aim (such an undertaking is probably impossible). Rather, the authors of the six chapters in this section selected studies that, in their judgment, have the best chance of being important to the emerging generalizations in both marine and terrestrial investigations.

In "Introduction to Spatial Statistics," Frank Davis presents a brief but useful collection of definitions and formulae for statistical quantities that apply to data distributed over two spatial dimensions. More important, however, Davis draws attention to the distinction between spatial heterogeneity—when variability depends upon absolute location—and spatial dependence—where an observation depends on nearby values, but not on absolute location. These are concepts that are often confused in the ecological literature. And the distinction between the two is important when one is attempting to infer the underlying cause of an observed spatial pattern.

Harold van Es, in "The Spatial Nature of Soil Variability and Its Implications for Field Studies," introduces geostatistics, presenting an appealing, intuitive discussion of a much-used quantity, the semi-variogram. The reader will appreciate the obvious connection to concepts introduced by Davis in the previous chapter (and, perhaps, to the reader's earlier exposure to "time series" quantities like auto-correlation). Van Es's chapter concludes with examples of these techniques applied in studies in soil science. The consideration of processes and techniques in soils is important to ecologists for two reasons. First, the heterogeneity of the soil setting mimics the heterogeneity that ecologists are faced with. And second, the cause of spatial pattern in terrestrial habitats, especially in plants, is often closely tied to the underlying soil distribution. And ecologists would like to know the cause and effect relationship, if any, between the two patterns.

Mark Abbott contributes a succinct summary of modern efforts to understand the growth rates of the autotrophs that exist at the "lowest" trophic level in oceanographic food webs ("Phytoplankton Patchiness: Ecological Implications and Observation Methods"). His aim is to emphasize those phenomena that are most dependent upon the non-uniform, non-random, and spatially variable character of the marine environment. Abbott introduces the intuitive notion of scale (more accurately, scale of variability) and uses it to show the need for simultaneous, coupled understanding of both physical and biological phenomena in the sea. Neither physics nor biology alone is sufficient to comprehend the mechanisms that lead to variability in marine populations and communities. He concludes with a brief review of measurement techniques used to observe varying populations of phytoplankton—from the largest to the smallest scales—a subject to which he has made numerous contributions in his own research.

Mimi Koehl and collaborators in "Measuring the Fate of Patches in the Water: Larval Dispersal" explore another coupled physical/biological issue at the interface between terrestrial and marine habitats: the transport of material to the sea from the shore (and vice-versa). Koehl et al. focus closely on how the magnitude of such transport can be quantitatively estimated in the difficult-to-measure region near the shore. They suggest how their findings might be readily applied to the important problem of larval transport to and from shore. Parenthetically, a substantial amount of our modern ecological wisdom has been derived from nearshore studies (e.g., Paine 1966). But we have proportionally much less quantitative information about the physical transport processes that are the source of food, recruits, etc. for these same shore communities about which we "know so much." Roughgarden et al. (1988) strongly champion this latter point.

Kirk Moloney, in "Determining Process Through Pattern: Reality or Fantasy?" presents a detailed analysis of one aspect of a terrestrial plant-herbivore system in Northern

California, USA—the distribution of gopher mounds, which are thought to be very important in determining the plant distribution in this community. The reader will be impressed that the blind use of different techniques can lead to different conclusions about what the "observed" patterns imply.

Finally, Graciela García-Moliner and co-authors in "Description and Analysis of Spatial Patterns" present a comprehensive comparison of techniques that are common in terrestrial vs. marine environments, with detailed examples from each. It is an appropriate conclusion to this first section of the complete volume. One should note the special emphasis given to the wavelet transform, a new technique, and its comparison to results that can be obtained from more traditional methods. Finally, García-Moliner et al. briefly mention fractals, fractal dimensions, etc. The true power of these ideas is still under study, but investigations using fractals promise to increase in the future.

REFERENCES

Cassie, R. M. 1959. Micro distribution of plankton. *N. Z. J. Science* 2:398-409.

Chatfield, C. 1984. *The Analysis of Time Series: An Introduction*. Chapman and Hall, New York.

Diggle, P. J. 1990. *Time Series: A Biostatistical Introduction*. Oxford University Press, Oxford.

Jassby, A. D., and T. M. Powell. 1990. Detecting changes in ecological time-series. *Ecology* 71:2044-2052.

Kolasa, J., and S. T. A. Pickett (eds.). 1991. *Ecological Heterogeneity*. Springer-Verlag, New York.

Mackas, D. L., K. L. Denman, and M. R. Abbott. 1985. Plankton patchiness: biology in the physical vernacular. *Bull. Mar. Sci.* 37:652-674.

Paine, R. T. 1966. Food web complexity and species diversity. *Amer. Nat.* 100:65-75.

Pickett, S. T. A., and P. S. White (eds.). 1985. *The Ecology of Natural Disturbance and Patch Dynamics*. Academic Press, Orlando, Florida.

Roughgarden, J., S. Gaines, and H. P. Possingham. 1988. Recruitment dynamics in complex life cycles. *Science* 241:1460-1466.

Shorrocks, B., and I. R. Swingland (eds.). 1990. *Living in a Patchy Environment*. Oxford University Press, Oxford.

Shumway, R. H. 1988. *Applied Statistical Time Series Analysis*. Prentice-Hall, Englewood Cliffs, New Jersey.

Shugart, H. H. 1975. *Time Series and Ecological Processes*. SIAM, Philadelphia.

1
INTRODUCTION TO SPATIAL STATISTICS

Frank W. Davis

INTRODUCTION

This chapter is intended to provide an overview of some basic theory and applications of spatial statistics. The student is assumed to be familiar with elementary statistical principles of probability, probability distributions, statistical moments, and significance tests. The emphasis here is on the ecological applications of spatial autocorrelation theory. These notes borrow heavily from textbooks by Cliff and Ord (1981) and Upton and Fingleton (1985) and from the monograph by Goodchild (1986). The reader should consult Anselin (1988) for a more advanced treatment of spatial autoregressive models. The review of geostatistics by Journel (1989) is also recommended. (See also the chapter in this volume on geostatistics.)

SPATIAL DATA

Spatial data consist of observations of a variable in a two-dimensional coordinate system, where each observation possesses a value and a location. There are many types of two-dimensional data that are distinguished by how space is represented or sampled as well as by how the variable is described and measured. For example, space may be represented as continuous or tessellated into a regular or irregular lattice (mosaic). The observations may consist of points, lines, or areas within that space, and, may range from categorical to ordinal to numerical in the values that they assume. Familiar examples include categorical point data in continuous space (e.g., mapped plant locations), categorical lines in continuous space (e.g., routes taken by different individuals of a species), numerical areal data on a regular lattice (e.g., digital imagery), and categorical data on an irregular mosaic (e.g., soil map). Point and line data will not be treated here, although many of the principles will apply.

SPATIAL HETEROGENEITY VERSUS SPATIAL DEPENDENCE

The objective of spatial statistics is not only to describe how things are distributed in space (e.g., random, clustered, uniform), but also to determine if spatial proximity is playing a causal role in the observed distribution. In this sense it is important to distinguish pattern caused by environmental or spatial heterogeneity from that caused by spatial dependence or autocorrelation. Spatial heterogeneity is expressed in data whose magnitude or variability depends on absolute location. For example, the composition of vegetation samples can vary from one rock type to another. Abundance of an animal species may vary from one elevation zone to another. In contrast, spatial dependence or *autocorrelation* is implied when the value of an observation depends on surrounding values but is independent of absolute location. Processes such as competition and diffusion produce spatial dependence of this kind.

Put another way, a variable can be spatially autocorrelated because of spatial interactions among neighbors or because that variable is determined by some other variable which is itself autocorrelated (Goodchild 1986). Cliff and Ord (1981) refer to these alternatives as *interactive* and *reactive* models of spatial dependence. The choice of statistical model for spatial data analysis hinges on whether spatial variation is a consequence of reaction to spatial heterogeneity or of true spatial dependence in the variable. Spatial data usually reflect a combination of the two, and it may be difficult to distinguish their effects.

Two Models of Spatial Heterogeneity

At one extreme, most or all variation in a spatial random variable Z occurs at the boundaries between uniform fields or patches. The resulting mosaic pattern can be formulated as a one-way analysis of variance (Burrough 1987):

$$Z = \bar{Z} + Z_i + \varepsilon, \qquad (1)$$

where $\bar{Z}$ is the mean value of Z over the sampling domain, Z_i is the systematic departure of Z from $\bar{Z}$ in region or location i, and ε is an error term. This model of environmental variation is the basis for cholopleth mapping and for much recent work in landscape eology.

At the other extreme, spatial heterogeneity is manifested as continuous, directional changes in Z, in which case a trend surface model may be appropriate. Let z_i be the value taken by continuous random variable Z at location i, where i is defined by spatial coordinates X_{1i} and X_{2i}.

$$E(z_i) = b_o + b_1 X_{1i} + b_2 X_{2i} + b_3 X_{1i}^2 + b_4 X_{1i} X_{2i} + b_5 X_{2i}^2 + \cdots \qquad (2)$$

The trend surface model is thus a linear regression model to predict Z as a function of spatial coordinates X_1 and X_2, where the regression coefficients are solved by ordinary least squares. The predicted surface will be linear, quadratic, cubic or higher depending on the number of terms on the right hand side of equation (2). There are a number of practical problems in fitting higher-order trend surfaces, as treated by Mather (1976). Edge effects can also be problematic (Ripley 1981).

In reality, spatial heterogeneity typically exhibits both discrete boundaries and continuous variation, depending on the sampling intensity (measurement scale) and the extent of the sampling domain. One solution is to model variation at multiple scales through a nested or hierachical design (e.g., trends within patches). The need for large samples limits the utility of this approach.

A General Model of Spatial Autocorrelation

Cliff and Ord (1981, p. 6) define spatial autocorrelation as "systematic spatial variation in [a mapped] variate." Hubert et al. (1981) describe a general cross-product statistic for quantifying such variation in geographical data:

$$r = \sum_i \sum_j W_{ij} C_{ij}, \tag{3}$$

where W_{ij} is a measure of spatial proximity between two locations and C_{ij} is a different measure of proximity based on observed values of Z at those locations. W_{ij} and C_{ij} are entries in the ith row and jth column of the square matrices **W** and **C**. Examples of proximity measures used in spatial analysis are provided in the next section.

SOME MEASURES OF SPATIAL AUTOCORRELATION

Join-Count Statistics

Several measures have been developed to quantify spatial dependence of a binary variable (say Black [B] or White [W]) mapped over a regular or irregular grid. A join-count refers to the number of cases in which adjacent grid cells occur in a certain condition. As described by Upton and Fingleton (1985), a binary map ($z_i = 1$ for black, $z_i = 0$ for white) can be described by the frequencies of BB, BW, and WW joins among adjacent cells out of a total of J joins. From equation (3), let $W_{ij} = 1$ if regions i and j are adjacent and $W_{ij} = 0$ otherwise. For gridded data, adjacency may be defined as sharing a common side ("rook's" definition, from chess), as sharing a corner ("bishop's" definition), or as either a side or corner ("queen's" definition). If in equation (3),

$$C_{ij} = (z_i - z_j)^2 \tag{4}$$

then $r/2$ is the number of times adjacent regions have different identities. If

$$C_{ij} = z_i z_j, \tag{5}$$

then $r/2$ is the number of BB joins. The number of WW joins is calculated as J — BB — BW.

It is also possible to extend join-count statistics to multicolored maps (Sokal and Oden 1978, Cliff and Ord 1981), although in general other measures, such as Moran's I or Geary's c, are recommended.

Moran's I and Geary's c

A widely used measure of spatial autocorrelation among mapped observations of a continuous or nominal variable is Moran's I statistic:

$$I = \frac{n}{S_0} \frac{\sum_i \sum_j W_{ij}(z_i - \bar{z})(z_j - \bar{z})}{\sum_i (z_i - \bar{z})^2}, \tag{6}$$

where z_i is the observed value at location i, $\bar{z}$ is the average of z_i over n locations, and S_0 is $\sum_i \sum_j W_{ij}(i \neq j)$. I is thus obtained from equation (3) by letting $C_{ij} = (z_i - \bar{z})(z_j - \bar{z})$ and dividing r by S_0 and the sample variance (Hubert et al. 1981). In the absence of autocorrelation, $I = -1/(n - 1)$. I approaches 1 with increasing positive autocorrelation.

The formula for Geary's c is:

$$c = \frac{(n-1)}{2S_0} \frac{\sum_i \sum_j W_{ij} (z_i - z_j)^2}{\sum_i (z_i - \bar{z})^2}. \tag{7}$$

In this case, $C_{ij} = (z_i - z_j)^2$ in equation (3). $c = 1$ in the absence of autocorrelation and approaches 0 with increasing positive autocorrelation. As demonstrated by Goodchild (1986), I and c provide very similar information. While c is more sensitive to absolute differences between paired observations, Cliff and Ord (1981) recommend I as a more reliable measure of autocorrelation.

Figure 1 shows varying levels of spatial autocorrelation (rook's condition) in a 50 x 50 grid. The grids were generated by initially assigning each cell a value sampled at random from a gaussian distribution (N(0,1)), then iteratively smoothing the grid until the pre-specified autocorrelation was attained.

The proximity matrix W_{ij} has been defined as binary for calculating join-count, I, and c statistics. However, proximity can be defined in any number of ways, including inverse euclidean distance, inverse city block distance, inverse resistance, and so on. In this sense, W_{ij} is a model of space, and its specification is critical to the appropriate measurement of spatial dependence in a process. Alternatively, one could explore the relationship between I (or some other measure) and W_{ij} to search for an appropriate spatial model, but there are practical limits to this approach for large data sets.

Hypothesis Testing

Several approaches have been taken to assess the probability that measured values of r in Equation 3 could occur due to chance in a spatial stochastic process. Cliff and Ord (1981) discuss sampling distributions for c and I, and show that under many circumstances and for moderately large samples these measures are asymptotically normally distributed. The first and second moments of r can be calculated as follows:

$$E(r) = S_0 T_0 / n(n-1) \tag{8}$$

$$var(r) = \frac{S_1 T_1}{2n(n-1)} + \frac{(S_2 - 2S_1)(T_2 - 2T_1)}{4n(n-1)(n-2)} + \frac{(S_0^2 + S_1 - S_2)(T_0^2 + T_1 - T_2)}{n(n-1)(n-2)(n-3)} - [E(r)]^2, \tag{9}$$

where: n is the number of locations;

$$S_0 = \sum_i \sum_j W_{ij} (i \neq j);$$

$$S_1 = 1/2 \sum_i \sum_j (W_{ij} + W_{ji})^2 \ (i \neq j);$$

$$S_2 = \sum_i (W_{i0} + W_{0i})^2;$$

$$W_{i0} = \sum_j W_{ij};$$

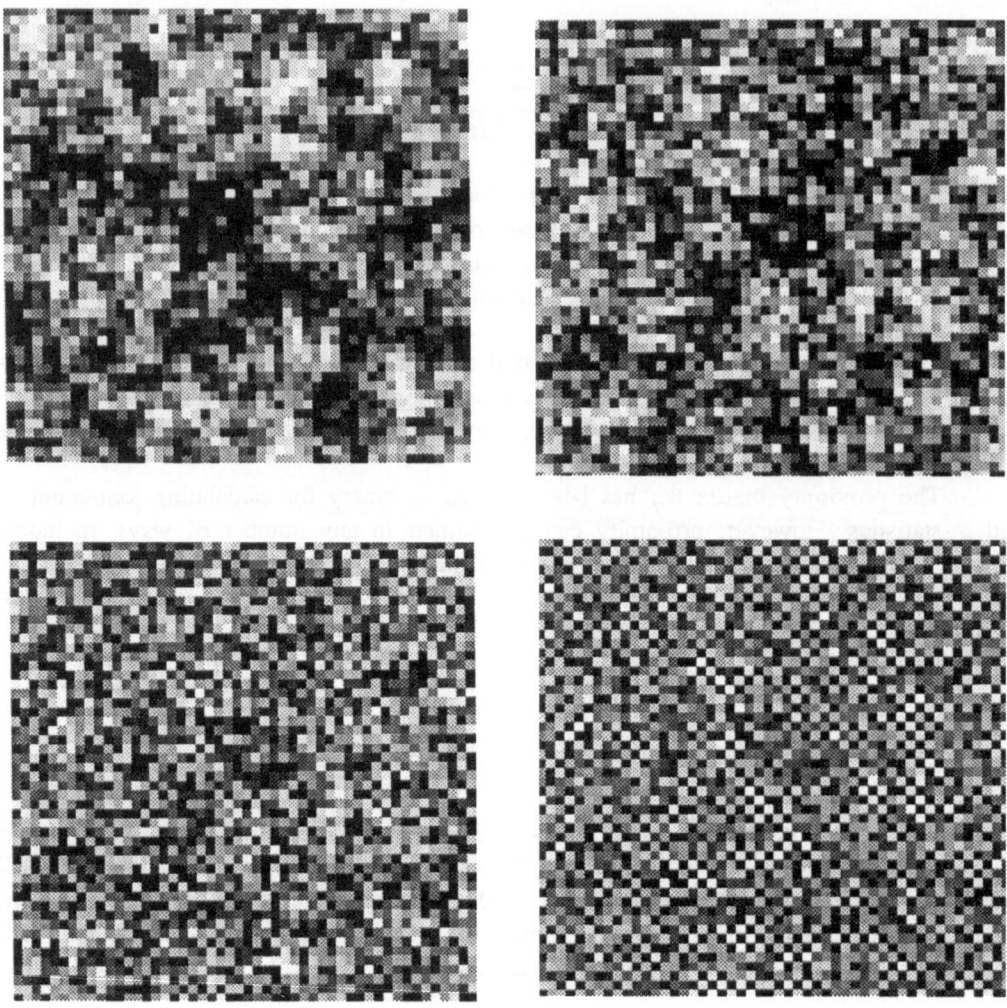

Figure 1. Four 50 x 50 grids of decreasing spatial autocorrelation (Moran's I, rook's condition) from I = 0.8 (upper left), I = 0.6 (upper right), I = 0.04 (lower left), to I = −0.8 (lower right).

$$W_{\alpha} = \sum_j W_{ji};$$

and T_0, T_1, and T_2 are calculated by substituting C_{ij} for W_{ij} in equations for S_0, S_1, and S_2, respectively.

Costanzo et al. (1983) propose the Pearson Type III approximation as preferable to the normal approximation due to the tendency for r to exhibit a skewed distribution under randomization experiments.

As an alternative to assuming an underlying probability distribution for r, one can permute the data to generate an empirical distribution. Goodchild (1986) discusses two alternative approaches termed "the randomization hypothesis" and "the resampling hypothesis." The randomization hypothesis proposes that the population from which the observed data were drawn consists of all possible ways in which the attributes z could be rearranged among the locations i. The resampling hypothesis proposes that the observed value of z assigned to location i is obtained from an infinitely large population by random, independent sampling at all locations. Thus the randomization hypothesis involves randomly re-assigning observed values of z across locations and calculating r, repeating this process to generate a probability distribution for r. The resampling hypothesis involves generating different sets of z from a pre-selected distribution (usually a normal distribution for interval or continuous data) to generate a distribution for r.

Random spatial distributions are the exception rather than the rule in ecology, and simply demonstrating that a variable exhibits significant spatial dependence may not be satisfying. A more practical concern is often whether or not the observed spatial auto-correlation is representative of the population as a whole, or, more specifically, what confidence intervals should be placed around estimates of r. Unfortunately, there is no obvious method for generating such confidence intervals for an observed pattern because that pattern is in effect a single sample of a spatial process. Repeated mapping through time offers one way of approaching this problem, assuming that the spatial process is not time-dependent.

Boundary Effects

Boundary effects occur when z_i is affected by z_j and location j falls outside the study area. This can lead to errors in estimating autocorrelation, especially for small data sets and those with high perimeter-to-area ratios. A number of methods have been proposed to deal with boundary effects, including buffer zones (impractical for small data sets), remapping observations to a torus, and downweighting observations as a function of their distance to a boundary. Griffith and Amrhein (1983) tested several alternatives and found that none adequately removed the bias introduced by boundary effects.

Stationarity

Measures of spatial dependence are interpretable and useful only to the extent that one can account for confounding effects of spatial heterogeneity. Put another way, one generally assumes that the process being investigated is spatially *stationary*; that is, that the expected value at a location, as well as the difference between that location and others at some distance and direction of separation, are independent of absolute location. More formally, a spatially continuous process $Z(x)$ defined at locations x is spatially stationary if:

$$E[Z(x)] = \mu, \tag{10}$$

$$var[Z(x)] = \sigma_Z^2, \tag{11}$$

and

$$cov[Z(x_i), Z(x_j)] = \sigma_Z^2 c(x_i - x_j), \tag{12}$$

where c is a correlation function depending of the relative locations of the two sites (Cliff and Ord 1981):

Non-stationarity in the mean value of Z may be removed by fitting regional means or by trend surface analysis, as described previously, prior to measuring autocorrelation. Alternatively, the increments, $Z(x + h) - Z(x)$, may be mean stationary with non-zero mean while the actual process is non-stationary (see below).

Non-stationarity in variance can be more difficult to correct. It is possible, though not easy, to develop a two-dimensional process that has a non-stationary variance while the increments of the process do not. When the mean is also non-stationary, variance may increase with the mean. A transformation such as a square root or log can help to make the variance more stationary.

For sufficiently large data sets such as those provided by remote sensing, segmentation of the study area into regions that are more homogeneous in their statistical moments and fitting separate statistical models to each region is probably the most straightforward, if cumbersome, way of dealing with non-stationarity.

Regionalized Variables

As mentioned previously, spatial variation in an ecological variable may be a function of environmental heterogeneity as well as spatial dependence in the variable. Often spatial heterogeneity produces broad patterns or large scale structure (large scale meaning over many observations) in the variable, spatial dependence is expressed locally over neighboring observations, and random (unexplained) variation occurs at or below the measurement scale (Burrough 1987). Treating these effects as additive:

$$Z = (\bar{Z} + Z_j) + \varepsilon(Z) + \varepsilon' \tag{13}$$

where:

$\bar{Z} + Z_j$ is deterministic or structural variation in Z as a function of the overall mean and the effect of location j (e.g., mosaic or trend surface variation);
$\varepsilon(Z)$ is spatially dependent or *autocorrelated*, but otherwise random, variation; and
ε' is residual unexplained error (usually assumed to be $N(0, \sigma^2)$).

A spatial variable described by equation (13) is known as a *regionalized variable* (Journel 1989).

CORRELOGRAMS AND VARIOGRAMS

The correlogram and variogram plot measured spatial autocorrelation as a function of distance and direction. Ignoring direction, for the general autocorrelation measure introduced in equation (3):

$$r(h) = \sum_i \sum_j W_{ij}(h)C_{ij}, \qquad (14)$$

where h is the distance of separation (*spatial lag*).

If W is structured so that $W_{ij} = 1$ if locations i and j are separated by distance h and 0 otherwise, and if $C_{ij} = [z_i - \mu][z_j - \mu]$, then $r(h)/(n-h)$ is a measure of spatial autocovariance, denoted $C(h)$. Note that $C(0)$ is the second moment or variance of Z, that is, $var(Z)$. The standard measure of autocorrelation, referred to here as $r'(h)$ to distinguish it from the general measure in equation (14), is:

$$r'(h) = \hat{C}(h)/\hat{C}(0), \qquad (15)$$

where $\hat{C}$ indicates sample estimators.

If W is structured as before and $C_{ij} = (z_j - z_i)^2$, then $r(h)/2(n-h)$ measures the *semivariance*, denoted $\gamma(h)$. The variogram is the plot of $\gamma(h)$ versus h.

If the mean, μ, is constant and $C(h)$ is constant over the region for a given h, then the process is considered to have *second-order stationarity*, in which case:

$$\gamma(h) = C(0) - C(h), \qquad (16)$$

and

$$r'(h) = 1 - \hat{\gamma}(h)/\hat{C}(h) \qquad (17)$$

Equation (17) states that under second-order stationarity, the variogram and correlogram provide virtually the same information. Semivariance is generally recommended in the presence of non-stationarity of the mean or variance, because it relies only on differences between locations (Burrough 1987). This advantage holds only where the differences are stationary and the process is not.

The estimated correlogram or variogram can be used to define a correlation length scale, which is a measure of the minimum distance between uncorrelated locations. This information can be useful for designing field surveys or obtaining rough estimates of the variance of spatial averages. The variogram is also used in kriging, a method of spatial interpolation based on the principle that interpolated data should exhibit the same spatial dependence as the sample data.

Periodicity in a process can also be detected as wavelike peaks in the correlogram with wavelength equal to the average size of the repeating pattern. However, spectral analysis is a more powerful method of modeling such variation.

SPATIAL EFFECTS IN INFERENTIAL STATISTICS

Ecologists typically apply statistical methods such as analysis of variance and regression to test for non-random association of variables measured over a number of locations. Significance tests assume that observations are independent, a condition that is frequently not met due to the presence of spatial autocorrelation in ecological variables. To the extent that any observation can be predicted from surrounding observations, there is redundant information in the sample and the effective sample size (n) may be considerably less than the number of observations.

Accounting for spatial effects in inferential statistics is not straightforward. Much has been published on this subject in the literature of econometrics and geography; however, there is little application of this work in ecology. At the very least, ecologists should record the location of samples and measure spatial autocorrelation before applying standard inferential tests. If significant autocorrelation is detected, the data can be "pre-whitened" to remove spatial dependence. Alternatively, a different set of test statistics can be applied that take account of autocorrelation.

As an example of spatial effects on statistical inference, consider the familiar regression model:

$$Y = Xb + e \qquad (18)$$

where:

Y is an N x 1 vector for the dependent variable;
X is a set of N x K matrices for explanatory variables;
b is regression parameters of the explanatory variables; and
e is an N x 1 vector of unobserved errors.

The Ordinary Least Squares (OLS) solution for the regression parameters assumes that errors are independent and normally distributed with mean 0 and variance σ^2. If errors are positively autocorrelated, estimates of t, F, and r^2 will be inflated and could lead to incorrect conclusions about the relationships between explanatory variables and Y. Spatial autocorrelation of errors can occur when the relationship between X and Y is non-linear, when important explanatory variables have not been included in the model, or when there is significant spatial interaction in Y. Thus spatial analysis of regression residuals is important not only to test whether the assumptions of OLS regression are met, but also to gain additional information that may improve the specification of the model. Cliff and Ord (1981) discuss at length the analysis of regression residuals and provide a test statistic based on Moran's I, as well as the statistical moments and tests of significance.

If there is spatial autocorrelation among residuals, there are many possible means of removing it, the simplest being a transformation or increase in explanatory variables. In general it may be more informative to model spatial autocorrelation explicitly. The simplest model of variation in Y is that all variation is due to spatial interaction, that is:

$$Y = \rho WY + e, \qquad (19)$$

where W is used as before to describe spatial proximity, ρ is a measure of spatial autocorrelation and e is uncorrelated error terms. Equation (19) is termed a *spatial autoregressive* model. Given W, OLS methods can be used to obtain a value for ρ that minimizes the squared differences between observed and predicted values of Y, although the least squares estimators have not proven very stable (Cliff and Ord 1981). A practical problem is that the OLS method requires obtaining the determinant of an N x N matrix, limiting its usefulness to relatively small data sets.

A more general model, similar to the model of a regionalized variable that was introduced earlier, combines both deterministic or explanatory variables and spatial interaction:

$$Y = \rho WY + Xb + e, \tag{20}$$

Once again, difficulties in estimating the model parameters limit the usefulness of this approach to relatively small data sets and to fairly simple systems involving a well-specified weight matrix and a few explanatory variables. The sensitivity of results to the choice of weight matrix is shown by Anselin (1986).

SOFTWARE FOR SPATIAL ANALYSIS

None of the major commercial statistical packages (e.g., SPSS, SYSTAT, BMDP, MINITAB) contains tests for spatial autocorrelation. There are a few packages for geostatistical analysis, including GEOPACK, which is available from the U.S. Environmental Protection Agency, Las Vegas, Nevada.

Anselin (1990) has recently developed SPACESTAT, a program that offers a range of techniques for spatial data analysis and is especially geared toward spatial effects in regression analysis. This software will be available through the National Center for Geographic Information and Analysis at the University of California, Santa Barbara, beginning Summer, 1992.

REFERENCES

Anselin, L. 1986. Non-nested tests on the weight structure in spatial autoregressive models: some Monte Carlo results. *J. Regional Science* 26: 267-284.

_____. 1988 *Spatial Econometrics: Methods and Models*. Kluwer Academic Publishers, Dordrecht.

_____. 1990. SPACESTAT: A Program for the Statistical Analysis of Spatial Data. Draft Manual, Department of Geography and Department of Economics, University of California, Santa Barbara.

Burrough, P. A. 1987. Spatial aspects of ecological data. In: R. H. G. Longman, C. J. F. ter Braak and O. F. R. van Tongeren (eds.). *Data analysis in community and lanscape ecology*. Pudoc, Wageningen. Pp. 213-246.

Cliff, A. D., and J. K. Ord. 1981. *Spatial Processes; models and applications*. Pion Limited, London.

Costanzo, C. M., L. J. Hubert, and R. G. Golledge. 1983. A higher moment for spatial statistics. *Geogr. Anal.* 15: 347-351.

Getis, A., and B. Boots. 1978. *Models of Spatial Processes*. Cambridge University Press, Binghamton, N.Y.

Goodchild, M. F. 1986. Spatial autocorrelation. *CATMOG (Concepts and Techniques in Modern Geography)* 47: 1-56.

Griffith, D. A., and C. G. Amrhein. 1983. An evaluation of correction techniques for boundary effects in spatial statistical analysis. *Geogr. Anal.* 15: 352-360.

Harvey, L. E., F. W. Davis, and N. Gale. 1988. The analysis of class disperion patterns using matrix comparisons. *Ecol.* 69: 537-542.

Hubert, L. J., R. G. Golledge, and C. M. Costanzo. 1981. Generalized procedures for evaluating spatial autocorrelation. *Geogr. Anal.* 13: 224-233.

Journel, A. G. 1989. Fundamentals of geostatistics in five lessons *Short Course in Geology: Volume 8.* American Geophysical Union, Washington, D.C.

Mather, P. M. 1976. *Computational Methods of Multivariate Analysis in Physical Geography.* J. Wiley, London.

Ripley, B. D. 1981. *Spatial Statistics.* J. Wiley, New York.

Sokal, R. R., and N. L. Oden. 1978. Spatial autocorrelation in biology 2. Some biological implications and four applications of evolutionary and ecological interest. *Biol. J. Linnean Soc.* 10: 229-249.

Upton, G. J. G. and B. Fingleton. 1985. *Spatial Data Analysis by Example.* J. Wiley and Sons, New York.

2

THE SPATIAL NATURE OF SOIL VARIABILITY AND ITS IMPLICATIONS FOR FIELD STUDIES

Harold M. van Es

INTRODUCTION

Recent quantifications of soil heterogeneity (e.g., Burgess and Webster 1980, Vieira et al. 1981, Yost et al. 1982, ten Berge et al. 1983, Wollum and Cassel 1984) have established the general validity of geostatistical principles for soil attributes. This means that soil and related properties (e.g., plant growth) are typically not independently distributed but characterized by autocorrelation. Besides the applicability of the use of geostatistical techniques such as kriging, this also has implications for other types of studies that are conducted in the landscape. This chapter addresses in general terms the nature of soil variability and its relation to soil-forming processes. In addition, it discusses sampling designs for various types of studies that are conducted in the landscape and are therefore affected by spatially dependent variance structures.

SOIL-FORMING FACTORS AND THEIR SPATIAL NATURE

Jenny (1941) proposed the concept of the five soil-forming factors. The spatial nature of these factors (parent material, climate, topography, organisms, and time) provide us with indications of the spatial variability structure for soils. All soil-forming factors, except the time factor, are characterized by regionality, i.e., they are geographically distributed, although often at different scales.

Parent materials, whether they are granitic saprolite, gravelly glacial outwash, lacustrine clays, or loess deposits (to name a few), all have distinct regionality. Their variability tends to be scale dependent and typically increases with distance. Although a few abrupt changes in parent material may occur with short distance, most lateral transitions in parent material are gradual.

Similar to parent material, climate shows strong regionality at various scale levels. At the scale of continents, climatological regions are typically related to their location with respect to water bodies, major orographic features, and/or latitude. At the micro-scale, climate may be affected by such features as slope orientation, topography, small water bodies, etc. At any scale, however, climates are geographically, not randomly, distributed, typically exhibiting gradual changes in space.

The third factor, topography, primarily affects hydrology and temperature, which in turn affect soil properties. For example, ridges tend to be better drained than foot slopes. In turn, drainage affects the weathering environment and thereby the physical and chemical makeup of the soil constituents in those locations. Clearly, topographic features also exhibit regionality, where they have a predictable rather than a random spatial distribution.

The fourth factor, organisms, refers to all biotic forces that affect soil. This includes all subsurface and supersurface animal and plant life. The spatial distribution of such organisms may be quite varied, but is typically regionalized. On a large scale, one may identify major plant ecological zones such as forest and grassland regions. On a small scale, soil variability may be influenced by single plants and/or animals.

The time factor influences soil bodies in opposite ways. First, the duration for which a soil-forming factor affects a soil body intensifies its effect. For example, warm humid climates acting upon soils for a long time period likely create more strongly weathered soil bodies compared to action for short time periods. Secondly, present soil properties reflect past soil-forming processes to a lesser extent if these processes occurred in the distant past compared to the recent past. In this manner, time de-intensifies the effect of a soil-forming process.

The concept of the five soil-forming factors, although not theoretically proven, provides us with insight into the structure of soil spatial variability. Due to the regionalized nature of the soil-forming factors, we also expect soil properties to be regionalized. It is important to recognize such *anticipated* spatial behavior, because this has important implications on how studies in the landscape, whether observational or experimental, should be conducted.

GEOSTATISTICAL ASSUMPTIONS

Geostatistics was initially conceived to improve the estimation of ore reserves. It uses the semivariogram function $\gamma(h)$ to describe the spatial structure of soil variability. This function quantifies the variance between pairs of observations as a function of their separation distance h:

$$\gamma(h) = 0.5 \ \text{Var}(Y_{i+h} - Y_i) \tag{1}$$

The theory of regionalized variables, on which geostatistics is based (Matheron 1963), recognizes that observations taken at short distances from each other are more alike than those taken further apart, i.e., observations are "regionalized." The semivariogram formalizes this spatial structure as a monotonically increasing function. Figure 1 shows an idealized semivariogram.

To understand geostatistical theory and its applicability to field observations, we must first recognize the concept of variance and scale. Any field observation y_i at point i is a realization of a random variable Y_i with a mean $E(Y_i)$, which is the expected value of Y_i, and a variance $\text{Var}(Y_i)$, which is the result of field variability as well as imprecision in sampling and analysis.

When two observations y_i and y_j from different locations in a field are compared, we encounter discrepancies between geostatistical assumptions and the assumptions made for conventional statistical analyses. The latter considers each observation to be a realization of the random variables Y_i and Y_j at points i and j. It also assumes that, if properly randomized, these variables are independently distributed with

(i) $\quad E(Y_i) = \mu$ for all i $\tag{2}$

and (ii) $\quad \text{Var}(Y_i) = \sigma_y^2$ for all i $\tag{3}$

On the other hand, geostatistics recognizes that the set of random variables Y_i at locations i is characterized by a structure in that observations are not independent, but are spatially correlated. A set of random variables with a spatial structure is called a *random function*, and a measured property with such a spatial structure is called a *regionalized variable*.

The intrinsic hypothesis in geostatistics (Journel and Huijbregts 1978) is:

(i) The expected value at any location in the field is the same:

$$E(Y_i) = \mu \qquad \text{for all i} \qquad (4)$$

(ii) A finite variance is defined for all observations at distance h from each other:

$$Var(Y_{i+h} - Y_i) = E(Y_{i+h} - Y_i)^2 = 2\gamma(h) \qquad (5)$$

Geostatistical theory thus recognizes the existence of a relationship between (co)variance and spatial separation. If observations separated by different distances are compared, as is done in studies conducted in the landscape, their expected variances may be unequal.

The semivariogram is useful for obtaining insight into the spatial structure of soil variability and optimizing estimation efforts (kriging). Another equally important contribution of the geostatistical concepts to studies conducted in the landscape is the realization that the

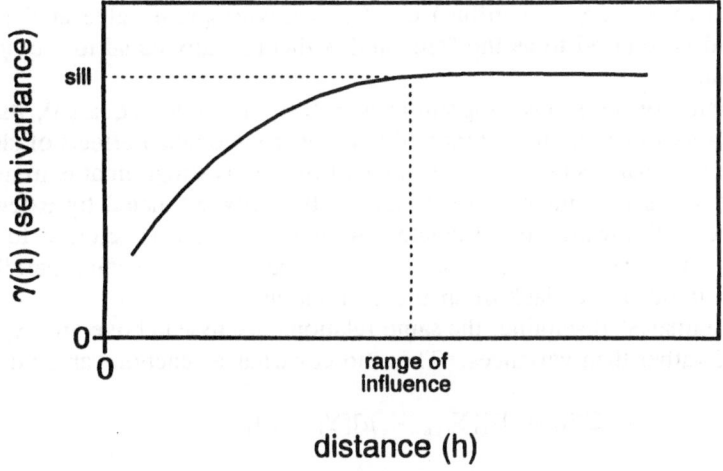

Figure 1. Idealized semivariogram.

sampling distance between observations is critically important in the research of regionalized variables.

SEMIVARIOGRAMS AND AUTOCORRELOGRAMS

Analysis of the γ-h relationship (equation 5) forms the basis for geostatistical studies.

Although spatial relations are defined in terms of the variance of the *difference* between observations (equation 5), the analysis is performed on half this quantity (γ(h) instead of 2γ(h)) in order to relate the semivariograms directly to the variance of observations as defined in conventional statistics (equation 3).

The semivariogram is estimated through

$$\gamma\ (h)\ =\ \frac{1}{2N(h)}\ \sum_{i=1}^{N(h)}\ [Y_{i+h}-Y_i]^2$$

(6)

where N(h) is the number of pairs of observations at a distance h from another. It is recommended that a sufficient number of observations are made so that N(h) equals 30 or more for all separation distances that are within less than half the width of the spatial domain. This insures a reasonably stable estimate of the semivariogram. Cressie and Hawkins (1980) proposed a robust estimator for [equation 6] as follows:

$$2\gamma(h)\ =\ \left[\frac{1}{N(h)}\ \sum_{i=1}^{N(h)}\ |Y_{i+h}-Y_i|^{1/2}\right]^4\ (0.457\ +\ 0.494/N(h))$$

(7)

The semivariogram (1) includes information on both the random and the spatially-structured aspects of regionalized variables. The "range of influence" for the semivariogram is the distance within which observations are spatially dependent, i.e., have lower variances. It is not uncommon for soil variables to show a twofold or higher increase from the shortest lag distance (h = 1) to the range of influence. The semivariogram value at the range of influence and beyond is referred to as the "sill" and is theoretically equal to the *apriori*, or non-correlated variance (σ_y^2).

Although the value of the semivariogram at h = 0 by definition equals 0, its value at extremely small distances may be much higher due to the compounded effects of short-scale variability and sampling/analysis errors. This value of the semivariogram at extremely short distances is referred to as the "nugget effect" and is typically estimated by extending the semivariogram curve back towards the ordinate. A semivariogram is referred to having a "pure nugget effect" if it shows no increase with distance, i.e., it remains parallel to the abscissa. This is an indication of lack of spatial correlation.

In the "spatial statistics" discipline, the same relations discussed above are expressed in terms of covariances rather than variances. The auto-covariance function can be defined as

$$C(h)\ =\ E\{[Y_{i+h}\ -\ \mu][Y_i\ -\ \mu]\}$$

(8)

This function is related to the semivariogram through (Hamlett et al. 1986)

$$C(h)\ =\ \sigma_y^2\ -\ \gamma(h)$$

(9)

The auto-covariance function is therefore negatively related to the semivariogram by subtraction from the *apriori* variance, σ_y^2. An idealized auto-covariance function is shown in Figure 2. For observations within the range of influence, auto-covariances are positive and observations are considered spatially correlated. Sometimes, this relation is depicted by standardizing the auto-covariance function with the *apriori* variance, which is referred to as the (auto)correlation function:

$$\rho(h) = \frac{C\ (h)}{\sigma_y^2}$$

(10)

where $\rho(h)$ equals 1 at h=0. $\rho(h)$ relates to the semivariogram as

$$\rho(h) = 1 - \frac{\gamma(h)}{\sigma_y^2}$$

(11)

The modeling of semivariograms and auto-covariograms is discussed by Journel and Huijbregts (1978) and by Isaaks and Srivastava (1989). A limited selection of semivariogram models is available, because they must be positive definite functions. This guarantees the existence and uniqueness of a solution to the kriging equations.

The auto-covariance function and the auto-correlation function are as good as or better than the semivariogram for description of spatial variability structures. For convenience, however, only the latter is used in this discussion.

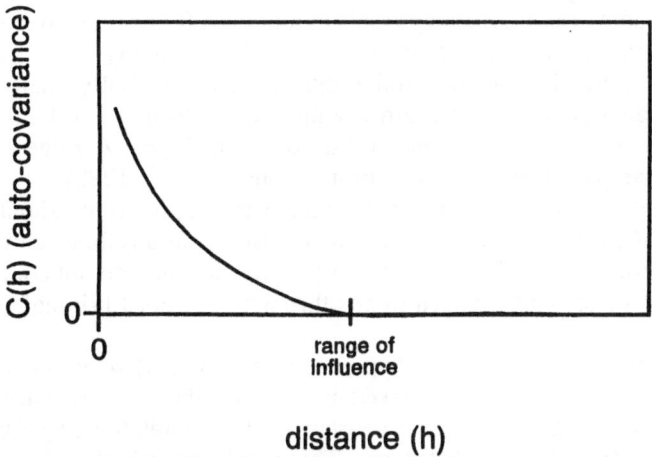

Figure 2. Idealized auto-covariance function.

KRIGING AND CO-KRIGING

The objective of most geostatistical studies is to optimize spatial estimation with information obtained from semivariograms. The kriging method uses a best linear unbiased estimator (B.L.U.E.) for this purpose. The distinguishing feature of (ordinary) kriging compared to other point estimation methods such as triangulation or inverse distance is that kriging minimizes the error variance. In addition, the kriging procedure yields an estimation variance (level of uncertainty) associated with each point estimate. Kriging is primarily applied to intensive mapping of regionalized variables and estimation of sparsely measured variables. For the latter purpose, it was first applied in mining engineering to improve the estimation of ore reserves.

Co-kriging involves point estimation with inclusion of a co-variable. The spatial cross-correlation between two variables, as expressed in the cross-semivariogram, is used to further optimize the estimation procedure. Co-kriging is particularly useful if a strongly correlated concomitant variable is available which is more intensely sampled. Detailed discussion on these procedures is beyond the scope of this chapter and the reader is referred to Journel and Huijbregts (1978) and to Isaaks and Srivastava (1989) for further explanation of kriging and co-kriging and similar procedures (block (co)kriging, universal kriging, etc.).

STATIONARITY, TRENDS, AND ANISOTROPY

In most cases involving soil and soil-related variables, spatial variability structures do not conform to the idealized depiction in Figures 1 and 2. The first component of the intrinsic hypothesis of geostatistics (equation 4) implies first-order stationarity, i.e., the same expected mean for the entire spatial domain. If first-order stationarity does not hold, then a field trend exists, often referred to as "drift" in geostatistical jargon. Trends are recognized in semivariograms through major deviations from the general pattern shown in Figure 1. David (1977) and Davidoff et al. (1986) have proposed quantitative tests for the presence of trends in field observations. Typically, trends generate a continuous increase in semivariance with lag distance, although in some cases they may result in increases followed by decreases. The latter is typically an indication of a spectral pattern. Trends commonly occur in fields and may be expected if the study site includes soil properties that gradually change over the spatial domain or if repeated patterns exist. An example of the former is a field that is laid out over a soil toposequence that spans from a ridge to a foot slope. A ridge-and-furrow-drainage pattern is an example of the latter (McBratney and Webster 1980).

Even if no predictable trend is apparent on the study site, its possibility should never be ignored in the analysis of field data. First-order stationarity is an absolute prerequisite for geostatistically based interpretations (i.e., (co)kriging). Field data that do not conform to this hypothesis therefore need to account for such trends through "universal kriging" (cf. Journel and Huijbregts 1978).

The second component of the intrinsic hypothesis (equation 5) does not imply strict second-order stationarity. This is further relaxed by the fact that the existence of $\gamma(h)$ is required only for a limited distance h, because geostatistical interpretations ((co)kriging) are based only on information from limited distances. This requirement is therefore referred to as "quasi-stationarity."

In some cases, nonstationarity can be dealt with through alternative description of the spatial variability structure. For example, *relative* semivariograms can be used to describe a spatial domain in which local variances are proportional to the local mean. In such case, the $\gamma(h)$ function can be adjusted by the local mean (Journel and Huijbregts 1978).

Spatial anisotropy, i.e., directional dissimilarities in the semivariogram, should also be anticipated when investigating soil variability, because soil-forming factors, especially parent material, climate, and topography, often exhibit patterns of directional dependence. For example, soils on a floodplain may show different spatial variability patterns in the perpendicular versus the parallel direction to the stream. Similarly, the structure of soil variability on a slope tends to be different along versus to across the contour. In general, the directional nature of processes that affect soil bodies, such as water and wind flow, and temperature/moisture gradients, typically results in an anisotropic spatial variability structure. Therefore, studies that consist of observations taken in the landscape should anticipate such anisotropy, and semivariograms must be constructed for various directions. Isaaks and Srivastava (1989) and Journel and Huijbregts (1978) discuss methods for incorporating anisotropy into semivariogram models.

SAMPLING STRATEGIES

The sampling strategy for studies in the landscape should depend on the ultimate use of the data. Three kinds of data uses can be distinguished, each with special consideration to the fact that observations in the field tend to be spatially correlated. If a study specifically aims at characterizing the spatial structure of a variable (e.g., to allow for (co)kriging), then observations should be made at close enough distance to allow for a good description of the semivariogram. In addition, one must be aware of the fact that the property of interest may not be stationary and regional biases may enter the semivariogram if the domain is not uniformly sampled. This requires a large number of observations. Conventionally, observations are made on a regular transect (one-dimensional) or grid (two-dimensional). Such sampling schemes, however, are inefficient, because they provide much larger numbers of observation pairs at short distances than at long distances. A nested sampling design is more efficient, in that it allows sample comparisons at short and long distance with approximately equal precision, while at the same time uniformly sampling the field.

Figure 3 shows an example of a nested triangular sampling scheme. The unit lag distance equals the distance between sampling locations within the smallest triangular cluster. The clusters in turn are separated by twice the unit lag distance. This scheme provides approximately equal precision for semivariogram estimation for up to 10 lags and allows for evaluation of anisotropy in three directions. If two-directional anisotropy is expected, nested squares rather than triangles may be used. The squares should be laid out parallel-perpendicular to the main axes of anisotropy. In a similar manner, nested schemes may be designed for transect sampling.

In a second type of study, the spatial correlation structure is of no interest per se, but one may wish to optimize sampling efficiency. Such is the case with many types of environmental-quality monitoring networks. When observations are taken at short distance from each other, they provide redundant information because they are spatially correlated. The spatial variability structure in this case is less relevant for the analysis of the data than for the design of the sampling scheme itself. To optimize the design and the number of true degrees of freedom, one needs to have knowledge of the range of spatial autocorrelation and assign sampling locations outside this range.

The third type of study is experimental (rather than observational) and involves the application of treatments. In such case, the intrinsic hypothesis is, by definition, violated because the treatments introduce nonstationarity of the mean (equation 4) as well as the spatial variance (equation 5). Due to the latter, the spatial variance structure is nonestimable because one cannot separate the treatment effects from their interaction with the spatial location (van Es et al. 1989). Experimental studies typically involve the replicated application of various treatments in "plots" whereby special attention is given to the spatial location of those plots. Conventionally, plot allocation is accomplished through the process of randomization. This insures that the observations obtained from the study are unbiased and independently distributed. However, when an experimental site is highly variable, the error term becomes inflated and the power to detect treatment differences is reduced. Traditionally, this problem has been addressed through the use of block designs, in which constraints are imposed on the imperfect randomization process by randomly allocating treatments within blocks only. Blocks are assumed to be relatively uniform and the block-to-block variability is removed from the error term, although with the loss of degrees-of-freedom. The Randomized Complete Block Design is now the most commonly used method for plot allocation in agronomic field trials. This design is a recognition of the spatial structure of soil variability, because variances are lower when treatments are grouped together in blocks. With our present knowledge of soil spatial variability, however, the block designs appear to have two inconsistencies: (i) blocks cannot be assumed to be uniform, and (ii) independence of observations within blocks rarely occurs, i.e., spatial correlation is still present.

Several alternatives to blocking have been developed to address the problem of spatial variability. Mendez (1970) evaluated six of these and found the methods of "nearest neighbor" and "trend analysis" to perform satisfactorily. The nearest neighbor method, suggested by Papadakis (1937) and discussed by Bartlett (1938, 1978) and Wilkinson et al. (1983), uses covariance adjustment from neighboring plots to reduce positional effects. This approach addresses soil variability by assuming a first-order autoregressive structure. Similar to geostatistics, it recognizes a variance structure which depends on spatial distance. However, the nearest neighbor method assumes that spatial dependence exists only for neighboring plots, and not for plots that are not directly adjacent.

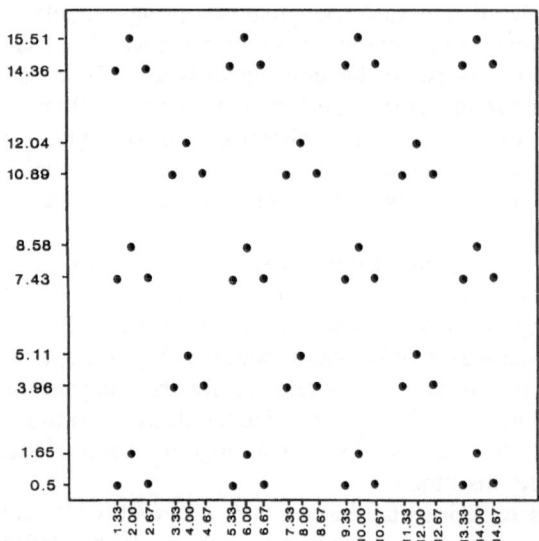

Figure 3. Nested triangular sampling scheme involving 54 observations. Numbers along axes indicate relative distances.

Trend analysis (Kirk et al. 1980, Tamura et al. 1988) removes the effects of spatial variability by fitting a response surface model using polynomial regression on the residuals. From a geostatistical perspective, this method removes field trends in the data. This may create problems, however, if the presence of field trends on the site cannot be validated. In addition, the choice of the number of terms to be included in the polynomial response surface is arbitrary and may cause overfitting and thereby excessive removal of variability from the error term.

Van Es et al. (1989) concluded that present experimental designs do not directly address the existence of a relationship between (co)variance and spatial separation (equations 5 and 8), as quantified in recent studies on soil heterogeneity. Based on this, treatment comparisons can be made with equal precision only if they are also made at equal distance. If this is not the case, then some treatment comparisons are made based on a higher variance (under the null hypothesis) than others. Van Es and van Es (1992) investigated the spatial nature of randomization and the effect of replication in insuring that treatment comparisons are, on the average, made at equal distance and therefore at equal precision. From theoretical derivations it was determined that conventional randomized designs do not insure equal precision of treatment comparisons, unless the experiment includes excessive replication. Based on simulated analyses of a uniformity trial through multiple randomizations, it was determined that some treatment comparisons conducted at the 0.05 error level were actually being conducted at only the 0.09 error level, because the spatial distance between plots for those treatment contrasts was longer than the average distance of treatment comparison upon which the error term is based. Two designs were proposed to address the problem of spatial inequity of treatment comparisons: incomplete blocks and spatially constrained randomization.

Small incomplete blocks, especially those involving only two plots, provide spatially balanced treatment contrasts, because all comparisons are made at equal distance *and* at the shortest distance (van Es et al. 1989). For experiments including a large number of treatments, however, blocks of two plots are impractical. Larger incomplete blocks, however, can then be used and still provide adequate insurance of spatially balanced treatment comparisons.

It is interesting to note that incomplete blocks were recommended over complete blocks for the purpose of further reducing error variances (Yates 1936). This constitutes a recognition of spatially correlated variability for soils, because error variances were reduced by decreasing block sizes. Nevertheless, despite this recognition, the randomization process was still recommended to insure independence among observations. In light of our current knowledge on soil variability, incomplete blocks are also a method for insuring spatially balanced treatment comparisons (van Es and van Es, in press).

Instead of using incomplete blocks, spatially balanced designs can also be obtained by constraining the plot allocations to insure (approximately) equal distances for all possible treatment comparisons. This may be accomplished through an iterative plot allocation process (van Es and van Es, in press).

CONCLUSION

Conventionally, studies on soil and related properties were based on the assumption that observations made in fields are independently distributed. It has now been firmly established that soil properties behave like regionalized variables and in most cases show strong spatial correlation. Based on this, field studies should anticipate and, if necessary, account for such variance structures.

Geostatistics is frequently applied to soils studies to analyze spatial variability patterns and to optimize interpolation and mapping through (co)kriging. Such studies, however, are very observation intensive and therefore not generally applicable. Other types of field studies, such as environmental monitoring networks or experiments involving treatment applications, are also affected by spatial correlation even though geostatistical techniques do not directly apply to such studies. Conventional designs based on random sampling do not directly address the relation between variance and distance. New design methods are now proposed to insure against adverse effects of spatial autocorrelation.

REFERENCES

Bartlett, M.S. 1938. The approximate recovery of information from field experiments with large blocks. *J. Agric. Sci.* 28:418-427.

Bartlett, M.S. 1978. Nearest neighbor models in the analysis of field experiments. *J. Royal Statist. Soc.* B 40:147-174.

Burgess, T.M., and R. Webster. 1980. Optimal interpolation and isarithmic mapping of soil properties I. The semivariogram and puntual kriging. *J. Soil Sci.* 31:315-331.

Cressie, N., and D. Hawkins. 1980. Robust estimation of the variogram, I. *Math. Geology* 12:115-125.

David. M. 1977. *Geostatistical Ore Reserve Estimation.* Elsevier Publishing Co., New York.

Davidoff, B., J.W. Lewis, and H.M. Selim. 1986. A method for verifying the trend in studying spatial variability of soil temperature. *Soil Sci. Soc. Am. J.* 50:1122-1127.

Hamlett, Y.M., R. Horton, and N.A.C. Cressie. 1986. Resistant and exploratory techniques for use in semivariogram analyses. *Soil Sci. Soc. Am. J.* 50:868-875.

Isaaks, E.H., and R.M. Srivastava. 1989. *Applied Geostatistics.* Oxford Univ. Press, New York.

Jenny, H. 1941. Factors of Soil Formation. McGraw-Hill, New York.

Journel, A.G., and Ch. Huijbregts. 1978. *Mining Geostatistics.* Academic Press, London.

Kirk, H.J., F.L. Haynes, and R.J. Monroe. 1980. Application of trend analysis to horticultural field trials. *J. Am. Hortic. Soc.* 105:189-193.

Matheron, G. 1963. Principles of geostatistics. *Econ. Geol.* 58:1246-1266.

McBratney, A.B., and R. Webster. 1981. Detection of ridge and furrow patterns by spectral analysis of crop yields. *Intern. Stat. Review* 49:45-52.

Mendez, I. 1970. Study of uniformity trials and six proposals to alternatives to blocking for the design and analysis of field experiments. Inst. of Statistics Mimeo Series 696. North Carolina State University, Raleigh, NC.

Papadakis, J.S. 1937. Methode statistique pour des experiences sur champ. *Bull. Inst. Amel. Plantes a Salonique* 23.

Tamura, R.N., L.A. Nelson, and G.C. Naderman. 1988. An investigation on the validity and usefulness of trend analysis for field plot data. *Agron. J.* 80:712-718.

ten Berge, H.F.M., L. Stroosnijder, P.A. Burrough, A.K. Bregt, and M.J. de Heus. 1983. Spatial variability of physical properties influencing the temperature of the soil surface. *Agric. Water Manag.* 6:213-226.

van Es, H.M., C.L. van Es, and D.K. Cassel. 1989. Application of regionalized variable theory to large-plot field experiments. *Soil Sci. Soc. Am. J.* 53:1178-1183.

van Es, H.M., and C.L. van Es. In press. The spatial nature of randomization and its effect on field experiments . *Agron. J.*

Vieira, S.R., D.R. Nielsen, and J.W. Biggar. 1981. Spatial variability of field-measured infiltration rate. Soil Sci. *Soc. Am. J.* 45:1040-1048.

Wilkinson, G.N., S.R. Eckert, T.W. Hancock, and O. Mayo. 1983 Nearest neighbor (NN) analysis of field experiments. *J. R. Stat. Soc.* B 45:151-211.

Wollum, A.G., II, and D.K. Cassel. 1984. Spatial variability of *Rhizobium japonicum* in two North Carolina soils. *Soil Sci. Soc. Am. J.* 48:1082-1086.

Yates, F. 1936. A new method arranging variety trials involving a large number of varieties. J. Agric. Sci. 26:424-455.

Yost, R.S., G. Uehara, and R.L. Fox. 1982. Geostatistical analysis of soil chemical properties of large land areas. I. Variograms. *Soil Sci. Soc. Am.* impressive. 46:1028-1032.

3
PHYTOPLANKTON PATCHINESS: ECOLOGICAL IMPLICATIONS AND OBSERVATION METHODS

Mark R. Abbott

INTRODUCTION

Phytoplankton ecosystems were once viewed as nearly static (or at least only slowly varying) systems on scales of roughly a few days and spatial scales of a few tens of kilometers. A notable example of research based on this view is the map of primary production by Koblentz-Mishke et al. (1970). As sampling techniques improved, variability was found to occur at smaller and smaller scales. Much of this variability was thought to be controlled by physical processes, such as turbulence, and these processes were included in models of the variance spectrum (Denman et al. 1977, Denman and Platt 1976). These ideas challenged the traditional notions of stability, suggesting that maps of "average" biomass and productivity were essentially meaningless; the structure of the variability of the planktonic ecosystem was thought to be more important.

The coupling of physical and biological processes as evidenced by temporal and spatial variability plays a key role in the dynamics and energy flow of both terrestrial and aquatic ecosystems. Although oceanic ecosystems were thought to be dominated by physical forcing, continued improvements in sampling and measurement of small-scale physiological processes suggested that biological processes might play an important role as well. For example, rapid uptake of nutrients and changes in photosynthetic characteristics led to the idea that physiological characteristics had evolved in the phytoplankton to take advantage of the specific frequency spectrum of environmental fluctuations (Harris 1978). This suggestion immediately leads to one of the main challenges of ecological theory: the distinction between proximate (present-day) and ultimate (evolutionary) causes. That is, is the present community structure shaped by events like competition, or by events that occurred in the distant past?

AN EXAMPLE FROM THE SOUTHERN OCEAN

The surface waters around Antarctica are frequently rich in nutrients, especially in the Antarctic Circumpolar Current (Martin et al. 1990). Other regions of the world ocean, notably the subarctic Pacific, also are nutrient-rich in terms of nitrate availability. However, primary productivity is not especially high, although nitrate never approaches limiting concentrations, and many hypotheses have been proposed, ranging from control by zooplankton grazers to light limitation. Recently, Martin et al. (1989, 1990) have proposed that iron is limiting primary production. As iron is essential for the production of nitrate reductase in phytoplankton, its absence in turn reduces nitrate uptake, hence limiting primary productivity. Such iron limitation appears to be a function of dust input from the atmosphere, which is the major source of iron to the upper ocean (Young et al. 1991, Duce et al. 1991). As regions such as the Southern Ocean and the subarctic Pacific are far removed from land masses, iron input is low, hence iron limitation of primary productivity.

Martin et al. (1990) used a series of incubations of phytoplankton samples collected in Antarctic waters to show that the addition of iron significantly stimulated productivity. Although these results have generated a fair amount of controversy (e.g., Banse 1991a, b; Martin et al. 1991; Dugdale and Wilkerson 1990; Broecker 1990), the results are compelling. However, in addition to stimulating phytoplankton growth, the iron additions also resulted in a dramatic shift in the species composition favoring large diatoms, which had previously been rare. Thus we need to explain both the increase in growth rates as well as changes in relative species abundances.

Several possible explanations have been put forward. First, diatoms need iron to take up nitrate, and then outcompete smaller phytoplankton when adequate iron is available. Second, the bottle experiments exclude grazers, which otherwise keep diatom populations under control. Third, the bottles are a dramatically different physical environment that favors diatoms over smaller forms.

Let us examine this last explanation. Research by Dugdale and Wilkerson (1989) suggests that there is a linear, positive relationship between nitrate uptake rates and the ambient nitrate concentration. Additional data by Dugdale and co-workers suggests that open ocean communities have a much weaker relationship between nitrate uptake rates and nitrate concentration than do coastal communities. (However, it should be noted that if uptake rates are normalized by the mean chlorophyll concentration, this difference is less pronounced.) It has been suggested that these differences may be the result of changes that have occurred over evolutionary time scales whereby some species are favored in stable, steady environments and others are favored in unstable, fluctuating environments. In particular, large species such as diatoms tend to be favored in unstable environments, whereas small species are favored in stable environments. The bottles used by Martin et al. (1990) may create an environment that favors larger species. It is interesting to note that the species shifts occurred after roughly 3 – 5 days, which is a long period at the phytoplankton scale. Thus the effects of iron may be confounded with changes in the physical environment that result in shifts in the phytoplankton community structure.

VARIABILITY AS A RESOURCE

It is well known in terrestrial ecology that environmental variability behaves as a "resource" that can be exploited. For example, Levins (1979) modeled such processes, showing that particular strategies could be employed to take advantage of environmental fluctuations. In a series of papers, Harris (1978, 1986) suggested that phytoplankton may have developed similar strategies, with particular physiological processes evolving to take advantage of the spectrum of environmental variability. Harris uses such a model to explain the succession of species in the mixed layer. As he points out, it is difficult (if not impossible) to use evolutionary mechanisms to explain proximate events (such as species composition) when we require that the mechanisms of ecology and evolution be consistent. For example, if we postulate that the coevolution of grazers and phytoplankton are largely responsible for present-day phytoplankton composition, then it would be suspect if we also argued that physiological responses to the changes in the nutrient/light regime are responsible for the seasonal changes in composition.

In examining phytoplankton dynamics, this chapter focuses on the interaction between environmental variability and the ability of species to cope with and adapt to this variability. The ecological and evolutionary response to these fluctuations in time and space will depend on the degree of environmental heterogeneity, on the way this heterogeneity is perceived by the organism, and on long-term trends. To illustrate: a short dash across the street when the temperature is −20°C may be possible without a sweater, whereas a two-block stroll is not; and if the temperature were −80°C, even the dash across the street might not be possible. Thus the focus here is on the scales and intensity of variability and the response of the organism.

This non-equilibrium view of phytoplankton dynamics emphasizes the variations in the environment, rather than simple averages. It also focuses on density-independent processes where competing species have different rates of increase, thus perceiving environmental variability differently. This view, as Harris (1986) points out, is counter to much of traditional biological oceanography, which has treated variability more as a sampling nuisance than as a critical component of the upper ocean environment.

As we examine how phytoplankton cope with environmental variability, we are confronted with a myriad of rates, species, etc. that can vary on a wide range of time and space scales. How do we decide which processes are important? As noted earlier, it is important to look at the scales of variability since environmental fluctuations are strongly scale-dependent. Such considerations lead to the "patchiness problem"; we must design our observation programs appropriately for the processes that we wish to study.

THE FLUID ENVIRONMENT

With the input of radiant energy at the surface, the ocean tends to be stably stratified, thus maintaining a state of minimum potential energy. Losses of energy through such processes as cooling or evaporative flux, and inputs of energy through wind mixing, act to reduce this stratification. As irradiance decreases exponentially with depth, phytoplankton growth is supported in the upper layers. However, this growth eventually reduces nutrient availability, and, coupled with stable stratification, eventually causes the upper layers to become nutrient-depleted. Thus physical processes that act to break down this stratification and increase the upward flux of nutrients should be important to phytoplankton ecology. However, these same processes also act to reduce light availability by dispersing phytoplankton over greater depths. The fundamental challenge for phytoplankton in the upper ocean is to maintain a balance between light availability (which is highest near the surface) and nutrient availability (which is highest at depth).

Although the most dramatic changes in the physical and biological environment occur in the vertical direction, processes that act in the horizontal, such as mesoscale eddies and fronts, are also important. However, such two-dimensional processes are usually associated with strong changes in the vertical dimension as well. For example, coastal fronts are often the site of strong vertical motions. Thus the predominant fluctuations can be described in terms of vertical processes.

These physical processes are never completely at rest, even during periods of extreme stratification. In addition, biological processes, such as excretion and grazing, are also taking place. The upper ocean environment is fluctuating at a variety of spatial and temporal scales. For an individual phytoplankter, the environment presents a complete spectrum of variability that in a sense must be "averaged." The action of this spectrum over evolutionary time scales has resulted in phytoplankton assemblages that can respond to changes that occur over a wide range of scales. From an oceanographic point of view, we must determine which scales are relevant for phytoplankton versus those that are simply "averaged." Using an anthropomorphic example, does a particular environmental fluctuation appear as a quick dash or a long walk through −20°C weather?

Scales of Variability

Several papers have appeared in recent years that describe the basic time and space scales associated with phytoplankton (e.g., Dickey 1991, Mackas et al. 1985, Harris 1986). In general, the evaluation of physical processes is based on their correspondence with the scales of biological processes. For example, internal waves with a period of 30 minutes may significantly affect the photosynthetic response, which varies on similar scales. This "scale matching" may not delineate all of the relevant processes, but it at least provides an initial framework to study biological/physical interactions. It should also be noted that the scales of horizontal and vertical processes are substantially different.

Denman and Powell (1984) provide a useful example of scale matching. Surface gravity waves have periods of several seconds in the upper few meters of the ocean. Thus a phytoplankter will be exposed to rapid variations in light as it moves up and down in the wave. Given that turnover times of phytoplankton are on the order of 0.5 per day, an individual phytoplankter will "average" thousands of waves. Internal gravity waves, such as those that occur on the thermocline, may move in rough synchrony with the daily cycle of solar irradiance. Hence the scale of physical forcing associated with internal waves is close to the time scale associated with phytoplankton growth, and we conclude that such waves are more important for phytoplankton growth than are surface gravity waves. Although the correspondence of scales does not ensure any causal relationship between physical and biological processes, it does provide an initial focus for research. The estimation of these scales based on characteristic values is discussed by Powell in this volume.

The biological response to variability is scaled to the size of the organism, its dispersal abilities, and its generation time. The sum of these responses is essentially the organism's "perception" of the environment. This perception will determine which fluctuations are important for study. Short time scale variations may be integrated by physiological mechanisms, whereas long time scale fluctuations may be avoided by resting stages or density-independent events (Harris 1986). For example, the onset of summer stratification may result in the sudden disappearance of diatoms from the upper ocean phytoplankton assemblage. Although such events are somewhat predictable, the presence of a broad range of scales in environmental variability essentially makes it impossible for an organism to optimize its response. Such reasoning led Harris (1978) to suggest that phytoplankton use the "noisy" signals of the environment as information. Changes in the nature of vertical mixing as stratification increases favor one assemblage over another; longer time scale changes in coastal upwelling may result in similar species shifts.

As noted by Denman and Powell (1984), biological processes control the time scales that should be considered. An analogy can be made to resonant or tuning effects; the resonant frequency is close to the frequency of forcing. Thus the biological time scales of growth (or generation time) set the temporal scales of interest. For example, tidal processes do not influence phytoplankton competition, as these scales are much longer than the tidal scale. However, phytoplankton division rates are sometimes near the time scale of the tides, so we might expect tides to influence growth rates. Denman and Powell (1984) make similar arguments concerning the control of the spatial scales of interest. As phytoplankton are largely passive in their distribution, spatial scales are largely set by the scales of the physical processes. Mesoscale eddies usually have scales of 50 – 200 km, and phytoplankton distributions usually have similar scales. Complications can arise due to nonlinear interactions between biological and physical processes, but in general this construct is useful for guiding our research.

Physiological Scale Processes

Thus far we have focused on the importance of physical forcing mechanisms on the scales of variability. However, it is apparent that the physiological responses are important as well. The separation of physics from physiology is arbitrary. However, the question remains: How important are the physiological scale processes? If we examine the environment at much finer scales, we discover that there is considerable patchiness in light and nutrients. For example, light availability is affected by small-scale vertical motions and turbulence, and nutrient availability is affected by grazing, excretion, and small-scale mixing events. As described by Harris (1986), most oceanographic studies assume that physiological processes are operating at equilibrium and do not take advantage of this small-scale variability. As we shall see in subsequent sections, this may not be the case. Rather, phytoplankton may have evolved specific strategies to exploit environmental variability at small scales. These strategies could include the ability to rapidly change nutrient and photosynthetic responses. Thus traditional sampling and analysis techniques that rely on steady-state, equilibrium dynamics may not be appropriate. For example, 0.5-day incubations to measure photosynthetic rate assume that the relevant processes are steady during the experiment. Large-scale surveys of chlorophyll fluorescence also assume that processes such as light utilization are fairly constant. Therefore, non-equilibrium processes may have a significant impact on the types of sampling and measurements that we can make.

Light utilization. Although the steps in the photosynthetic process are well known, there are many time-varying steps that are not understood. A basic description can be found in any number of oceanographic texts, such as Kirk (1983) and Falkowski (1980). The process has been described by Falkowski and Kiefer (1985, p. 716) as "the controlled production and dissipation of an electrochemical gradient where oxidation of water provides a source of free electrons and the initial driving energy is free energy released by the de-excitation of an excited pigment molecule." The pigment molecules that initially capture photons can either pass this energy to the photosynthetic system, re-emit the energy as fluorescence, or dissipate it as heat. In addition to the processes of photosynthesis, it should be noted that phytoplankton can significantly affect the light environment, in both amount and spectral composition, through the processes of absorption and scattering. Sathyendranath et al. (1991) and Lewis et al. (1990) suggest that light absorption by phytoplankton traps solar energy in the upper ocean, hence affecting air/sea heat flux. This would tend to reinforce the stability of the upper ocean, perhaps further reducing the upward vertical flux of nutrients. This would suggest that high biomass layers near the surface of the ocean would be inherently unstable, as their source of nutrients would effectively be shut off.

The standard method to examine light utilization is through the use of photosynthesis/irradiance (P vs. I) curves. This method relies on measuring photosynthesis at several different light intensities. Early measurements involved incubations over several hours, but recent techniques allow incubations to be completed in less than one hour, thus minimizing photoadaptation and species changes. Several functional forms are used to fit the data (e.g., Platt et al. 1980), but most rely on estimating three parameters: α, β, and P_{max}. The initial slope of the P vs. I curve is defined by α, which is a measure of the photosynthetic efficiency. It was thought that α would be relatively constant from species to species, but field measurements show that it varies by at least a factor of 20. α also appears to depend on cell size and pigment composition and varies on time scales of one day to several days. β is a measure of photoinhibition that occurs at high light intensities. Such inhibition appears to be the result of damage to the reaction centers of photosystem II caused by ultraviolet radiation (Harris 1986). As expected, the recovery time of photosynthesis depends strongly on the length of exposure to damaging levels of irradiance. β also appears to vary diurnally, as noted by Hood et al. (1991). Lastly, P_{max} is closely linked with α and is not strictly independent. That is, P_{max} is essentially set by the initial slope of the P vs. I curve.

These three photosynthetic parameters have had broad application in describing light utilization by phytoplankton. However, it should be obvious that each parameter covers a wide range of specific physiological processes and does not describe some inherent process; each is merely a mathematical convenience. For example, α depends strongly on chloroplast size and orientation and the number photosynthetic units. Pigment composition can affect photoinhibition. Each of these processes is associated with its own time scale of variation, thus confounding our attempts to estimate phytoplankton time scales based on relatively simple measures of P vs. I. However, this is not to say that such an approach is hopeless; rather we must exercise caution in our interpretations.

One outgrowth of P vs. I research was the prediction that α could be used to estimate the maximum quantum yield (ϕ_{max}) of photosynthesis (defined as the number of moles of carbon dioxide fixed by one absorbed photon). Since the maximum quantum yield varies only between about 0.1 and 0.125, and given that ϕ_{max} and α are linearly related, there were expectations that α should not vary markedly from species to species or from environment to environment. However, the relationship between ϕ_{max} and α depends on the details of light absorption by the phytoplankton. As these can (and do) vary over a broad range of values, α is not constant. A set of measurements by Hood et al. (1991) showed that α (along with the other P vs. I parameters) can vary over small spatial scales. These scales are on the order of several meters in the vertical and a few kilometers in the horizontal. In large part, these changes are driven by changes in the species composition, but physical forcing (such as subduction of surface water masses) also plays a critical role.

Lande and Lewis (1989) presented a model of photoadaptation in a turbulent mixed layer, based on time scales estimated from P vs. I experiments. Building on work by Cullen and Lewis (1988), this model relied on simple first-order kinetics to model photoadaptation. The model showed that for slow changes in irradiance, linear changes with depth in the P vs. I parameters would occur in the mixed layer. That is, photoadaptation scales were much smaller than mixing scales. However, more rapid changes in irradiance seriously complicated these results, and nonlinear dynamics would be more important.

Nutrient uptake. Several decades of limnological and oceanographic research have noted the rapid turnover of the soluble pools of nutrients in the surface waters of lakes and oceans. These observations have led to the concept of "limiting nutrients," which has been the foundation of much ecological research. As discussed by Harris (1986), researchers have often assumed a simple relationship between nutrient concentration and phytoplankton growth rate. That is, a state variable (nutrient concentration) can be related to a dynamic variable (growth rate).

This assumption has been based on two observations. First, analysis of phytoplankton culture data showed that phytoplankton growth rates decreased over time as nutrients became depleted. However, the initial supply of nutrients represents the total available supply; processes affecting nutrient input such as vertical mixing are not relevant to laboratory cultures. Second, field studies showed large-scale correlation between nutrient abundance and growth rates over both time and space. Studies of the spring phytoplankton bloom in the North Atlantic "confirmed" the view that growth rates were governed by nutrient availability.

On small temporal and spatial scales, we must consider other processes that are not relevant to either large-scale or laboratory studies. Specifically, processes involved with nutrient recycling such as pool sizes, uptake rates, and resupply rates become critical. A simple analogy to one's personal finances can be made. On an annual time scale, our income correlates well with our expenses. However, such large-scale budgets do not tell us much about dynamics, especially cash flow. For example, if we are paid only once a year, our spending patterns will be considerably different than if we are paid daily. In the former case, a large internal storage pool must be used carefully, but its presence will allow us to make large purchases fairly easily. In the latter case, we must more carefully regulate our daily spending patterns in order to build up a storage pool that may be required to meet large monthly bills, such as mortgage payments. Thus our checkbook balance may tell us something about our financial situation, but without information on cash flow, the picture is woefully incomplete. It is also apparent that the time scales of the cash flow (daily versus annual paychecks) will greatly affect our strategies both for using money and for observing the relevant processes.

In many regions of the ocean, nutrient concentration is not a useful predictor of growth rates. Earlier, we discussed nutrient-rich regions (for example, the Southern Ocean) where growth rates are lower than expected, but large areas of the ocean are nutrient-poor yet support relatively high growth rates. Such oligotrophic oceans (such as the central gyres) were thought to be nutrient-limited, with primary productivity dominated by regenerated production (Dugdale and Goering 1967). Such conditions imply a low f-number (Eppley and Peterson 1979), which is the ratio of new production (supported by nitrate supplied from outside the euphotic zone) to total production.

About 15 years ago, it was suggested that phytoplankton were growing much more rapidly in oligotrophic oceans than had been expected. Goldman et al. (1979) noted that growth rates were nearly always at their maximum, suggesting that nutrients were not limiting. Several explanations have been proposed, but the present view is that there is a tight coupling between small grazers and the phytoplankton. This results in rapid recycling of nutrients on very small scales, so that, while growth is maintained, the ambient nutrient concentrations as measured on larger scales remain low or undetectable. Large particles, such as marine snow, may enhance the close physical coupling necessary to maintain this rapid recycling. Thus, the overall system is at steady state in a statistical sense, but small micropatches or particulates may not be at equilibrium. Recent measurements have shown large concentrations of micrograzers and small autotrophs, leading to the concept of the microbial loop. Thus the oligotrophic oceans may be likened to a small, rapidly spinning wheel.

There has been considerable controversy concerning the relative magnitudes of new and regenerated production in the oligotrophic oceans (e.g., Shulenberger and Reid 1981, Platt 1984, Reid and Shulenberger 1986, Craig and Hayward 1987). Platt et al. (1989) have suggested that the f-number is considerably higher than previously supposed. Much of the controversy revolves around the use of "bulk" measurements of state variables in comparison with estimates of dynamic quantities such as primary productivity. As noted by Platt et al. (1989), bulk measurements generally have much longer time scales and their interpretation often relies on steady state assumptions. After an analysis, Platt suggests that intermittent pulses of nitrate are supporting the higher values of new production. Such pulses are difficult to resolve in a statistically robust manner, and bulk estimates of ambient nitrate concentrations are generally low or undetectable. These pulses may be forced by mesoscale eddies or storms. (It should be noted that Hayward (1987) pointed out the dilemma in the central Pacific where the nitricline is significantly deeper than the euphotic zone, making the mechanism supporting upward flux of nutrients into the euphotic zone more problematic.) As these pulses would be more common at the base of the euphotic zone (just above the nitricline), we would expect to see vertical structure in phytoplankton response. The upper layer, where nutrients are always low, should be dominated by phytoplankton that have a high nutrient utilization efficiency and are closely coupled with small grazers for efficient nutrient recycling. The lower layer, which is subjected to occasional pulses of nutrients, should be dominated by less efficient phytoplankton that are more loosely coupled with grazers.

Harris (1986) suggests that phytoplankton are largely opportunistic and grow rapidly under favorable conditions and merely persist under poor conditions. This implies that recycling rates at various trophic states should be roughly correlated with varying growth strategies. Harris further suggests that those species that are actively growing are growing rapidly, at rates close to their maximum relative growth rates. However, only a small portion of the phytoplankton community may be growing at any one time. In oligotrophic oceans, as nutrients become depleted, the phytoplankton community shifts to species favored under conditions of rapid nutrient regeneration and grazing as nutrients become depleted. Such shifts in species composition can happen in the vertical dimension (as suggested by Platt 1984) as well as seasonally.

Variability in the environment calls into question our assumptions of steady state behavior and, as we shall see in the next section, significantly affects our observation strategy. Levins (1979) modeled the competition between species in a variable environment, showing that a nonlinear growth response to a change in resource levels would allow species to coexist. Turpin and Harrison (1979, 1980), in a set of laboratory experiments, showed that phytoplankton species composition could be manipulated by varying the periodicity of nutrient pulses. Frequent additions of small pulses favored small species, whereas infrequent, large pulses favored larger species. It was also shown that phytoplankton can track pulses of ammonia with surprising precision, even down to nearly undetectable levels. This tracking consisted of changes in storage capabilities and growth rates to suit environmental variability. For example, addition of a single ammonia pulse per day to a diatom culture soon resulted in a culture with a much higher nutrient uptake velocity, a decreased half-saturation constant, and biphasic uptake kinetics (rapid uptake following a pulse followed by slower uptake). As shown by the model developed by Turpin et al. (1981), species composition is a complex function of the spectrum of environmental fluctuations and the spectrum of physiological responses. It should also be noted that the physiological response is not fixed, but can adapt (within some limits) depending on the patterns of environmental variability.

The work by Turpin and Harrison also showed that cell size plays a critical role in the biological response. Large cells tend to be favored where pulses are large and infrequent. Although their half-saturation constant is larger, they have a larger maximum nutrient uptake velocity and larger storage capacities. Small cells are favored by constant environments, or where nutrient pulses are small and frequent. Thus we would expect Platt's (1984) two-layer system in oligotrophic oceans to have small cells near the surface in the "recycling" regime, where nutrients occur in small, frequent pulses as a result of grazer excretion; and large cells at depth, where nutrient pulses are driven by physical forces of eddies or vertical overturns. Although available data are consistent with this view, there is still considerable controversy over the magnitude of primary productivity in oligotrophic oceans, especially in regard to the fraction of total production that is supported by upwelled nitrate (new production). If this new production is indeed supported by isolated events, then it suggests that global scale productivity needs to be studied at smaller scales than previously supposed.

If our hypotheses depend on specific processes, then we must tailor our observations accordingly. It has been suggested that the oceans play a much less important role in global carbon cycling than previously supposed (Tans et al. 1990), but this result is based on noisy and incomplete data. We must exercise great caution, as poorly designed sampling methods (or measuring the wrong variable) can bias our conclusions. In our discussion of oligotrophic oceans, it appears that bulk observations of nutrient levels may be quite misleading. If primary productivity (and especially new production) are driven by rare events, then our sampling must take this into account. Bottle experiments where large amounts of nitrate are added may change both the physiological responses as well as the species composition, thus confusing the results. As noted by Harris (1986), our assumptions concerning steady-state responses are likely not to be correct, and we must conduct our experiments accordingly.

OBSERVATION METHODS

The study of phytoplankton patchiness began with direct microscopic counts of water samples (e.g., Cassie 1959). Although such measurements provide detailed information concerning species distributions, it is nearly impossible to collect a sufficient number of samples and process them in a reasonable amount of time. In many instances, the species data behave as a point process and are thus more difficult to interpret. In an attempt to avoid these difficulties, various bulk measurements were developed, such as chlorophyll extractions, ATP, displacement volume, etc. All of these give "reasonable" estimates of biomass with increased resolution of spatial patterns (but at the cost of less detailed information).

In the early 1970s, oceanographers and limnologists began the development of various optical and acoustical techniques to study plankton distributions. The goal was to increase the spatial resolution of the measurements, as it was becoming clear that plankton were patchy on all scales. The optical and acoustical techniques offered the promise of increased resolution without the enormous manpower needs of discrete water sample analysis (whether for species counts or for bulk quantities such as chlorophyll).

Light absorbed by chloroplasts can be used in either photosynthesis or production of heat, or it can be re-emitted as fluorescence. The standard fluorometric techniques rely on fluorescence by photosystem II and were adapted for use in oceanography nearly 30 years ago (Lorenzen 1966, Holm-Hansen et al. 1965). Chlorophyll fluorescence was used as an indicator of chlorophyll content, which was then used to infer biomass concentration. The technique could be used either for discrete samples or with a flow-through system for rapid, under way sampling. These fluorescence surveys stimulated increased interest in phytoplankton patchiness, especially the relation between biological patterns and physical forcing (e.g., Platt 1972, Powell et al. 1974, Denman 1976).

There are several pitfalls in the use of fluorescence data. First, fluorescence is not a constant function of chlorophyll. Fluorescence can vary as a function of species composition, light level, physiological state, and time of day. This makes the conversion from fluorescence to chlorophyll problematic without the frequent collection of calibration samples. Second, chlorophyll itself is not a constant function of biomass, as chlorophyll content per cell can change dramatically. The fundamental problem is that fluorescence is a physiological process and can vary on physiological scales; its use as a "state" variable to study ecological processes that vary on much longer time scales must be approached with caution.

In an attempt to tease more information out of the basic fluorescence signal, several advanced fluorometric techniques have been developed in the last ten years. Flow cytometry has been adapted from the medical sciences and used to study individual cell sizes and backscattering, in addition to fluorescence (Yentsch and Yentsch 1984). Spectral fluorescence has been used to measure pigment distributions besides chlorophyll a. Laser-induced fluorescence allows the study of detailed fluorescence excitation/response features with fine spectral resolution (Cowles et al. 1990). Sun-stimulated fluorescence, in addition to being used to estimate chlorophyll a concentration, has been used to estimate primary productivity on a nearly instantaneous basis (Kiefer et al. 1989, Chamberlin et al. 1990). Lastly, "double flash" fluorometers use sequences of high and moderate light flashes to estimate primary productivity (Falkowski and Kiefer 1985).

Beam transmissometers are used to estimate the scattering and absorption properties of water over a fixed distance. In general, the wavelength of the light source is chosen appropriately so that the data can be used to estimate particulate concentrations (Spinrad 1986, Bishop 1986). Various studies have been made to develop conversion factors to estimate suspended material concentrations as well as particulate organic carbon concentrations. Siegel et al. (1989) used a time series of beam transmissometer data to estimate primary productivity.

Other optical instrumentation, notably spectroradiometers, have also been used to study phytoplankton distributions. This method relies on the spectrally dependent absorption of light by various pigments. This process lies at the heart of the various bio-optical algorithms that are used in conjunction with satellite remote sensing (Smith et al. 1991, Gordon et al. 1988).

Although the focus of this chapter has been on phytoplankton, it should be noted that advances have been made in the automated counting and sorting of zooplankton. Optical techniques have been used, but acoustic techniques have been the primary method. Using various frequencies, it is possible to estimate size classes, as the backscattering signal depends on the size and other characteristics of the zooplankter. However, these techniques are useful only for larger forms; acoustical methods cannot resolve microzooplankton.

Dickey (1991) gives a thorough overview of the measurement methods and the in-water platforms that have been used to support the various instrument packages. These platforms range from traditional ship-based towed, profiling, and pumping systems to moored and drifting instrument packages. While in-water techniques are essential for studying certain processes at particular temporal and spatial scales, satellite remote sensing has also expanded in the last decade to provide a view of processes occurring over much larger scales.

Remote sensing can be divided into two types: active, which involves the transmission and reception of electromagnetic radiation by the satellite; and passive, which involves only reception. Active remote sensing of the ocean is confined to the use of radars. These are altimeters, which are used to sense small changes in sea surface elevation, and scatterometers, which are used to study wind stress at the sea surface. Synthetic aperture radars are used primarily in sea ice research, but some physical features of the ocean, notably internal waves and ocean fronts, can be detected by such radars. An overview of these methods can be found in Fu et al. (1990).

Passive remote sensing, in the visible, thermal infrared, and microwave portions of the electromagnetic spectrum, has also been used extensively in studies of ocean processes. Visible remote sensing has been used primarily to study chlorophyll abundance, as an indicator of phytoplankton. Sensing in the thermal infrared provides data on sea surface temperature. Passive microwave remote sensing can be used to study a variety of ocean processes, including sea surface temperature, wind speed, latent heat flux, and sea ice. Abbott and Chelton (1991) provide a detailed review of the methods and applications. Satellite remote sensing began in earnest in the late 1970s with a series of "technology demonstration" missions. The success of these missions resulted in plans for a series of satellites focused on oceanographic research, rather than merely on testing technology. These programs have begun with the launch of the European ERS-1 satellite and will continue through the 1990s with a series of missions by the U.S., France, Japan, and the European Space Agency. In the late 1990s, the U.S., in collaboration with Japan and Europe, will begin a series of comprehensive missions to study the Earth as a system.

The scientific evolution of satellite remote sensing started with simple validation and verification of the satellite signals in relation to ocean processes. Satellite data were used to provide the large-scale "context" for ship studies that could cover only a limited region. As longer time series became available, more attention was paid to using satellite data in and of itself, primarily through statistical analysis. Recent research is now concentrating on assimilating satellite data into numerical models to provide a four-dimensional view of the ocean.

When coupled together, advances in satellite remote sensing and in-water measurements of the ocean show a bewildering degree of variability on all scales. The days of the smooth, well-behaved map of the world ocean are no more. More important, this variability is a key element of ocean dynamics, both physically and biologically. We cannot hope to understand, let alone predict, the ocean without an adequate picture of this variability. Although the tools of oceanographic research have significantly improved our understanding, they may also serve to hinder this process. If we step back and look at the total picture of ocean observation, we can discern two trends. First, more attention is being paid to large-scale surveys from both satellite sensors and in-water autonomous samplers, such as moorings and drifters. Second, measurement and sampling techniques have given us the ability to measure processes at finer and finer scales, such as nutrient uptake, vertical distributions, and pigment structure. Both trends include state (standing stock, abundances) and dynamic (growth, nutrient uptake) variables; but in general, small-scale surveys focus on intensive rate measurements and large-scale surveys focus on extensive, standing stock measurements. As Harris (1986) has pointed out, we have generally assumed that processes occurring at the two ends of the spectrum are in equilibrium. However, it is clear that this is not the case.

The danger is that the two trends in measurement techniques may become more separated. Increasingly detailed physiological measurements may improve the accuracy and precision of our observations, but this is only part of the overall error of the measurements. The other source of error arises from undersampling (or, worse, biased sampling) and missing key components of variability. The large-scale surveys sacrifice this technical precision by gathering large quantities of data at a wide range of temporal and spatial scales. The initial experiences with fluorometers generated an enormous body of literature detailing the failures and pitfalls of the use of fluorescence to infer phytoplankton biomass. However, these studies generally ignored the advantages of the fluorescence method in collecting data at scales that traditional bottle samples could not hope to resolve. Somehow we must make the linkage between large- and small-scale processes. We must decide which processes must be resolved and at what scales. It appears that modeling may provide a way out of this dilemma; rather than develop models after collecting data, more attempts must be made to use models to guide sampling and measurement strategies. Such an integration of modeling and observation is only beginning in physical oceanography; it is even more critical in biological studies.

CONCLUSIONS

We have now come full circle, after beginning with a case study of the productivity paradox of the Southern Ocean. It is clear that small-scale dynamics play a critical role in larger-scale processes. The response by the phytoplankton is not merely passive; rather, the spectrum of environmental fluctuations can influence the types of responses utilized by phytoplankton. Thus the proximate causes influencing present-day distributions are consistent with the ultimate causes that occur on evolutionary scales. Variability is clearly ubiquitous at all scales; the challenge is to apply our observation techniques appropriately, striking a balance between extensive, less accurate measurements and intensive, more accurate measurements. Patchiness is more than a mere nuisance; it is a vital resource in the planktonic ecosystem.

ACKNOWLEDGMENTS

Many of the ideas presented here have been developed over the years through discussions with my colleagues in oceanography and limnology. I especially wish to acknowledge Ken Denman, Marlon Lewis, Graham Harris, and Tom Powell. This work was supported by grants from the National Aeronautics and Space Administration and the Office of Naval Research.

REFERENCES

Abbott, M.R., and D.B. Chelton. 1991. Advances in passive remote sensing of the ocean. *Rev. Geophys.*, suppl., *U.S. Nat. Rept.*, 571–589.

Banse, K. 1991a. Iron availability, nitrate uptake and exportable new production in the subarctic Pacific. *J. Geophys. Res.* 96:741–748.

_____. 1991b. Iron nitrate uptake by phytoplankton, and mermaids. *J. Geophys. Res.* 96:20,701.

Bishop, J.K.B. 1986. The correction and suspended material calibration of Sea Tech transmissometer data. *Deep Sea Res.* 33:121–134.

Broecker, W.S. 1990. Comment on "Iron deficiency limits phytoplankton growth in Antarctic waters" by John H. Martin et al. *Global Biogeochem. Cycles* 4:3–4.

Cassie, R.M. 1959. Micro distribution of plankton. *N. Z. J. Science* 2:398–409.

Chamberlin, W.S., C.R. Booth, D.A. Kiefer, J.H. Morrow, and R.C. Murphy. 1990. Evidence for a simple relationship between natural fluorescence, photosynthesis, and chlorophyll in the sea. *Deep Sea Res.* 37:951–973.

Cowles, T.J., R.A. Desiderio, J.N. Moum, M.L. Myrick, and S.M. Angel. 1990. Fluorescence microstructure using a laser/fiber optic profiler. *Proc. SPIE-Int. Soc. Opt. Eng.* 1302:336–345.

Craig, H., and T. Hayward. 1987. Oxygen supersaturation in the ocean: Biological versus physical conditions. *Science* 235:199–202.

Cullen, J.J., and M.R. Lewis. 1988. The kinetics of algal photoadaptation in the context of vertical mixing. *J. Plankton Res.* 10:1039–1063.

Denman, K.L. 1976. Covariability of chlorophyll and temperature in the sea. *Deep Sea Res.* 23:539–550.

Denman, K.L., and T. Platt. 1976. The variance spectrum of phytoplankton in a turbulent ocean. *J. Mar. Res.* 34:593–601.

Denman, K.L., and T.M. Powell. 1984. Effects of physical processes on planktonic ecosystems in the coastal ocean. *Oceanogr. Mar. Biol. Ann. Rev.* 22:125–168.

Denman, K.L., A. Okubo, and T. Platt. 1977. The chlorophyll fluctuation spectrum in the sea. *Limnol. Oceanogr.* 22:1033–1038.

Dickey, T.D. 1991. The emergence of concurrent high-resolution physical bio-optical measurements in the upper ocean and their applications. *Rev. Geophys.* 29:383–413.

Duce, R.A., et al. 1991. The atmospheric input of trace species to the world ocean. *Global Biogeochem. Cycles* 5:193–259.

Dugdale, R.C., and J.J. Goering. 1967. Uptake of new and regenerated forms of nitrogen in primary productivity. *Limnol. Oceanogr.* 12:196–206.

Dugdale, R.C., and F.P. Wilkerson. 1989. Regional perspectives in global new production. In: M.M. Denis (ed.). *Oceanologie Actualite et Prospective.* Centre d'Ocean. Marseille, France, p. 289.

_____. 1990. Iron addition experiments in the Antarctic: A reanalysis. *Global Biogeochem. Cycles* 4:13–19.

Eppley, R.W., and B.W. Peterson. 1979. Particulate organic matter flux and planktonic new production in the deep ocean. *Nature* 282:677–680.

Falkowski, P. (ed.). 1980. *Primary Productivity in the Sea.* Plenum, New York, 531 pp.

Falkowski, P., and D.A. Kiefer. 1985. Chlorophyll a fluorescence in phytoplankton: Relationship to photosynthesis and biomass. *J. Plankton Res.* 7:715–731.

Fu, L.-L., W.T. Liu, and M.R. Abbott. 1990. Satellite remote sensing of the ocean. In: B. Le Mehaute and D.M. Hanes (eds.). *The Sea,* vol. 9. Wiley-Interscience, New York, pp. 1193–1236.

Goldman, J.C., J.J. McCarthy, and D.G. Peavey. 1979. Growth rate influence on the chemical composition of phytoplankton in oceanic waters. *Nature* 279:210–215.

Gordon, H.R., O.B. Brown, R.H. Evans, J.W. Brown, R.C. Smith, K.S. Baker, and D.K. Clark. 1988. A semi-analytic radiance model of ocean color. *J. Geophys. Res.* 93:10,909–10,924.

Harris, G.P. 1978. Photosynthesis, productivity and growth: The physiological ecology of phytoplankton. *Arch. Hydrobiol. Beih. Ergeb. Limnol.* 10:1–171.

_____. 1986. *Phytoplankton Ecology: Structure, Function, and Fluctuation.* Chapman and Hall, New York, 384 pp.

Hayward, T.L. 1987. The nutrient distribution and primary production in the central North Pacific. *Deep Sea Res.* 34:1593–1627.

Holm-Hansen, O., C.J. Lorenzen, R.W. Holmes, and J.D.H. Strickland. 1965. Fluorometric determination of chlorophyll. *J. Cons. Perm. Int. Explor. Mer.* 30:3–15.

Hood, R.R., M.R. Abbott, and A. Huyer. 1991. Phytoplankton and photosynthetic light response in the coastal transition zone off northern California in June 1987. *J. Geophys. Res.* 96:14,769–14,780.

Kiefer, D.A., W.S. Chamberlin, and C.R. Booth. 1989. Natural fluorescence of chlorophyll a: Relationship to photosynthesis and chlorophyll concentration in the western South Pacific gyre. *Limnol. Oceanogr.* 34:868–881.

Kirk, J.T.O. 1983. *Light and Photosynthesis in Aquatic Ecosystems.* Cambridge Univ. Press, New York, 401 pp.

Koblentz-Mishke, O.J., V.V. Volkovinsky, and J.G. Kabanova. 1970. Plankton primary production of the world ocean. In: W.S. Wooster (ed.). *Scientific Exploration of the Southern Pacific*. National Academy of Science, Washington, D.C., pp. 183–193.

Lande, R., and M.R. Lewis. 1989. Models of photoadaptation and photosynthesis by algal cells in a turbulent mixed layer. *Deep Sea Res.* 36:1161–1175.

Lewis, M.R., M.-E. Carr, G.C. Feldman, W. Esaias, and C. McClain. 1990. Influence of penetrating solar radiation on the heat budget of the equatorial Pacific Ocean. *Nature* 347:543–545.

Levins, R. 1979. Coexistence in a variable environment. *Amer. Nat.* 114:765–783.

Lorenzen, C.J. 1966. A method for the continuous measurement of in vivo chlorophyll concentration. *Deep Sea Res.* 13:223–227.

Mackas, D.L., K.L. Denman, and M.R. Abbott. 1985. Plankton patchiness: Biology in the physical vernacular. *Bull. Mar. Science* 37:652–674.

Martin, J.H., R.M. Gordon, S.E. Fitzwater, and W.W. Broenkow. 1989. VERTEX: Phytoplankton/iron studies in the Gulf of Alaska. *Deep Sea Res.* 36:649–681.

Martin, J.H., S.E. Fitzwater, and R.M. Gordon. 1990. Iron deficiency limits phytoplankton growth in Antarctic waters. *Global Biogeochem. Cycles* 4:5–12.

_____. 1991. We still say iron deficiency limits phytoplankton growth in the subarctic Pacific. *J. Geophys. Res.* 96:20,699–20,700.

Platt, T. 1972. Local phytoplankton abundance and turbulence. *Deep Sea Res.* 19:183–187.

_____. 1984. Primary productivity in the central North Pacific: Comparison of oxygen and carbon fluxes. *Deep Sea Res.* 31:1311–1319.

Platt, T., C.L. Gallegos, and W.G. Harrison. 1980. Photoinhibition of photosynthesis in natural assemblages of marine phytoplankton. *J. Mar. Res.* 38:687–701.

Platt, T., W.G. Harrison, M.R. Lewis, W.K.W. Li, S. Sathyendranath, R.E. Smith, and A.F. Vezina. 1989. Biological production of the oceans: The case for a consensus. *Mar. Ecol. Prog. Ser.* 52:77–88.

Powell, T.M., et al. 1974. Spatial scales of current speed and phytoplankton biomass fluctuations in Lake Tahoe. *Science* 189:1088–1090.

Reid, J.L., and E. Shulenberger. 1986. Oxygen saturation and carbon uptake near 28°N, 155°W. *Deep Sea Res.* 33:267–271.

Sathyendranath, S., A.D. Gouveia, S.R. Shetye, P. Ravindran, and T. Platt. 1991. Biological control of surface temperature in the Arabian Sea. *Nature* 349:54–56.

Shulenberger, E., and J.L. Reid. 1981. The Pacific shallow oxygen maximum, deep chlorophyll maximum, and primary productivity, reconsidered. *Deep Sea Res.* 28:901–919.

Siegel, D.A., T.D. Dickey, L. Washburn, M.K. Hamilton, and B.G. Mitchell. 1989. Optical determination of particulate abundance and production variations in the oligotrophic ocean. *Deep Sea Res.* 36:211–222.

Smith, R.C., K.J. Waters, and K.S. Baker. 1991. Optical variability and pigment biomass in the Sargasso Sea as determined using deep-sea optical mooring data. *J. Geophys. Res.* 96:8665–8686.

Spinrad, R.W. 1986. A calibration diagram of specific beam attenuation. *J. Geophys. Res.* 91:7761–7764.

Tans, P.P., I.Y. Fung, and T. Takahashi. 1990. Observational constraints on the global atmospheric CO_2 budget. *Science* 247:1431–1438.

Turpin, D.H., and P.J. Harrison. 1979. Limiting nutrient patchiness and its role in phytoplankton ecology. *J. Exp. Mar. Biol. Ecol.* 39:151–166.

_____. 1980. Cell size manipulation in natural marine, planktonic, diatom communities. *Can. J. Fish. Aquat. Sci.* 37:1193–1195.

Turpin, D.H., J.S. Parslow, and P.J. Harrison. 1981. On limiting nutrient patchiness and phytoplankton growth: A conceptual approach. *J. Plankton Res.* 3:421–431.

Yentsch, C.M., and C.S. Yentsch. 1984. Emergence of optical instrumentation for measuring biological properties. *Oceanogr. Mar. Biol. Ann. Rev.* 22:55–98.

Young, R.W., K.L. Carder, P.R. Betzer, D.K. Costello, R.A. Duce, G.R. DiTullio, N.W. Tindale, E.A. Laws, M. Uematsu, J.T. Merrill, and R.A. Feely. 1991. Atmospheric iron inputs and primary productivity: Phytoplankton responses in the North Pacific. *Global Biogeochem. Cycles* 5:119–134.

4
MEASURING THE FATE OF PATCHES IN THE WATER: LARVAL DISPERSAL

Mimi A. R. Koehl, Thomas M. Powell, and Geoff Dairiki

INTRODUCTION

The aquatic contributions to this volume have focused on the dynamics of either patches on the substratum (benthic communities) or patches in the water column (pelagic communities). This contribution couples these two foci, exploring the importance to benthic organisms of the dynamics of patches in the water. We will outline benthic processes that depend on the transport of materials in the water column. We will describe an empirical technique for quantifying the mixing and transport of patches in nature, and will present some examples of our measurements for wave-swept rocky shores (habitats important in ecological research but difficult to study hydrodynamically). Then we will discuss how future modeling efforts might incorporate these findings.

IMPORTANCE OF THE DYNAMICS OF PATCHES IN THE WATER COLUMN TO BENTHIC ORGANISMS

Benthic organisms often depend on the water moving around them for the transport to their vicinity of dissolved substances (e.g. gases, nutrients) and particulate food, for the removal of wastes, for the transport of gametes (of spawners), and for the dispersal and subsequent return to the shore of propagules (e.g., planktonic larvae, spores). Therefore, if a patch of water is depleted of resources by a benthic organism, or is filled with larvae, gametes, or waste products released by the creature, the fate of that patch can have important consequences for the organism and its neighbors. Such mass transport for organisms on wave-swept rocky shores is poorly understood, in spite of the extensive use of rocky-shore communities for basic ecological research and of the role water movement can play in the ecologically important processes listed above.

Rather than address all of the above processes, let us focus on larval transport. A number of recent reviews of the topic are available (e.g., Norcross and Shaw 1984, Scheltema 1986, Levin 1990, Okubo 1992), so we only mention a few examples here. The supply of larvae is an important factor affecting benthic community structure in some coastal sites (e.g., Bernstein and Jung 1979, Connell 1985, Gaines and Roughgarden 1985, Roughgarden et al. 1987, 1988). Water movement can have important effects on the population dynamics of benthic species with planktonic larval stages, as described by, e.g., Jackson (1986) and Possingham and Roughgarden (1990). Moreover, the transport of larvae by moving water can have significant evolutionary consequences because it is an important determinant of gene flow between populations of marine organisms (e.g., Jackson 1974, Burton and Feldman 1982, Burton 1983, Hedgecock 1979, 1982; Hedgecock et al. 1982, Palumbi 1992). The relationship of larval dispersal to the evolution of life-history strategies of marine organisms has also been considered (e.g., Strathmann, 1974, 1980, 1985; Obrebski 1979; Jackson and Strathmann 1981; Palmer and Strathmann 1981; Jablonski and Lutz 1983).

The role of water motion in transporting marine larvae has been investigated over large and small spatial scales. The effects of oceanic currents on global distributions of species with long-lived larvae have been studied (e.g., Scheltema 1971, 1975, 1986; Scheltema and Carlton 1984). The consequences of coastal circulation (on spatial scales of one to tens of kilometers) upon geographic distributions of various organisms with planktonic larvae and on their retention in bays and estuaries have also been investigated (e.g., Wood and Hargis 1971, Rothlisberg 1982, Cronin and Forward 1982, Sulkin and van Heukelem 1982, DeWolf 1983, Rothlisberg et al. 1983, Levin 1983, Tegner and Butler 1985, Bakun 1986, Emlet 1986, Johnson et al. 1986, Roughgarden et al. 1987, 1988; Boicourt 1988). In addition, the role of internal waves and fronts in cross-shelf transport (spatial scale of tens of meters to kilometers) of larvae has also been considered (Zeldis and Jillet 1982, Shanks 1983, 1986; and Kingsford and Choat 1986). Moreover, water flow on very small spatial scales (mm's to cm's) has been shown to be important to larval settlement (e.g., Crisp 1955, Eckman 1983, 1987; Hannan 1984; Jumars and Nowell 1984; Nowell and Jumars 1984; Butman 1987; Pawlick 1991).

Water motion in nearshore environments on the spatial scale of meters (i.e., 0.1 m to 10 m) connects effects at large and small scales, providing the transport between the small-scale near-substratum flow and the larger-scale circulation patterns whose effects on larval distributions have been studied already. Nonetheless, mass transport on the meter scale along wave-swept rocky shores, and its consequences for larval transport, have received little attention from biologists. Similarly, physical studies of mechanisms of mass transport in the ocean (such as dispersal of pollutants (e.g., Myers and Harding 1983), or transport of beach sand (e.g., Bascom 1980, Basco 1983) have not focused on the spatial scale of meters near wave-beaten rocks.

It is not surprising that mass transport on the spatial scale of meters has not been studied for wave-swept rocky shores, in spite of the ecological importance of such habitats. Water flow at such sites is complicated by the interaction of waves, tidal currents, and complex topography (e.g. Koehl 1977, 1982, 1984, 1986; Denny 1988), and waves crashing onto rocks provide a very hostile environment for current meters, drifters, and other oceanographic instrumentation. We will present here a simple dye-tracking technique that can be used to measure mass transport in this challenging habitat.

TECHNIQUE FOR MEASURING MIXING AND MOVEMENT OF A PATCH OF WATER

All sufficiently small deformable bodies, including parcels of fluid, have only three possible modes of motion (Sommerfeld 1964). First, the body can translate (i.e. the center of mass of the body moves). Second, the body can rotate (e.g. about its center of mass), and third, the body can expand or contract in three mutually orthogonal directions. One can observe directly these three modes of motion in a fluid by labeling a parcel of the fluid with some marker, such as a visible dye like flourescein, and recording the evolution of the "dye blob" with, for example, a sequence of photographs. The rate of horizontal movement of the center of mass of the fluid blob is commonly called advection by oceanographers and meteorologists. The rate of rotation of the fluid parcel is half its vorticity. The spread of the fluid parcel as it mixes with the adjacent environment is called diffusion -- in this case turbulent diffusion because the molecular processes causing diffusion are much smaller than the mixing effects of fluid turbulence. Such labeling with dye and quantitative tracking (say, via photography) has a venerable history in engineering and oceanography (e.g. Pritchard and Carpenter 1960, Okubo 1971) and provides information about fluid motion similar to that obtained from studies using instrumentation like drifters or current meters (e.g. Lemmin 1991). Note that the tracking of dye is not different in concept from the tracking of features on satellite images to obtain motions of fluid at much larger spatial scales (e.g. Emery et al. 1986, Strub and Powell 1987, and Tokmakian et al. 1990).

Advection, vorticity, and turbulent diffusion rates (on spatial scales of meters to tens of meters) can be measured using the photographic technique reported by Koehl et al. (1987, 1988). A volume of flourescein dye is released at one instant into the water, and photographs of the dye are taken from a fixed reference point at timed intervals. Dye solution (made using water from the habitat) can be released via a port affixed to the substratum, by gentle pouring, or via rupture of a balloon containing dye; alternatively, a small amount of flourescein powder can be released directly into the water. The expanding blob of dye is photographed at timed intervals with a camera mounted above the water's surface. The tripod on which the camera is mounted is outfitted with shaft encoders that indicate the horizontal and vertical angles of the camera. These angles, as well as the time that each photograph is taken, are recorded on a lap-top computer. [For those without such equipment, the angle of the camera with the horizontal can be measured using a line level and protractor (Fig. 1,A). The camera must remain unmoved during the sequence of photographs, and the times at which each photo are taken must be recorded by hand. The photographs should include a reference landmark, such as a rock or post, that does not move so that successive photographs can be superimposed.] A photograph must also be taken from each camera location of a size scale on the surface of the water.

The photographs of the dye blobs are projected onto a digitizing tablet and the perimeter of each successive blob is traced. The size scale, reference landmark, and camera angle data are used to correct the coordinates of each point in the blob for parallax (Figs. 1,B and 1,C). By making the simplifying assumption that the dye is evenly dispersed within the blob (see below), we calculate the position of the centroid of the blob at each time (t_1, t_2,..., t_n). If Δs_n is the distance the centroid has traveled in the nth time interval (i.e., between t_{n-1} and t_n), then the advective speed (U_n) for the nth time interval is calculated as:

$$U_n = (\Delta s_n)/(t_n - t_{n-1}) \tag{1}$$

as illustrated in Figure 2.

Using Figure 2 as a concrete example, we see that a blob not only spreads, but also rotates about its centroid. If one calculated the rate of spread of the blob with respect to fixed (i.e., non-rotating) axes, then the effects due to expansion could not be separated from those due to rotation. We need a measure of the rate of expansion of the blob that is not confounded by any rotations the blob may have undergone. One method of removing the effects of rotation is to calculate the spread about the two principal axes of the blob. The first principal axis, which we denote by "a", is the direction of greatest spread; the second principal axis, "b", is perpendicular to the first. The identification of principal axes of a solid is a common calculation in rigid body dynamics, described in all physics and engineering mechanics texts (e.g. Marion 1967). A simple and practical way to perform these calculations is to use principal component analysis (PCA; e.g. Chatfield and Collins 1980). The algorithms for PCA are widely available in statistical packages for personal computers. Preisendorfer (1988, Chapter 2) presents a very

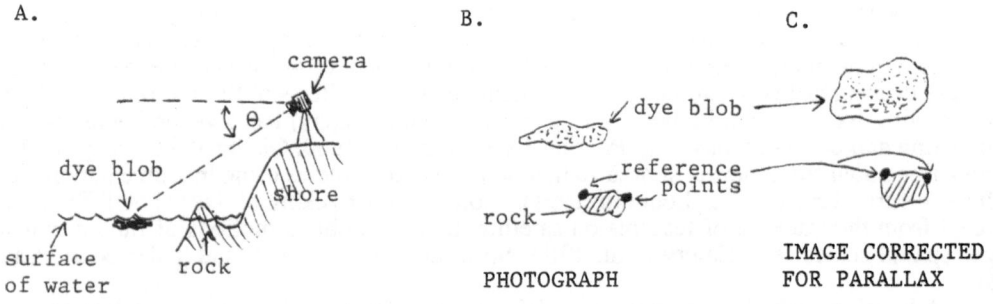

Figure 1. A. Diagram of a camera on the shore used to photograph a blob of dye in the water. The angle between the line of view of the camera and the horizontal (θ) is used in calculating the corrections for parallax. B. Diagram of the photograph taken by the camera in A. Points fixed with respect to the shore (indicated by black dots) are used to align photographs of the same blob taken at later times. C. Diagram of the blob shown in B after correction for parallax.

53

clear discussion of the use of the technique in a simple two-dimensional problem like the one we consider here, and his results can be readily adapted to our dye blob calculation. Using the terminology of PCA, the eigenvalues associated with these two directions (i.e. the principal axes) measure the variance (i.e. the spread) along these two axes. We call these two eigenvalues σ_a^2 and σ_b^2; σ_a is the best description (in a statistical sense) of the linear dimension of the blob along its direction of greatest spread, and σ_b is that for the perpendicular direction. We further define instantaneous diffusion coefficients (K, diffusivities) for the spread of dye along the first (K_a) and second (K_b) principal axes, respectively, by:

$$K_a = (1/2)d\sigma_a^2/dt$$

$$K_b = (1/2)d\sigma_b^2/dt.$$

(2)

In our data analysis we approximate the instantaneous time derivatives above with finite differences:

$$d\sigma^2/dt \approx (\sigma^2_n - \sigma^2_{n-1})/(t_n - t_{n-1})$$

(3)

as illustrated in Fig. 3,C. In addition, we determine overall diffusivities (K_A and K_B) for the entire sequence of photographs for a dye blob by calculating linear regressions for plots of σ_a^2 and σ_b^2 versus time (Fig. 3,A). Note that K_a and K_b (or K_A and K_B) relate the flux of some quantity (in this case, dye) to the concentration gradients of that quantity (in this case, again, gradients in dye concentration). Symbolically, one expresses this relationship (called Fick's Law) by

$$F_a = K_a \, \partial C/\partial x_a$$

(4)

F_a is the flux of dye in the "a" direction (units of F_a: grams of dye per unit area per unit time), and C is the concentration of dye (units of C: grams of dye per unit volume). Note that the units of K_a are (length^2)/(time); see Okubo (1980, Chapter 2).

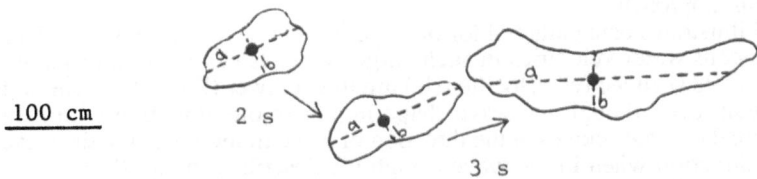

Figure 2. Tracings (corrected for parallax) of the perimeters of a dye patch at three successive times at Toad Point, Tatoosh Island, WA, a wave-swept rocky shore site. The time intervals between tracings are indicated next to the arrows between the tracings. The dark dots indicate the position of the centroid of the dye patch at each time, and the dashed lines indicate the first principal axis ("a", the longer line) and the second principal axis ("b", the shorter line perpendicular to "a" at each time). Note that the blob translates, expands, and rotates slightly.

Our description of the dye-tracking technique has been brief, and we cannot in the space allotted in this volume address all the methodological concerns a reader might have about the reliability of the approach. However, a few questions do demand at least cursory attention. First, the water flows in the rocky shore environments we have studied are decidedly three-dimensional, while the photographic dye-tracking technique gives only a two-dimensional view of the transport. By measuring the concentration of dye in water samples collected at various depths at the same times that photographs of dye blobs were taken, we found that vertical mixing of the dye was very rapid at a shallow site (depth = 1 to 2 m, at Hopkins Marine Station, Pacific Grove, CA; unpubl. data gathered in collaboration with M. Denny). There was no detectable difference between the dye concentrations near the surface and near the bottom within a few seconds of the time of dye release. Accordingly, the 2-D aerial views recorded via photography give a good description of the temporal evolution of the fully three-dimensional patch at such a shallow, wave-swept site. At deeper sites where vertical mixing is of concern, an anchored SCUBA diver or camera can photograph dye blobs from the side, thereby recording the vertical and (one direction of) the horizontal transport and spread of the blob. Such photographs can be analyzed using the technique described above, as we have done for some coral reef sites near Discovery Bay, Jamaica (Koehl and Jackson unpubl. data). A second concern about the dye-tracking technique is whether the transport of a dissolved substance (dye) is a good indicator of the transport of particles in the water, such as larvae, spores, or food. We have addressed this question by simultaneously releasing dye and particles of various sinking or rising velocities in a wave-swept surge channel on Tatoosh Island, WA. Concentrations of dye and particles in water samples taken at stations various distances (0 to 30 m) from the release point show that dye and particles travel together and that their rates of change of concentration at a point are indistinguishable. Hence, the movement of dye is a good marker for the passive movement of particles suspended in the water column, like larvae. At sites with more gentle water motion, however, the dye-tracking technique can only describe the transport of water and dissolved substances, but not that of organisms that sink, rise, or swim rapidly compared with the mixing and advection of the water. Finally, although we have mentioned the possible application of these techniques to calculation (and interpretation) of the changes in vorticity in the fluid as the patch evolves, we defer any discussion of this potentially interesting topic to later articles.

EXAMPLES OF MEASUREMENTS OF WATER TRANSPORT AT WAVE-SWEPT ROCKY SHORES

We have used the technique described above to measure advective transport and turbulent mixing at a variety of coastal sites. We will report here a few examples of our data for wave-swept rocky shores to illustrate the type of information about patch dynamics that can be obtained using this approach.

Figure 3 illustrates data gathered for one dye run in a wave-swept surge channel. Even though instantaneous water velocities in such surge channels can be high (often exceeding 5 m/s, Koehl 1977, 1984) advective velocities (U) are much lower (Fig. 3,B). The water motion in waves is orbital; even though the wave shape moves in one direction with respect to the substratum, a parcel of water moves in the direction of wave motion when in the wave crest, but in the opposite direction when in the wave trough (as described in e.g. Bascom 1980, Koehl 1984, 1986; Denny 1988). The flow over wave-swept rocks and in surge channels is also basically oscillatory when these environments are subjected to breaking waves, that is, water washes up the shore and then returns seaward (Koehl 1977). Although water moves back and forth, there is some net pumping of water in the direction of wave motion (Stokes drift, as described in e.g. LeMehaute 1976). Water motion at wave-swept shores is further complicated by flow oscillations of longer period, such as edge waves (e.g. LeBlond and Mysak 1978, Holman 1983) and tidal currents. A pattern develops of slow net movement of water towards, along, and away from the shore. Since the pattern of this water transport depends on local topography, the dye-tracking technique is a very useful tool for quantifying net water transport to and from particular sites where benthic ecology is being studied. Although the details of water movement vary from one wave-swept site to another, a general conclusion we can draw from our dye studies is that advective velocities (U) are much lower than instantaneous water velocities. Therefore, although wave-swept habitats appear to be characterized by rapid water movement, this impression is very misleading when considering the net transport of material in such environments.

Our dye blob data also revealed that mixing in these shallow wave-swept habitats is not isotropic. As illustrated in Figure 3,A, the mixing coefficient (K_A) in the direction parallel to the direction of water flow is much greater than the mixing coefficient at right angles to this (K_B). The mechanism responsible for this anisotropy is most likely shear dispersion (for details see e.g. Bowden 1965; Okubo 1968; 1980). Friction with solid boundaries slows nearby water motion, hence velocity gradients (boundary layers) develop in the water between the substratum and the free-stream flow (for further discussion see e.g. Nowell and Jumars 1984). Figures 4,A and B illustrate how such shear in the water can lead to enhanced dispersion along the axis parallel to the freestream velocity. Similarly, pockets of slowly-moving water (e.g. trapped in pools, canopies of organisms, or other topographic irregularities characteristic of complex rocky shores) can also cause anisotropic mixing and slowed advection of patches of water-borne material, as diagrammed in Figures 4,C and D, and described by Okubo (1973).

Our dye studies at a variety of coastal sites have not only shown that assumptions of high transport and isotropic diffusion are not correct for wave-swept shores, but have also revealed that U and K for a particular spot are very sensitive to time in the tidal cycle and to weather (Koehl et al. 1987, 1988). In addition, we have found that U's and K's for different sites within the same kilometer stretch of shoreline can differ from each other by orders of magnitude due to the influence of local topography and macrophyte canopies (Koehl et al. 1987; 1988).

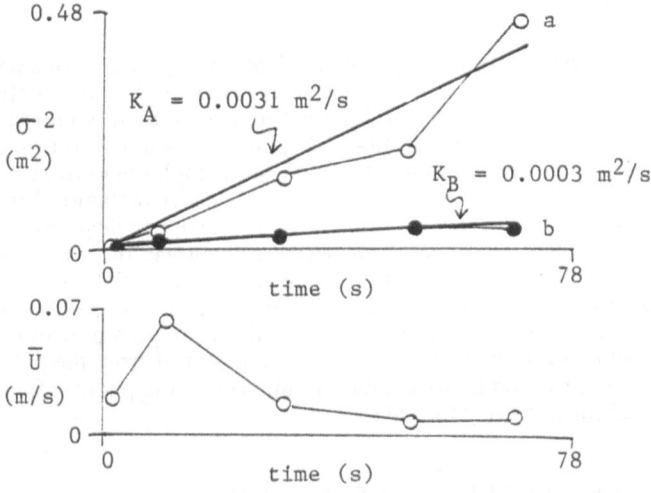

Figure 3. Example of measurements of mass transport in a surge channel on Tatoosh Island, WA, using the dye-tracking technique described in the text. A. The eigenvalues, σ_a^2 and σ_b^2, (measures of spread of the patch) along the first (open circles) and second (dark circles) principal axes of the dye blob, plotted as functions of time. The straight lines represent the least-squares, best-fit, straight lines (i.e., linear regressions) calculated through these points. The slopes give the overall diffusivities (K_A and K_B) along the two principal axes (parallel to and perpendicular to the direction of greatest spread, respectively, of the blob). B. The speed of advection of the centroid of the blob, plotted as a function of time.

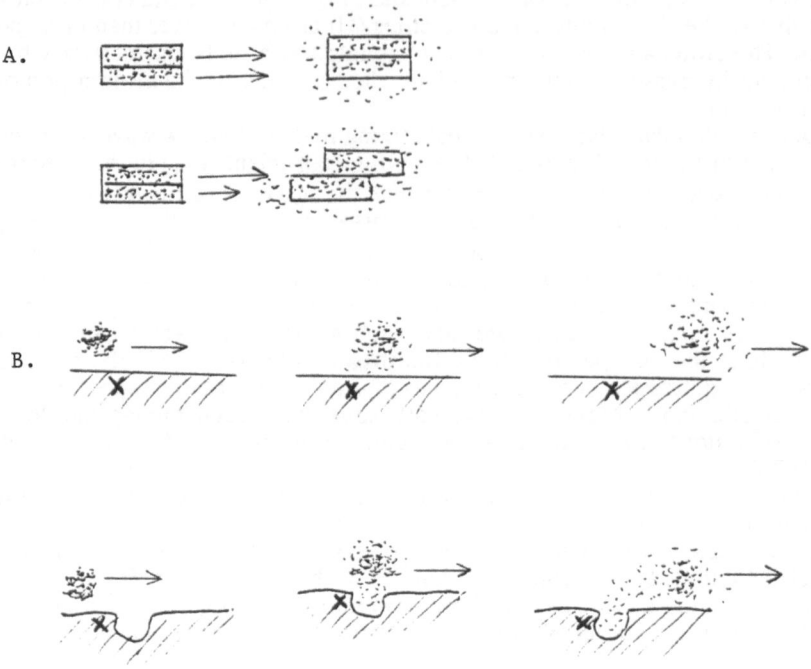

Figure 4. Diagrams of two mechanisms by which anisotropic spreading of a patch of water can occur near a shore. Both mechanisms result in slowed advection of the centroid of the patch because the rear end of the patch is retained for a longer time than it would be if there were no interaction of the patch with the substratum. A patch of water-borne material (e.g., dye molecules) is indicated by the dots, and the direction of the ambient current is indicated by the arrow. A. Diagram of two adjacent slabs of water (indicated by rectangles) whose positions at time 1 are shown on the left, and at time 2 on the right. The upper diagram illustrates adjacent layers of water between which there is no shear, whereas the lower diagram illustrates a layer of water moving more slowly than the layer above it (as would occur, for example, near the substratum). B. Diagram of a patch of material moving past a shore (hatched area) that is smooth (upper diagram) versus a shore with an irregularity (e.g., a pocket of slowly-moving water in a hole, a plant canopy, etc., - lower diagram). Material from the patch mixes into the pocket as the patch moves by, and then continues to mix out of the pocket after the leading edge of the patch has moved further along the shore.

MODELS OF THE DYNAMICS OF PATCHES OF LARVAE

A number of mathematical models have been developed to address various ecological questions that depend on the transport of patches of larvae, as reviewed by Okubo (1992). Most of these deal with populations of pelagic organisms (e.g. fish), and incorporate advection, isotropic diffusion, growth, and sometimes mortality terms. Several models have also been developed for benthic populations. For example, Jackson and Strathmann (1981) considered the role that mixing played in determining the recruitment of larvae into the benthic stages. They did not consider advection or anisotropic diffusion. Further, Possingham and Roughgarden (1990), in an advective-diffusive model of larval dispersal and settlement, allowed absorbing boundaries (to simulate settlement) and/or reflective boundaries at the shore, in addition to mortality and space limitation on the shore. Diffusion was again assumed to be isotropic. In a model that simulated details of turbulence in the surf zone, Denny and Shibata (1989) examined the effect that vertical turbulent mixing has upon the delivery of the larvae of benthic organisms to the substratum, but they ignored horizontal mixing and advection. However, using Csanady's (1973) turbulent plume formulation, they did consider three-dimensional transport in their model of the fertilization success of eggs and sperm spawned by benthic organisms. Finally, in studies of the dispersal of coral larvae along barrier reefs (Sammarco and Andrews 1988, 1989), Andrews et al. (1989) used a detailed numerical circulation model to study the role that transport plays over both ecological and evolutionary time scales.

In an earlier section of this article we noted that the role of larval transport via water movement has been shown to be important in at least four ecological/evolutionary considerations: 1) the structure of benthic communities; 2) the dynamics of benthic populations that have a planktonic stage; 3) gene flow between such populations of marine organisms; and 4) the evolution of (benthic) life-history strategies. This list is certainly not exhaustive. The models we cited in the previous paragraph (and the many more that we did not cite, e.g. Okubo 1992) are the initial attempts to assess how the details of water flow affect these important ecological and evolutionary considerations. [N.b., although the effects of water motion on population structure have been modeled, so far as we can ascertain no model exists that incorporates the role that water flow plays in determining benthic community structure, though there can be no doubt of its importance. See, e.g., Roughgarden et al. 1985, or Roughgarden et al. 1988.] All of the model studies noted above, though by no means an exhustive survey, concluded that transport of any kind -- advective, diffusive, turbulent, etc. -- played a substantial role, often the dominant role, in the biological processes under investigation. But few of the (benthic) models incorporated even moderately realistic advection and diffusion characteristics [Andrews's (1989) use of a detailed numerical model is, perhaps, an exception]. One reason for this retreat from realism is the complication that it brings to the modeler's efforts. However, another more important reason is the sad recognition that very few measurements of the relevant transport parameters exist. This state of affairs could change, and should change, rapidly. The dye release techniques we sketched in this article are so simple that they could readily be incorporated into many biological field investigations. In this way important estimates of transport (coarse estimates, to be sure) could accompany the detailed biological data collected by investigators. Furthermore, many such studies in the nearshore, both observational and modeling, could benefit substantially from the addition of realistic transport estimates.

CONCLUSION

The shore environment is a complex mosaic with great variability in transport from site to site over small spatial scales, and from time to time, (even over hours, and certainly over a tidal cycle). Our measurements lead us to the "common sense" conclusion that such spatial and temporal variability can be of considerable ecological and evolutionary importance. We argue that the dye tracking technique provides a simple, quantitative method of classifying various habitats as to their actual transport characteristics. Moreover, the results can be incorporated into the present generation of models of benthic populations and communities.

ACKNOWLEDGMENTS

This research was supported by NSF grant #OCE-8717028 (to M.A.R.K.) and NSF grant #OCE-8717678 (to T.M.P.). We are grateful to T. L. Daniel and M. W. Denny for collaborating on some aspects of this research, to F. Dardis, S. Distefano, J. Ishimoto, K. Lynn, and S. Zaret for technical assistance, and to A. Okubo for helpful discussions. We thank R. T. Paine for facilitating our work on Tatoosh; research at Tatoosh is possible only with permission of the Makah Tribal Council and the U. S. Coast Guard.

REFERENCES

Bakun, A. 1986. Local retention of planktonic early life stages in tropical demersal reef/bank systems: The role of vertically structured hydrodynamic processes. *Workshop on Recruitment in Tropical Coastal Demersal Communities*. Cuidad del Carmen, Mexico.

Basco, D. R. 1982. Surf-zone currents. *Coastal Eng.* 7:331-355.

Bascom, W. 1980. *Waves and Beaches*. Anchor Press/Doubleday, Garden City, NJ.

Bernstein, B. B., and N. Jung. 1979. Selective pressures and coevolution in a kelp canopy community in southern California, *Ecol. Monogr.* 49:335-355.

Boicourt, W. C. 1988. Estuarine larval retention mechanisms on two spatial scales. In: V. S. Kennedy (ed.). *Estuarine Comparisons*. Academic Press, New York, pp. 445-458.

Bowden, K. F. 1965. Horizontal mixing in the sea due to a shearing current, *J. Fluid Mech.* 21:83-95.

Burton, R. S. 1983. Protein polymorphisms and genetic differentiation of marine invertebrate populations. *Mar. Biol. Letters* 4:193-206.

Burton, R. S., and M. W. Feldman. 1982. Population genetics of coastal and estuarine invertebrates: Does larval behavior influence population structure? In: V. S. Kennedy (ed.). *Estuarine Comparisons*. Academic Press, New York, pp. 537-551.

Butman, C. A. 1987. Larval settlement of soft-sediment invertebrates: The spatial scales of pattern explained by active habitat selection and the emerging role of hydrodynamic processes. *Oceanogr. Mar. Biol. Ann. Rev.* 25:113-165.

Chatfield, C., and A.J. Collins. 1980. *Introduction to Multivariate Analysis*. Chapman and Hall, New York

Connell, J. H. 1985. The consequences of variation in initial settlement vs. post-settlement mortality in rocky intertidal communities, *J. Exp. Mar. Biol. Ecol.* 93:11-46, 1985.

Crisp, D. J. 1955. The behavior of barnacle cyprids in relation to water movement over a surface. *J. Exp. Biol.* 32:569-590.

Cronin, T. W., and R. B. Forward. 1982. Tidally timed behavior: Effects on larval distributions in estuaries. In: V. S. Kennedy (ed.). *Estuarine Comparisons*. Academic Press, New York, pp. 505-521.

Csanady, G. T. 1973. *Turbulent Diffusion in the Environment*. Reidel, Boston.

Denny, M.W. 1988. *Biology and the Mechanics of the Wave-swept Environment*. Princeton Univ. Press, Princeton, N.J.

Denny, M. W., and M. Shibata. 1989. Consequences of surf-zone turbulence for settlement and external fertilization. *Amer. Nat.* 134:859-889.

DeWolf, P. 1983. Ecological observations on the mechanisms of dispersal of barnacle larvae during planktonic life and settling. *Neth. J. Sea Res.* 6:1-129.

Eckman, J. E. 1983. Hydrodynamic processes affecting benthic recruitment. *Limnol. Oceanogr.* 28:241-257.

Eckman, J. E. 1987. The role of hydrodynamics in recruitment, growth, and survival of *Argopectin irradians* (L.) and *Anomia simplex* (D'Obrigny) within eel grass meadows. *J. Exp. Mar. Biol. Ecol.* 106:165-192.

Emery, W. J., A. C. Thomas, M. J. Collins, W. R. Crawford, and D. L. Mackas. 1986. An Objective Method for Computing Surface Velocities from Sequential Infrared Satellite Images. *J. Geophys. Res.*: 91:12865-12878.

Emlet, R. B. 1986. Larval production, dispersal, and growth in a fjord: A case study on larvae of the sand dollar *Dendraster excentricus*. *Mar. Ecol. Prog. Ser.* 31:245-254.

Gaines, S., and J. Roughgarden. 1985. Larval settlement rate: A leading determinant of structure in an ecological community of the marine intertidal zone. *Proc. Nat. Acad. Sci. U.S.A.* 82:3707-3711.

Hannan, C. A. 1984. Planktonic larvae may act like passive particles in the turbulent near-bottom flows. *Limnol. Oceanogr.* 29:1108-1116.

Hedgecock, D. 1979. Biochemical genetic variation and evidence of speciation in *Chthamalus* barnacles of the tropical eastern Pacific Ocean. *Mar. Bio.* 54:202-214.

Hedgecock, D. 1982. Genetic consequences of larval retention. In: V. S. Kennedy. *Estuarine Comparisons*. Academic Press, New York, pp. 553-568.

Hedgecock, D., M. L. Tracey, and K. Nelson. 1982. Genetics. In: *Biology of Crustacea, Volume 2*. Academic Press, New York, pp. 284-403.

Holman, R. 1983. Edge Waves and the configuration of the shoreline. In: P. D. Komar (ed.). *CRC Handbook of Coastal Processes and Erosion*. CRC Press, pp. 21-34.

Jablonski, D., and R. A. Lutz. 1983. Larval ecology of marine benthic invertebrates: Paleobiological implications. Biol. Rev. 58:21-89.

Jackson, G. A., and R. R. Strathman. 1981. Larval mortality from offshore mixing as a link between precompetent and competent periods of development. *Amer. Nat.* 118:16-26.

Jackson, J. B. C. 1974. Biogeographic consequences of eurytoppy and stenotropy among marine bivalves and their evoutionary significance. *Amer. Nat.* 105:542-560.

Jackson, J. B. C. 1986. Modes of dispersal of clonal benthic invertebrates: Consequences for species' distributions and genetic structure of local populations. *Bull. Mar. Sci.* 39:588-606.

Johnson, D. E., L. W. Botsford, R, D, Methot, and T. Wainwright. 1986. Wind stress and cycles in Dunginess crab (*Cancer magister*) catch off California, Oregon, and Washgton, *Can. J. Fish. Aquat. Sci.* 43:838-845.

Jumars, P. A., and A. R. M. Nowell. 1984. Fluid and sediment dynamic effects on marine benthic community structure. *Amer. Zool.* 24:45-55.

Kingsford, M. J., and J. H. Choat. 1986. Influence of sruface slicks on the distribution and onshore movements of small fish. *Mar. Biol.* 91:161-171.

Koehl, M. A. R. 1977. Effects of sea anemones on the flow forces they encounter. *J. Exp. Biol.* 69:87-105.

Koehl, M. A. R. 1982. The interaction of moving water and sessile orgnaisms. *Sci. Am.* 247:124-132.

Koehl, M. A. R. 1984. How do benthic organisms withstand moving water? *Amer. Zool.* 24:57-70.

Koehl, M. A. R. 1986. Form and function of macroalgae in moving water. In: T. J. Givnish (ed.). *On the Economy of Plant Form and Function.* Cambridge University Press, Cambridge, pp. 291-314.

Koehl, M. A. R., T. M. Powell, and T. L. Daniel. 1987. Turbulent transport near rocky shores: Implications for larval dispersal. *EOS* 68:1750.

Koehl, M. A. R., T. M. Powell, and T. L. Daniel. 1988. Turbulent transport of marine larvae near rocky shores. *Amer. Zool.* 28:113A.

LeBlond, P.H., and L.A. Mysak. 1978. *Waves in the Ocean.* Elsevier, Amsterdam.

LeMehaute, B. 1976. *An Introduction to Hydrodynamics and Water Waves.* Springer-Verlag, New York.

Lemmin, U. 1989. Dynamics of horizontal turbulent mixing in a nearshore zone of Lake Geneva. *Limnol. Oceanogr.* 34:420-434.

Levin, L. A. 1983. Drift tube studies of bay-ocean water exchange and implications for larval dispersal. *Estuaries* 6:363-371.

Levin, L. A. 1990. A review of methds for labeling and tracking marine invertebrate larvae. *Ophelia* 32:1115-144.

Marion, J.B. 1970. *Classical Dynamics of Particles and Systems.* Academic Press, New York.

Myers, E. P., and E. T. Harding. 1983. *Ocean Disposal of Municipal Wastewater: Impacts on the Coastal Environment.* Seagrant College Program, MIT, Cambridge, MA.

Norcross, B. L., and R. F. Shaw. 1984. Oceanic and estuarine transport of fish eggs and larvae: A review. *Trans. Amer. Fish. Soc.* 113:153-165.

Nowell, A. R. M., and P. Jumars. 1984. Flow environments of aquatic benthos. *Ann. Rev. Ecol. Syst.* 15:303-328.

Obrebski, S. 1979. Larval colonization strategies in marine benthic invertebrates. *Mar. Ecol. Prog. Ser.* 1:293-300.

Okubo, A. 1968. Some remarks on the importance of the "shear effect" on horizontal diffusion. *J. Oceanogr. Soc. Japan* 24:20-29.

Okubo, A. 1971. Oceanic diffusion diagrams. *Deep-Sea Res.* 18:789-802.

Okubo, A. 1973. Effects of shoreline irregularities on streamwise dispersion in estuaries and other embayments. *Neth. J. Sea Res.* 6:213-224.

Okubo, A. 1980. *Diffusion and Ecological Problems: Mathematical Models.* Springer-Verlag, Berlin.

Okubo, A. 1992. The role of diffusion and related physical processes in dispersal and recruitment of marine populations. In: P.W. Sammarco and M. Heron (eds) Marine larval dispersal and recruitment: an interdiscipinary approach. Springer-Verlag, Berlin.

Palmer, A. R., and R. R. Strathmann. 1981. Scale of dispersal in varying environments and its implications for life history of marine invertebrates. *Science* 225:1478-1480.

Palumbi, S. 1992. Marine speciation on a small planet. *TREE* 7:114-118.

Pawlick, J. R., C. A. Butman, and V. R. Starczak. 1991. Hydrodynamic facilitation of gregarious settlement of a eef-building tube worm. *Science* 251:422-424.

Possingham, H., and J. Roughgarden. 1990. Spatial population dynamics of a marine organism with a complex life cycle. *Ecology* 7:973-985.

Preisendorfer, R. W. 1988. *Principal Component Analysis in Meteorology and Oceanography. Developments in Atmospheric Science, Volume 17.* Elsevier, Amsterdam.

Pritchard, D. W., and J. H. Carpenter. 1960. Measurement of turbulent diffusion in estuarine and inshore waters. *Bull. Int. Assoc. Sci. Hydrol.* 20:37-50.

Rothlisberg, P. C. 1982. Vertical migration and its effect on dispersal of penaeid shrimp larvae in the Gulf of Carpenteria, Australia. *Fish. Bull.* 80:541-554.

Rothlisberg, P. C., J. A. Church, and A. M. G. Forbes. 1983. Modelling the advection of vertically migrating shrimp larvae. *J. Mar. Res.* 41:511-538.

Roughgarden, J., S. Gaines, and S. Pacala. 1987. Supply-side ecology: The role of physical transport processes. In: P. Giller and J. Gee (eds.). *Organization of Communities Past and Present. Proc. Brit. Ecol. Soc. Symp., Aberystwyth, Wales.* Blackwell, Oxford.

Roughgarden, J., S. Gaines, and H. P. Possingham. 1988. Recruitment dynamics in complex life cycles. *Science* 241:1460-1466.

Scheltema, R. S. 1971. The dispersal of the larvae of shoal-water benthic invertebrate species over long distances by oceanic currents. In: D. J. Crisp (ed.). *Fourth European Marine Biology Symposium.* Cambridge University Press, Cambridge, pp. 7-28.

Scheltema, R. S. 1975. Relationship of larval dispersal, gene-flow and natural selection to geographic variation of benthic invertebrates in estuaries and along coastal regions. *Estuarine Res.* 1:372-391.

Scheltema, R. S. 1986. On dispersal and planktonic larvae of benthic invertebrates: An eclectic overview and summary of problems. *Bull. Mar. Sci.* 39:290-322.

Scheltema, R. S., and J. T. Carlton. 1984. Methods of dispersal among fouling orgnaisms and possible consequences for range extension and geographic variation. In: J. D. Costlow and R. C. Tipper (eds.). *Marine Biodeterioration: An Interdisciplinary Study.* Naval Institute Press, pp. 127-133.

Shanks, A. L. 1983. Surface slicks associated with tidally forced internal waves may transport pelagic larvae of benthic invertebraates and fishes shoreward. *Mar. Ecol. Prog. Ser.* 13:311-315.

Sommerfeld, A. 1964. *Mechanics of Deformable Bodies.* Academic Press, New York.

Strathmann, R. R. 1974. The spread of sibling larvae of sedentary marine invertebrates. *Amer. Nat.* 108:29-44.

Strathmann, R. R. 1980. Why does a larva swim so long? *Paleobiology* 6:373-376.

Strathmann, R. R. 1985. Feeding and nonfeeding larval development and life history evolution in marine invertebrates. *Ann. Rev. Ecol. Syst.* 16:339-361.

Strub, P. T., and T. M. Powell. 1987. Surface temperature and transport in Lake Tahoe: inferences from satellite (AVHRR) imagery. *Cont. Shelf Res.* 7:1001-1013.

Sulkin, S. D., and W. vanHeukelem. 1982. Larval recruitment in the crab *Callinectes sapidus* Rathbyn: An amendment to the concept of larval retention in estuaries. In: V. S. Kennedy (ed.). *Estuarine Comparisons.* Academic Press, New York, pp. 459-475.

Tegner, M., and R. A. Butler, 1985. Drift-tube study of the dispersal potential of green abalone (*Haliotis fulgen* larvae in the Southern California Bight: Implicaitons for recovery of depleted populations. *Mar. Ecol. Prog. Ser.* 26:73-84.

Tokmakian, R.T., P. T. Strub, and J. McClean-Padman. 1990. Evaluation of the Maxiumum Cross-Correlation Method of Estimating Sea-Surface Velocities from Sequential Satellite Images. *J. Atmos. Oceanic. Tech.* 7:852-865.

Wood, L. and W. J. Hargis, 1971. Transport of bivalve larvae in a tidal estuary. In: D. J. Crisp (ed.). *Fourth European Marine Biology Symposium.* Cambridge University Press, Cambridge, pp. 459-475.

Zeldis, J. R., and J. B. Jillet. 1982. Aggregation of pelagic *Mundia gregaria* (Fabricus)(Decapodia, Anomura) by coastal fronts and internal waves. *J. Plankton Res.* 4:839-857.

5

DETERMINING PROCESS THROUGH PATTERN:
REALITY OR FANTASY?

Kirk A. Moloney

MOTIVATION

Terrestrial ecologists are becoming increasingly aware of the need to understand the dynamics
of ecological systems across a range of spatial and temporal scales (e.g., Wiens 1989, Denslow
1990, Chesson 1990). Part of the reason for this interest is a growing awareness of the large-
scale ecological problems impacting society. For example, global warming, acid rain,
deforestation, and the development of the antarctic ozone hole all involve processes acting over
a very broad range of spatial scales. Unfortunately, we are faced with a fundamental problem in
understanding the relationship between these broad-scale environmental problems and basic
ecological processes. This is due in large part to the approaches traditionally employed by
ecologists.
        In the past, most ecological studies have considered a limited set of conditions—both
spatially and temporally—in characterizing ecological processes. Field studies and experiments
in general have been designed to factor out what is perceived to be "the problem" of spatial and
temporal heterogeneity. Modeling work to a large extent has also been constrained to a
consideration of the dynamics of systems whose underlying parameters are time and space
invariant. Clearly our interests must be broadened to include an explicit consideration of spatial
and temporal heterogeneity in the dynamics of ecological systems if we are to expand our
understanding to encompass processes occurring over a range of scales. Indeed, a widespread
interest in formulating these approaches has developed over the last decade or so (see Wiens
1989; Tuljapurkar 1984; Moloney 1988, 1990; Levin 1989; Chesson 1990; Hobbs and
Mooney 1991; Moloney et al., in press for a few recent examples). In fairness, it should be
pointed out that earlier research exploring the importance of spatial and temporal heterogeneity
in ecological systems has been done (e.g., Cowles 1899, Watt 1948, Davis 1968), however
this approach must be more widely adopted in the future.
        The approach I outline below illustrates some of the concerns mentioned above, and
suggests a possible methodology for exploring the innate spatial variability observed in most
ecological systems. The work discussed characterizes a particular system—an annual grassland
found on serpentine soils at Jasper Ridge, California. This provides a logical framework for
exploring the relationships coupling local ecological dynamics to pattern evolution at a broader
scale, and allows comparisons between model predictions and field data. A comparative
approach yields a more powerful research program than modeling or field studies taken alone
(Moloney et al., in press). (See Hobbs and Mooney (1985) and McNaughton (1968) for a
general introduction to the vegetation structure of the Jasper Ridge grassland).

APPROACH

A major force affecting the distribution of plants at Jasper Ridge, and consequently landscape
pattern, is the widespread activity of gophers within the grassland community (Hobbs and
Mooney 1985, Moloney et al. 1991). Ten to twenty percent of the landscape is affected by the
formation of gopher mounds every year (Hobbs and Mooney 1985; see Figure 1). Mounds

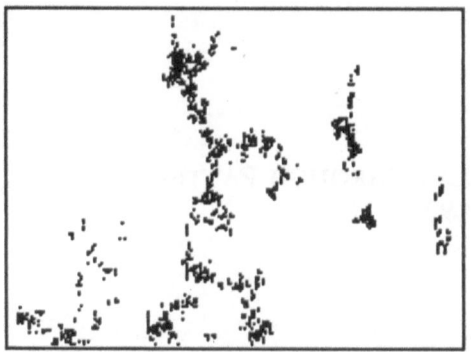

Figure 1. Digitized image of the distribution of newly formed gopher mounds within the serpentine grassland at Jasper Ridge, CA. The image was hand digitized from a false infrared image obtained by low-level flyby during the summer of 1989. The area shown is approximately 45 m x 64 m (1800 x 2544 pixels) in the vertical and horizontal directions, respectively.

destroy all of the local vegetation, and recolonization can occur only through dispersal of seed from an undisturbed site. Clearly, an understanding of the dynamics of the system must revolve, at least to some extent, around an understanding of the spatial and temporal distribution of the disturbance regime caused by gopher activity.

A first attempt at understanding the disturbance regime at Jasper Ridge might involve some form of pattern analysis; especially as there is a growing interest among ecologists in using pattern analysis techniques in the characterization of ecological landscapes (e.g., Krummel et al. 1987, Milne 1988, Moloney et al. 1991). One technique that is commonly used in exploring spatial relationships is autocorrelation analysis. There are a number of other techniques that can be used, but I will limit the discussion to autocorrelation analysis as a means of illustrating the approach.

Autocorrelation analysis depicts the average degree of spatial coherence in a process as a function of the distance, or lag, separating points in the system. At any lag k, the autocorrelation function $\rho(k)$ is derived from the more general autocovariance function $\gamma(k)$. The sample function for determining the autocovariance from a one-dimensional set of contiguous sample points is given in equation (1), where x_i is the measured value of some state variable at position i, $\bar{x}$ is the sample mean for x, and N is the total number of sample points:

$$\gamma(k) = [\sum_{i=1}^{N-k} (x_i - \bar{x})(x_{i+k} - \bar{x})] / [N - k]. \tag{1}$$

The autocorrelation at any lag is then easily determined through the relationship $\rho(k) = \gamma(k)/\gamma(0)$, with the value $\gamma(0)$ being equivalent to the measured variance in the system.

Autocorrelation analysis might be viewed as a suitable method for determining the critical scales of the spatial relationships involved in producing the distribution of individual gopher mounds, but what exactly can be determined from autocorrelation relationships? The answer to this question is not as straightforward as it would at first seem and warrants a detailed analysis.

The autocorrelation relationship shown in Figure 2a was obtained through an analysis of the digitized image of newly formed gopher mounds shown in Figure 1. The image contained 1800 horizontal scan lines of 2544 pixels each. A separate autocorrelation analysis was conducted on each scan line, and the results were then averaged over all 1800 lines. There is a clear shift in the slope of the autocorrelation function at a scale smaller than the average width of an individual mound as measured in the field (37.6 ± 1.7 cm; Hobbs and Mooney 1985). Below 20 to 30 cm, the decay rate of the autocorrelation relationship appears to be approximately linear, whereas at larger scales the rate appears to be exponential. It is tempting to conclude that the scale at which there is a shift in the decay rate corresponds to the approximate scale of the process producing the pattern. However, this would be an incorrect inference, especially as an underestimate of the scale of the pattern-forming process would be expected due to the use of a one-dimensional analysis in examining a two-dimensional process.

An even more suggestive view of the pattern-forming process can be obtained through a log transform of the autocorrelation relationship (Figure 2b). Three general regions are observed: (1) from 0.0 m to 0.2 m there is a region of rapid exponential decay; (2) from 0.2 m to 1.8 m there is a region of slower exponential decay; and (3) above 1.8 m the autocorrelation flattens and then drops off sharply towards zero. How might these relationships be interpreted? A general argument might run as follows: At the smallest scales the pattern is dominated by noise in the image. The change in slope at around 0.2 m corresponds to the smaller spatial scales involved in the pattern producing process—the off-center intercept of individual mounds by transects. At intermediate scales—0.2 m to 1.8 m—there is a high degree of spatial coherence in the image corresponding to the fact that mounds are not formed individually, but tend to occur in clusters representing regions of intense activity by gophers (cf. Figure 1). The size of the disturbances intercepted by transects across the image vary from partial intercept of an isolated mound to the intercept of relatively large clusters—the largest being approximately 1.8 m across (Figure 2b). At the longest length scales—greater than 1.8 m—there are few if any regions of contiguous disturbance, and as a result the autocorrelation rapidly drops off to zero.

A)

B)

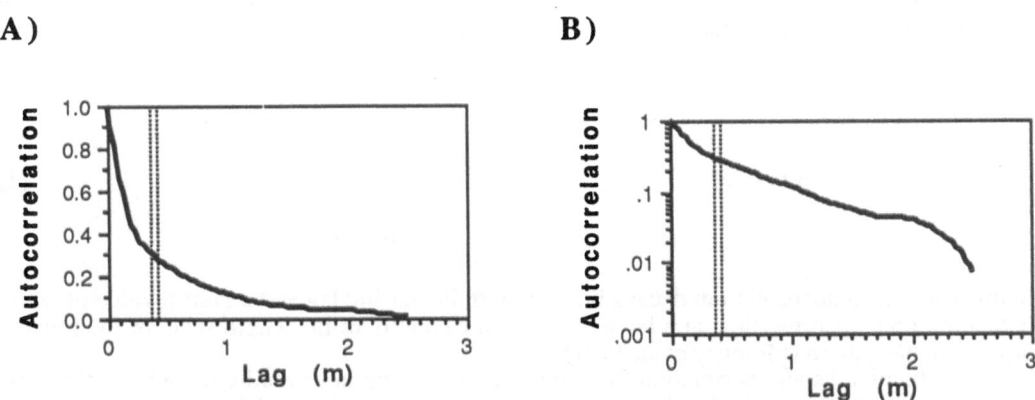

Figure 2. Autocorrelation analysis of the horizontal scanlines of the digitized map of newly formed gopher mounds shown in Figure 1. The broken vertical lines indicate the 95% confidence interval for the diameter of newly formed gopher mounds (Hobbs and Mooney 1985). A) Linear scale. B) Log Scale.

How precise are these results in depicting the critical scales of the spatial relationships involved in producing pattern in the system? At the very least, we know that a downward bias exists in identifying the length scales of the smallest scale processes involved. Although straightforward allometric transformations may be available for correcting this downward bias, other unknown factors may complicate the ability to interpret the relationships shown by the autocorrelation function.

The interpretation presented above was guided in large measure by *a priori* knowledge of the disturbance regime, both through studies in the field and through visual inspection of the pattern shown in Figure 1. Figure 1 is relatively easy to interpret, at least at a qualitative level, because only two states are present in the image—disturbed versus undisturbed sites. A more complex image—one having multiple states—would be much harder to interpret in an *ad hoc* fashion through simple visual inspection. In the latter case, one would be tempted to draw stronger conclusions on the basis of pattern analysis alone.

Although pattern analysis may prove useful in developing inferences about the important scales at which ecological processes operate within ecological landscapes, there is an inherent, and very real, danger that analytical techniques, such as autocorrelation analysis, will be used primarily as a tool for verifying preconceived ideas regarding the relationships coupling pattern to process (cf. Krummel et al. 1987). To avoid this problem, it is necessary to develop a more critical understanding of how the spatial dynamics of pattern formation are interpreted through pattern analysis. One approach is to explore the dynamics of simple pattern-forming models and examine how they are interpreted through pattern analysis techniques such as autocorrelation analysis.

In exploring pattern formation by disturbance, the simplest one-dimensional model we can consider would be an additive process that produces individual disturbances of equal size— in this case, length—and that has an equal probability of occurrence at any point in space (see Moloney et al. 1991 for a more detailed development of the model). In this case, the disturbance model can be characterized as a simple moving average process:

$$X_s = Z_s + Z_{s-1} + ... + Z_{s-m} ,$$ (2)

where X_s is the state of the system at point s, Z_s is a binomially distributed random variable, and m + 1 equals the width of an individual disturbance.   The autocorrelation for the model in equation (1) can be easily derived and yields a very simple expression:

$$\rho(k) = 1 - \frac{k}{[m+1]} \quad \text{for} \quad 0 < k < m+1$$

$$= 0 \quad \text{elsewhere.}$$ (3)

In this case, the autocorrelation decays linearly with increasing lag and attains a value of zero, indicating no autocorrelation, at a lag corresponding exactly to the length scale of the process producing the pattern (Moloney et al. 1991).

Although the autocorrelation function is quite striking in its ability to pick out the scale of individual disturbances in the simple moving average model, the model is not very realistic in depicting a natural disturbance regime, due to a number of assumptions that are explicit in the model. By relaxing some of these assumptions, we can begin to explore more realistic scenarios. For example, the assumption that all disturbances are of the same size is clearly not true in the case of the disturbance regime at Jasper Ridge. Even if all of the mounds producing the disturbance pattern were of the same size, linear transects would intercept mounds at varying distances from their centers, leading to a mixture of patch lengths that contribute to the autocorrelation function. How good, then, is the autocorrelation function in determining the relative contribution of patches of different size to the formation of an ecological pattern?

We can approach an answer to the above question by writing down an extension of equation (1) that accounts for a mixture of patches, assuming that the process again is additive and that patches of different size (m_i) are characterized by different patch initiation probabilities p_i (see Moloney et al. 1991). If we let $\sigma_i^2 = p_i (1 - p_i)$ represent the variance in the initiation probability of a patch of size m_i—assuming a binomial distribution for the patch initiation process—we find that the autocorrelation function is fully specified by knowing the initiation probabilities and patch sizes for all of the patch types involved in the process (see Moloney et al. 1991 for full development of the model). In fact, we can recover the structure of the system through autocorrelation analysis by examining the rate of change in the slope of the autocorrelation function, since as shown in Moloney et al. (1991):

$$\sigma_i^2 = \text{Var}(X_S)[\ \rho(m_i + 1) + \rho(m_i - 1) - 2\,\rho(m_i)\ ]\ . \qquad (4)$$

This relationship allows us to solve for the initiation probability of each patch type, as we have assumed a binomial distribution, and therefore

$$p_i = \frac{1 - \sqrt{1 - 4\sigma_i^2}}{2} \qquad (5)$$

The theoretical results presented in equations (4) and (5) are quite interesting, but can any insight into the structure of the disturbance regime at Jasper Ridge be gained using these techniques? An estimate of initiation probabilities for patches of different size can be obtained through an analysis of the autocorrelation relationship shown in Figure 2, and can then be compared to a more direct estimate of patch initiation probabilities obtained through a direct count of the number of disturbed sites of different size occurring in the digitized map in Figure 1 (see Figure 3).

Disturbed sites, for the purposes of the comparative analysis, were defined to be any horizontal array of contiguous pixels in Figure 1 that were completely occupied by newly formed disturbances. A direct estimate of patch initiation probabilities was then obtained by dividing the number of linear clusters of size m_i by the total number of pixels in the image (in this case, 1800 x 2544 pixels).

There was very close agreement in the magnitude of the estimates of patch initiation probabilities using the direct count and autocorrelation techniques (Figure 4). However, the estimate obtained from the autocorrelation relationship predicts a greater proportion of small-size patches and a lesser proportion of intermediate-size patches than does the estimate obtained through a direct count of clusters. What is the source of this disagreement? One assumption in the mixed patch model presented above is that the pattern-forming process is additive. This assumption is clearly violated, as it is impossible to distinguish between a patch formed by the overlap of multiple disturbances and a patch formed by one large disturbance. Because of the non-additivity of the process, the autocorrelation function decays more rapidly than it would if the process were additive (Moloney et al. 1991), which leads to an overestimate of the importance of small scale processes in forming the pattern. However, the direct estimate is also biased, but in the opposite direction. Large clusters were implicitly assumed to be the result of a single patch-forming event, although several individual disturbances were clearly involved in forming the larger disturbances in the Jasper Ridge system. The implication here is that,

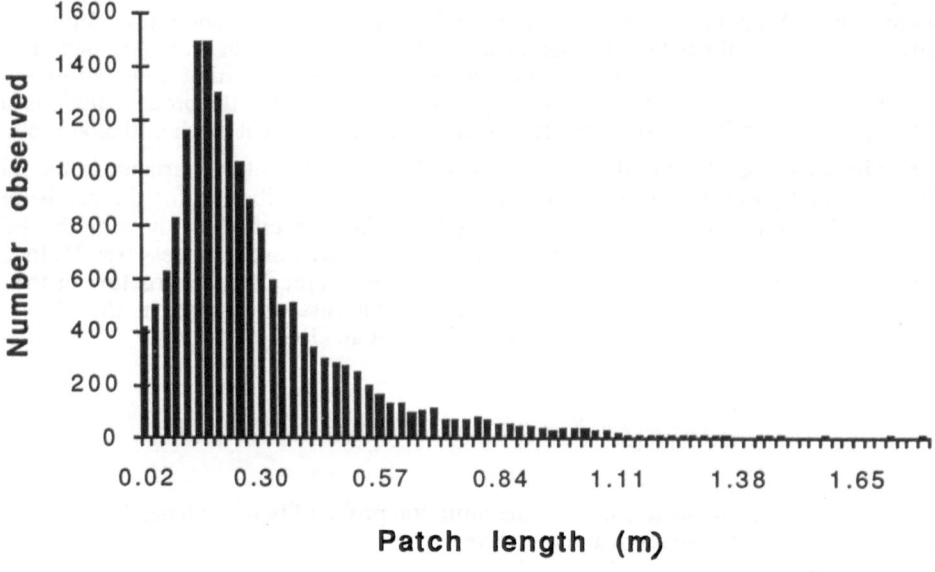

Figure 3. The number of disturbed sites of varying size in the digitized gopher mound map (Figure 1) as classified by patch length. Patch length was defined to be the number of pixels contiguously occupied by newly formed gopher mounds along horizontal scan lines in Figure 1. Disturbances were excluded from the analysis if they began or ended at the edge of the digitized image. Each scanline was composed of 2544 pixels, with each pixel being approximately 0.025 m wide.

although we can tally up the number of clusters of any given size, we still cannot determine the scale of the processes producing the ecological pattern of interest. Is there any hope that pattern analysis techniques can provide quantitatively satisfying insights into the structure of ecological systems? Perhaps.

CONCLUSION

Clearly, neither the direct count nor the autocorrelation methodologies presented above provide a quantitatively exact understanding of the scale of the processes producing the disturbance pattern observed at Jasper Ridge. There are a number of reasons for this failure, including the non-additivity of the disturbance process. Perhaps the most important reason is that all of the approaches presented above assume that disturbances are distributed independently of one another, even though there is a distinct tendency for mounds to occur in clusters due to the territorial behavior of gophers (Reichman et al. 1982). In fact, spatial substructuring is common to most ecological systems (e.g., see Powell 1989, Wiens 1989) and can be expected to introduce a strong bias in the development of pattern through ecological processes. This will directly affect the view of the system provided through pattern analysis techniques. We cannot know the exact nature of this bias unless more complex models are developed, particularly ones that integrate an element of spatial autocorrelation into the pattern-forming process. Once this is done, we can begin to explore the response of the autocorrelation function—or other suitable pattern analysis techniques—to the degree of spatial autocorrelation involved in producing pattern.

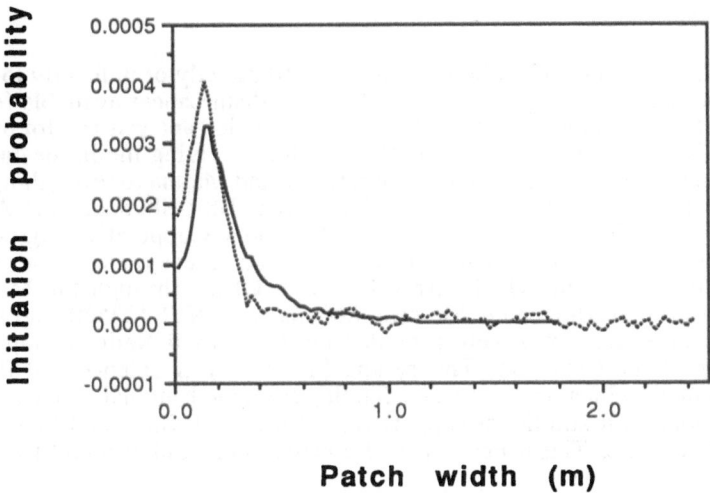

Figure 4. Patch initiation probabilities at Jasper Ridge as determined through two techniques applied to the digitized map of newly formed gopher mounds shown in Figure 1. The stippled line represents initiation probabilities determined through autocorrelation analysis, and the solid line represents probabilities determined through direct count methods. (See the text for a detailed description of these two methods of determining initiation probabilities.)

An argument might also be made that the approach taken in this paper would produce better results if a two-dimensional analysis and two-dimensional modeling techniques were employed. Although one could draw on the growing literature regarding two-dimensional pattern analysis techniques in designing a study similar to this one (e.g., Ford 1976, Ripley 1981), there would be a significant cost incurred in developing suitable, two-dimensional analytical models that could be solved explicitly for their autocorrelation structure, if indeed this could be done. In fact, it is unclear whether further insight would even be gained into the issues being considered here through a two-dimensional approach. The only clear advantage would arise if the pattern or underlying process were inherently anisotropic. In that case, the nature of the directional dependency would be revealed through a two-dimensional analysis, but the same problems of interpretation would exist as in the one-dimensional case; it would still be quite difficult to obtain a quantitatively exact understanding of the underlying process if it were either non-additive or if individual realizations of the process were coupled spatially through an underlying dependency. In fact, a greater understanding of how to interpret a two-dimensional analysis in the anisotropic case would be most easily attained by first developing a better understanding in one dimension.

Perhaps a deeper insight into the relationship between pattern and process will evolve through studies such as the one presented here, allowing us to use pattern analysis to learn a great deal about the structure of ecological systems; or perhaps we will learn that pattern analysis can tell us little, because of the smearing effect of the overlap of ecological processes across a number of scales. Although it seems clear that the potential for pattern analysis is great, extreme care must be taken in the rush to infer process from ecological patterns, at least until more is learned about the exact nature of the relationship between pattern and process.

ACKNOWLEDGMENTS

I would like to thank several people who have contributed directly or indirectly to this paper: Nona Chiariello for making the digitized image of gopher disturbances available for analysis; Simon Levin for suggesting a log transform of the autocorrelation analysis and for providing an enjoyable and stable work environment; Hal Mooney for involving me in the Jasper Ridge work; Tom Powell and John Steele for editorial comments; and Martha for everything else.

The research was supported in part by NSF grant BSR-8806202, the Andrew W. Mellon Foundation, the Ecosystems Research Center under Cooperative Agreement CR-812685-03 from the United States Environmental Protection Agency, Department of Energy grant DE-FG02-90ER60933, the U. S. Army Research Office through the Mathematical Sciences Institute of Cornell University, McIntire-Stennis grant NYC183550, and Hatch grant NYC183430. Computer resources were provided by the Cornell National Supercomputer Facility, a resource of the Center for Theory and Simulation in Science and Engineering (Cornell Theory Center), which receives major funding from the National Science Foundation and IBM Corporation, with additional support from New York State and members of the Corporate Research Institute. This is Ecosystems Research Center Publication ERC-248.

REFERENCES

Chesson, P. 1990. The changing role of equilibrium in ecological models. *Bull. of the Ecological Society of America* 71(2):118.

Cowles, H. 1899. The ecological relations of the vegetation on the sand dunes of Lake Michigan. *Botanical Gazette* 27:95-117, 167-202, 281-308, 361-391.

Davis, M. 1968. Climatic changes in Southern Connecticut recorded by pollen deposition at Rogers Lake. *Ecology* 50:409-422.

Denslow, J. 1990. Tropical communities as non-equilibrium assemblages. *Bull. of the Ecological Society of America* 71(2):137.

Ford, E. 1976. The canopy of a scots pine forest: Description of a surface of complex roughness. *Agricultural Meteorology* 17:9-32.

Hobbs, R. J., and H.A. Mooney. 1985. Community and population dynamics of serpentine grassland annuals in relation to gopher disturbance. *Oecologia* 67:342-351.

Hobbs, R. J., and H. Mooney. 1991. Effects of rainfall variability and gopher disturbance on serpentine annual grassland dynamics. *Ecology* 72:59-68.

Krummel, J., R. Gardner, G. Sugihara, R. O'Neill, and P. Coleman. 1987. Landscape patterns in a disturbed environment. *Oikos* 48:321-324.

Levin, S. A. 1989. Challenges in the development of a theory of community and ecosystem structure and function. In: J. Roughgarden, R. M. May, and S. A. Levin (eds.). *Perspectives in Ecological Theory*. Princeton University Press, Princeton, NJ, pp. 242-255.

McNaughton, S. 1968. Structure and function in California grasslands. *Ecology* 49:962-972.

Milne, B. 1988. Measuring the fractal geometry of landscapes. *Applied Mathematics and Computation* 27:67-79.

Moloney, K. 1988. Fine-scale spatial and temporal variation in the demography of a perennial bunchgrass. *Ecology* 69:1588-1598.

Moloney, K. 1990. Shifting demographic control of a perennial bunchgrass along a natural habitat gradient. *Ecology* 71:1133-1143.

Moloney, K., A. Morin, and S. Levin. 1991. Interpreting ecological patterns generated through simple stochastic processes. *Landscape Ecology* 5:163-174.

Moloney, K., S. Levin, N. Chiariello, and L. Buttel. In press. Pattern and scale in a serpentine grassland. *Theoretical Population Biology* .

Powell, T. 1989. Physical and biological scales of variability in lakes, estuaries, and the coastal ocean. In: J. Roughgarden, R. M. May, and S. A. Levin (eds.). *Perspectives in Ecological Theory*. Princeton University Press, Princeton, NJ, pp.157-177.

Reichman, O., T. Whitham, and G. Ruffner. 1982. Adaptive geometry of burrow spacing in two pocket gopher populations. *Ecology* 63:687-695.

Ripley, B. 1981. *Spatial Statistics*. John Wiley and Sons, New York, NY.

Tuljapurkar, S. 1984. Demography in stochastic environments. I. Exact distributions of age structure. *Journal of Mathematical Biology* 19:335-350.

Watt, A. S. 1948. Pattern and process in a plant community. *Journal of Ecology* 35:1-22.

Wiens, J. 1989. Spatial scaling in ecology. *Functional Ecology* 3:385-397.

6
DESCRIPTION AND ANALYSIS OF SPATIAL PATTERNS

Graciela García-Moliner, Doran M. Mason, Charles H. Greene, Agustín Lobo,
Bai-lian Li, Jianguo Wu, and G. A. Bradshaw

INTRODUCTION

Spatial pattern is a conspicuous characteristic of any ecosystem and has received much
attention from researchers over the last decade and a half (e.g., Steele 1978, Pickett and
White 1985, Kolasa and Pickett 1991), most recently under the name of landscape ecology
(Forman and Godron 1986, Turner and Gardner 1991). Description of patchiness in marine,
freshwater, and terrestrial systems presents different problems, particularly in terms of
mechanisms of patch formation (e.g., Wiens 1976 and Pickett and White 1985 in terrestrial
systems; and Legendre and Demers 1984, Mackas et al. 1985, Hamner 1988, Barry and
Dayton 1991, and Downing 1991 in aquatic systems). Deutschman et al. (this volume)
supply specific examples of mechanisms of patch formation.

In this chapter we attempt to describe patchiness in ecosystems and a set of analytical
tools useful for the study of spatial pattern across different scales and systems. Specifically,
we address three topics: (1) the description of patches in relation to scale, (2) sampling a
patchy environment, and (3) the analysis and interpretation of spatial data. Additionally,
marine and terrestrial ecosystems will be contrasted and compared to elucidate the inherent
differences between the two very different systems.

PATCHES AND PATCHINESS

Description of a "Patch:" The Terrestrial Concept

When the distribution of a phenomenon on land becomes discontinuous, the system is
considered to be "patchy." A patch at the landscape scale may contain a certain number of
within-community patches consisting of aggregates of physical entities and biological
individuals, which in turn may exhibit patchiness on finer scales—a "hierarchical scheme"
(Wiens 1976). At the community level, a patch may consist of a tree gap or a gopher
mound. Although all patches are realistically three-dimensional, humans generally perceive
patches in lower dimensionality. The term *patch* typically tends to emphasize spatial
discontinuity and internal homogeneity, but patches may vary greatly in geometry, content,
heterogeneity, and boundary characteristics.

Patchiness: The Marine Concept

The terrestrial phenomenon of a discontinuous patch has dominated the landscape literature. However, ecological processes in the ocean associated with spatio-temporal scales of water movement reveal "patchiness" rather than patches. These are dealt with through the Stommel diagrams (Haury et al. 1978) revised by Marquet et al. (this volume). The inherent fluid-dynamic nature of the environment necessitates a different set of criteria (Legendre and Demers 1984). Patches in the sea are best defined by the purpose of the study, the processes that create the patch, and the scale of the sampling. This is an indication of basic differences in the terrestrial and marine systems; or between substrate-dependent (intertidal and benthic) and substrate-independent (pelagic) systems. From this perspective, the intertidal and benthic marine ecosystems are viewed in a similar manner to terrestrial systems, defined in terms of the absence of an entity or open space where the boundaries are often discrete. Intertidal zones where high-energy wave activity removes assemblages of organisms and leaves behind a barren patch open for colonization (Paine and Levin 1981) are similar to gaps in forest stands.

Pelagic organisms and plants are often aggregated through physical, chemical, and behavioral processes. Physical phenomena such as fronts and Langmuir circulation can produce linear concentrations of planktonic organisms (e.g., Denman and Powell 1984, Mackas et al. 1985).

SAMPLING

One of the main problems in describing patchiness is the large rate of spatial change in the ocean, compared to terrestrial systems. As a result, there have to be major differences in the sampling design. The physical factors include: three-dimensional motion; absence of physical barriers to vertical and horizontal migration (Hamner 1988); low requirement of structural support for biota due to high specific gravity of water (Parsons 1976); and damping of short-term thermal variability due to high specific heat content of water (Steele 1985). Differences between the two systems' biology include: dominance of poikilotherms at the higher trophic levels; life history strategies (Steele 1985); and the small size of primary producers vs. the large size of primary producers of terrestrial ecosystems (Issacs 1977). It is these differences that present distinct sampling challenges in marine and terrestrial ecosystems.

The results of these differences are evident in the data obtained and, in consequence, in the methods used in analysis. Thus, it is relatively easy to run long transects on a ship sampling continuously for certain physical and biological parameters. This allows the use of data-intensive analytical methods such as Fourier analysis. On the other hand, there are problems in the temporal resolution of these time series: Are they truly synoptic?

On land it is usually extremely difficult to get long transects, and so methods such as geostatistics have been developed (see van Es, this volume). However, the problems in reconciling spatial and temporal variability are very much less on land.

In both sectors, satellite imagery can provide an entirely new, and different, view of each environment. Thus, in the ocean, color data has provided details of the two-dimensional fine structure and has radically altered our perceptions of smaller scale structure. On land, the opposite is true, and remote sensing has opened up the potential for large-scale analysis and modeling.

ANALYSIS TECHNIQUE, DATA SETS, AND EXAMPLES

Conventional statistical techniques, such as ANOVA, are not well suited for use on spatial data sets due to the assumptions of independence and normality. In spatially derived data sets, observations are generally not independent but rather are spatially correlated at some scale of resolution. Thus, other techniques must be used when analyzing spatial data sets. The analytical techniques for the study of spatial patterns or the variability in patchiness treated in this chapter include autocorrelation function (ACF), power spectra, wavelets, multivariate analysis, empirical orthogonal functions (EOF), and fractals.

A general review on description and analysis of the determination of spatial pattern on land can be found, for example, in Legendre and Fortin (1989) and references therein (see also van Es, this volume). In the oceans, power spectral analysis is probably the most widely used technique for pattern recognition with transect data (e.g., Platt and Denman 1975, Barale and Trees 1987, Barale 1987). Deutschman et al. (this volume) have analyzed krill data using both spectral analysis and wavelet transforms (see also the section on wavelet analysis later in this chapter) and in addition have developed a model (discussed in Grünbaum 1991) to describe the interactions between physical phenomena and behavioral interactions in Antarctic krill. Overall, the coupling of biotic/abiotic phenomena is a major factor in determining pattern and in differentiating pattern and process (Abbott, Denman, this volume).

Analyses of Time-Series and Spatial Series Data: The Autocorrelation Function and the Spectral Density Function

The autocorrelation function (ACF) and the spectral density function (power spectrum) are two of the most widely used diagnostic tools in signal processing (Jassby and Powell 1990). The power spectrum is the Fourier transform of the autocovariance function. The power spectrum decomposes the total variance in the data set into component contributions associated with each wave number in the spatial range of interest. The estimated spectrum may exhibit peaks at those wave numbers that contribute most to oscillations of the recorded variable in space.

For shorter series, analysis of the ACF often provides more reliable results (Jassby and Powell 1990). The ACF is used to quantify autocorrelations among points in a time series or one-dimensional spatial series separated by different temporal or spatial lags.

These techniques are useful only for analyzing one-dimensional serial data (Chatfield 1984). For two-dimensional data sets there are two-dimensional Fourier methods and geostatistics, but these essentially convert the data to one spatial scalar. The problem of pattern recognition, which the human eye does so easily, is still analytically intractable (e.g., Mandelbrot 1983).

Examples from biological oceanography. Spectral analysis has a relatively long history of use in physical oceanography, a field in which periodic phenomena are commonly encountered (e.g., Stommel 1963). The method was introduced to biological oceanography in the mid-1970s (Platt and Denman 1975) and has yielded a variety of interesting results for theoretical ecologists to ponder (e.g., Steele 1985, Powell 1989).

Recently, Levin (1990) compared the spectra for Antarctic krill, temperature, and chlorophyll data from the Southern Ocean (Figure 1). Temperature and chlorophyll spectra were similar at all scales, but the krill spectrum was flatter at higher wave numbers (smaller spatial scales: less than 20 km). Levin hypothesized that this finding might be indicative of physical processes driving the large-scale distributions of all three variables and of biological processes, specifically active aggregation behavior, driving the krill distributional patterns at smaller spatial scales.

With Levin's hypothesis in mind, we set out to analyze the fine-scale (meters to 10s of meters) patterns of krill in a data set collected by Greene et al. (in press) from three different sound-scattering layers (SSLs) in the Gulf of Maine (Figure 2). The species of krill (*Meganyctiphanes norvegica*) studied in the Greene et al. investigation was different from the species (*Euphausia superba*) discussed by Levin (1990) and by Deutschman et al. (this volume). Although *Meganyctiphanes norvegica* is rarely found to school or actively aggregate as intensively as *Euphausia superba*, we had some evidence of patchiness in the observed fine-scale patterns (Table 1; Greene et al., in press). Therefore, we were interested in determining whether or not the ACF could improve our characterization of the observed fine-scale patchiness.

The results of the analyses did little to improve our characterization of the observed fine-scale patchiness. The ACF was found to drop off rapidly with increasing spatial lag in data from all three SSLs. In only four out of the 15 cases analyzed were the autocorrelations significant at one lag interval (Table 2), and in no cases were they significant at two or more lag intervals. These results suggest that the characteristic length scales of patches were less than 1.5 - 3.0 m, and thus unresolvable by our sampling methods. The power spectra from the three SSLs were found to be flat (Figure 3), indicating no obvious periodicities in the observed fine-scale patchiness.

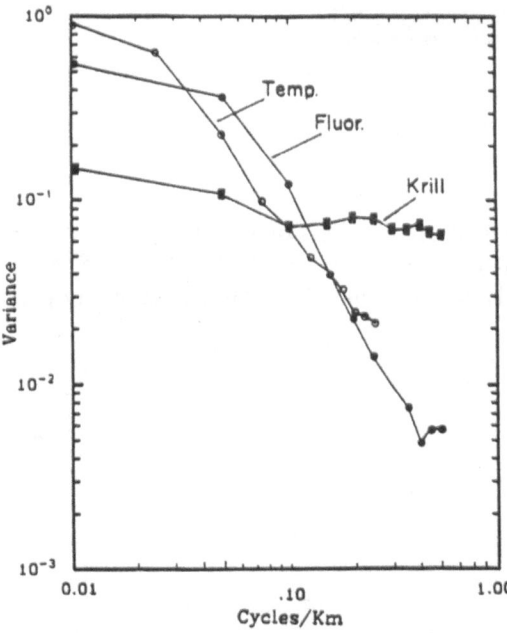

Figure 1. Mean spectral plots for Antarctic krill, chlorophyll fluorescence, and temperature from Levin (1990), as reprinted from Weber et al. (1986).

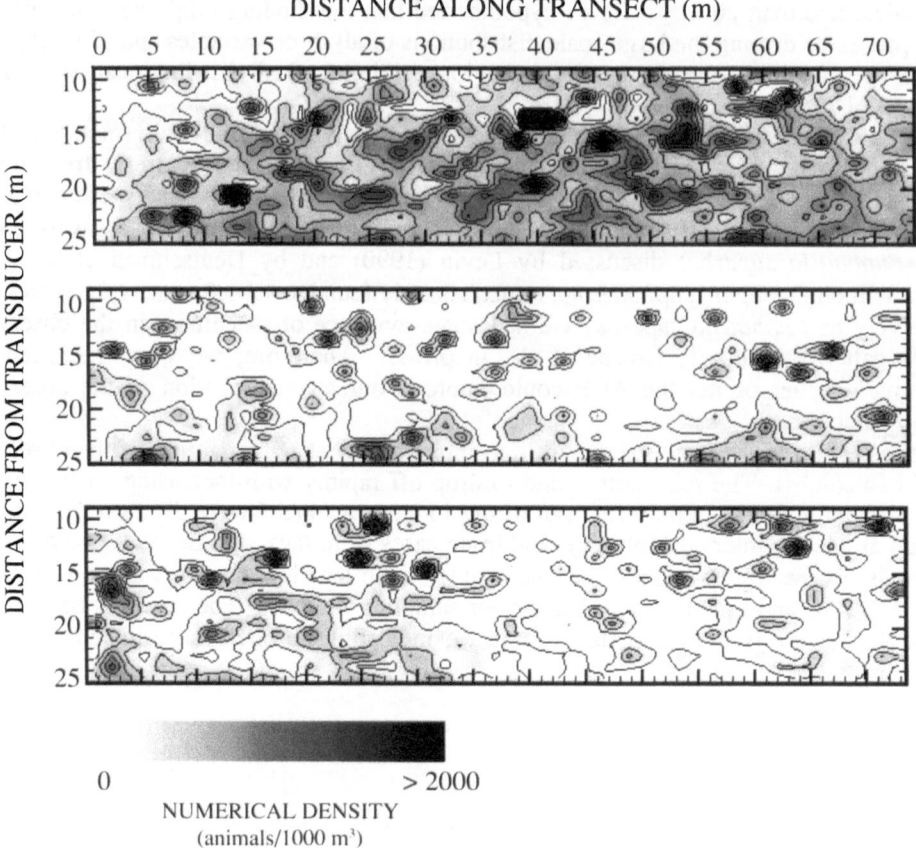

Figure 2. Fine-scale horizontal distributions of krill in a) daytime deep SSL (DDSSL), b) nighttime deep SSL (NDSSL), and c) nighttime shallow SSL (NSSSL). Numerical densities are given in units of animals per 1000 m3 (from Greene et al., in press).

Table 1. Numerical density and spatial aggregation statistics for krill in daytime deep SSL (DDSSL), nighttime deep SSL (NDSSL), and nighttime shallow SSL (NSSSL) from Gulf of Maine. Estimates of mean numerical density and Lloyd's (1967) index of mean crowding are given in units of animals per 1000 m³; Lloyd's index of patchiness is nondimensional.

	Mean Numerical Density (+2 SE)	Index of Mean Crowding	Index of Patchiness (+2 SE)
DDSSL	594 (26)	820	1.38 (0.04)
NDSSL	200 (18)	491	2.46 (0.24)
NSSSL	234 (18)	503	2.15 (0.16)

Index of Mean Crowding $= \bar{x} = (s^2/\bar{x} - 1)(1 + s^2/n\bar{x}^2)$

Patchiness $= [\bar{x} + (s = s^2/\bar{x} - 1)(1 + s^2/n\bar{x}^2)/x$

where $\bar{x}$ = mean, s^2 = variance,
n = number of samples,
and $1 + s^2/n\bar{x}^2$ is the sampling bias correction factor.

Comparisons of Mean Numerical Densities Among SSL's (ANOVA and Duncan's Multiple Range Test): DDSSL > NSSSL > NDSSL

Table 2. Autocorrelation at one spatial lag (1.5 m) for krill nighttime deep SSL (NDSSL), and nighttime shallow SSL (NSSSL) from Gulf of Maine. An asterisk indicates statistically significant autocorrelation; NS indicates nonsignificant autocorrelation.

Transect Distance from Transducer (m)

	21	22	23	24	25
			Autocorrelation		
DDSSL	0.227 NS	0.216 NS	0.247 NS	0.116 NS	0.419 *
NDSSL	0.096 NS	0.136 NS	0.410 *	0.391 *	0.119 NS
NSSSL	0.182 NS	0.250 NS	0.247 NS	0.160 NS	0.527 *

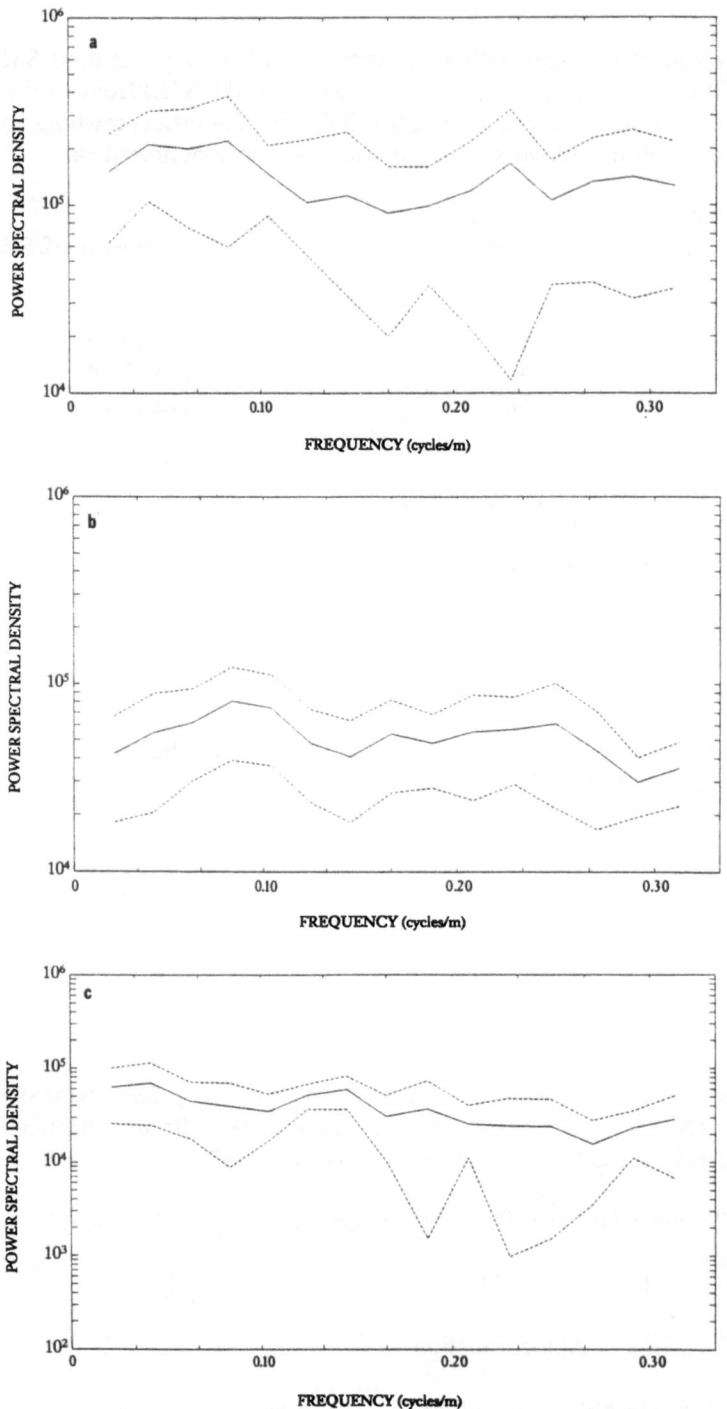

Figure 3. Power spectra for three SSLs: a) DDSSL, b) NDSSL, and c) NSSSL. Dashed lines correspond to 95% confidence intervals.

We infer from these results that, although *Meganyctiphanes* individuals exhibited significant fine-scale patchiness in the three SSLs examined (Table 1), the krill aggregations themselves tended to be relatively small and distributed more or less randomly in the environment. Greene et al. (in press) concluded that this type of fine-scale patchiness would be particularly difficult for predators to exploit in an efficient manner.

Wavelet Analysis

Wavelet analysis is a technique that has been employed for the analysis of spatial pattern and time series (Mallat 1988, Argoul et al. 1989, Bradshaw 1991). Three functions comprise wavelet analysis: the wavelet transform, wavelet variance, and the wavelet cross-covariance. The wavelet transform is defined in one dimension in the continuous form as:

$$\omega(a,x) = 1/a \int_{-\infty}^{\infty} f(x)g(x-b/a)dx \qquad (1)$$

where f(x) is the data function, g(x) is the analyzing wavelet, a is the scale, and b is the point around which the wavelet is centered.

Similar to Fourier spectral analysis, the transform effects a scale-by-scale decomposition of the data. In contrast to spectral analysis, the transform is a local filter and, as a function of both scale, a, and position, x, it retains location information. Because of this property, the wavelet transform can be used to examine the relationship of spatial pattern across scales. This is particularly useful in cases where the data is hierarchical or multi-scalar in structure or the data is non-uniformly distributed along the transect (Bradshaw and Spies, in press). Thus, the presence and intensity of fine-scale features may be related to higher-order structures at coarser scales.

As an example of the method, the wavelet transform was calculated for a transect selected from a 1 m resolution digital image of a Douglas fir forest canopy in western Cascades, Oregon (Figure 4). The wavelet transform provides a graphical display of the hierarchical structure of the overall pattern: i.e., the pattern of the individual tree crowns at one scale nested in a higher-order pattern at which the individual trees are clustered (Figure 5).

The analyzing wavelet may be selected from several possible functions given certain admissibility requirements (e.g., Mexican hat and Haar wavelets, Daubechies 1988). For example, if one is interested in the detection of gradients or edges, a step function such as the Haar wavelet would be a suitable choice (Gamage 1990; see also next section). The Mexican hat function was used in the previous canopy example.

The wavelet variance is a function derived from the wavelet transform to facilitate the identification of dominant scale(s) (e.g., patch sizes) among two or more data sets. It is defined as

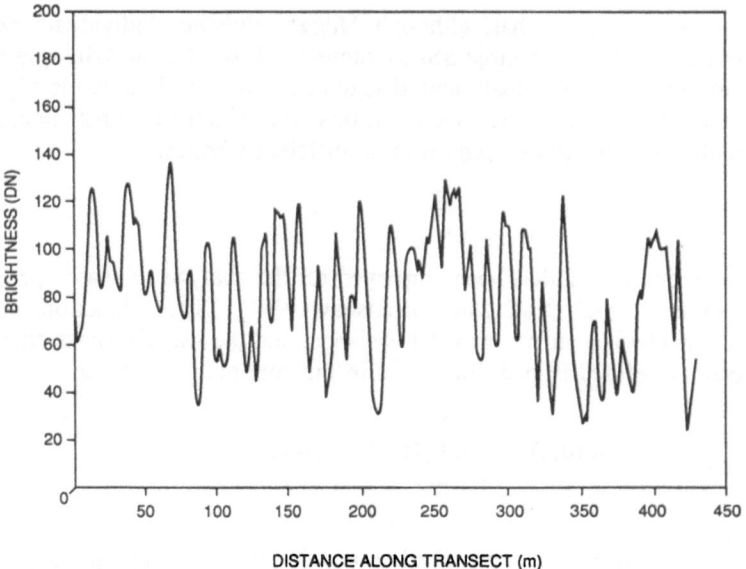

Figure 4. Transect of canopy brightness taken from a Douglas fir old growth stand using low-altitude videography (1 m resolution). The peaks and troughs in the data correspond to bright, illuminated tree crowns alternating with dark gaps in the canopy, respectively.

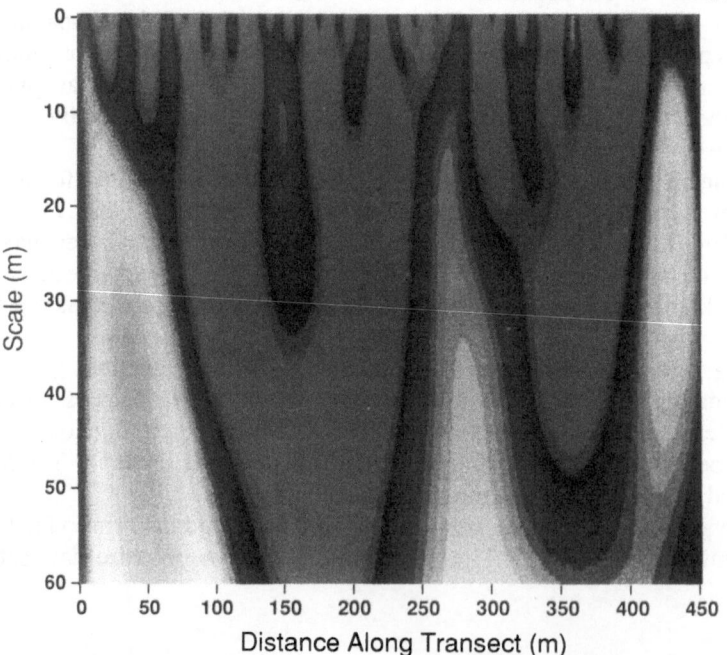

Figure 5. The wavelet transform calculated for the transect in Figure 4. The x-axis corresponds to location along the transect (meters). The y-axis indicates the scale of pattern. High (bright) values of the transform correspond to peaks, while low (dark) values correspond to troughs in the data. Note there are two dominant scales of pattern: a fine scale nested within the larger domains.

$$w(a) = \int_{-\infty}^{\infty} \omega^2(a,x)dx \qquad (2)$$

The wavelet variance was calculated for the transect in Figure 5 and for four other transects taken from four forest canopies at distinct stages of development: young, mature, mature-old growth mixture, and a young-mature mixture. The peak and amplitude of the wavelet variance for each transect distinguish the canopy texture of each stand (Figure 6).

In terms of signal processing, the wavelet variance provides a measure of the average energy contributed by each scale (or frequency) to the overall signal. Although both Fourier power spectra and wavelet variances generally appear to provide similar information (Bradshaw 1991), in many cases the wavelet variance has the advantage of being used in concert with the hierarchical information provided by the wavelet transform.

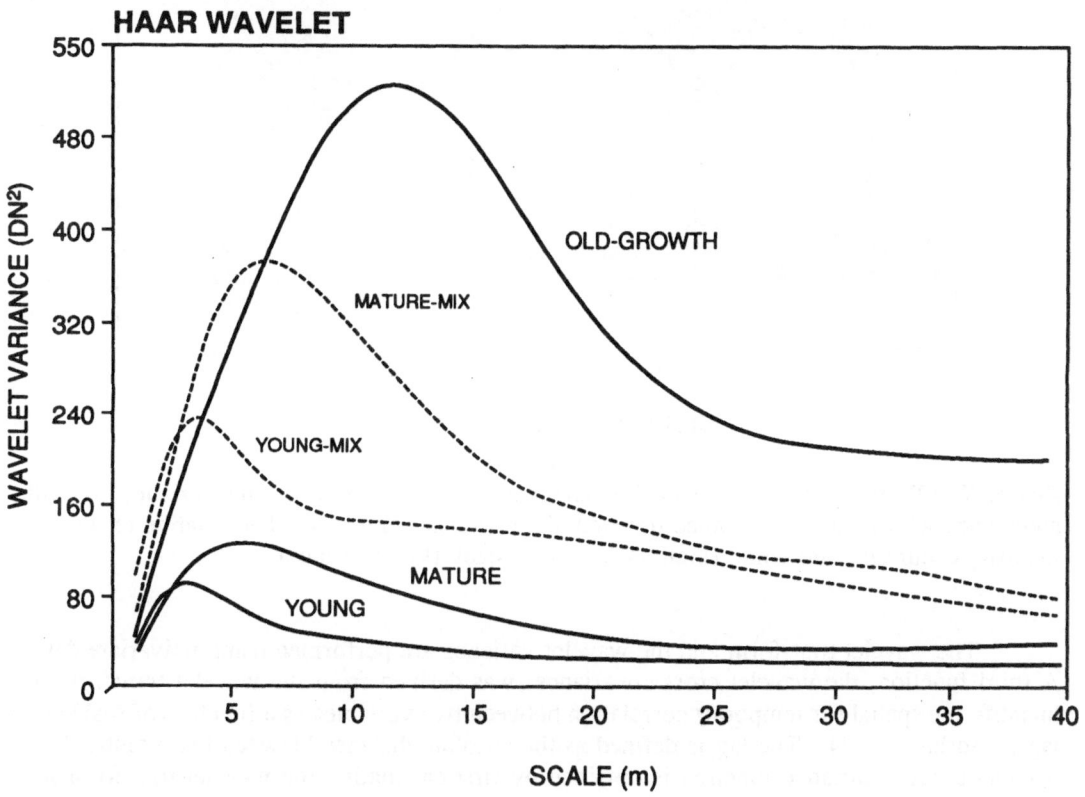

Figure 6. The wavelet variance calculated for canopy data as in Figure 4 for five forest age-class stands: young, mature, old growth, mature-old growth mixture, and young-mature mixture. The peak indicates the scale at which the pattern is dominated. The amplitude is proportional to the contrast in the data signal.

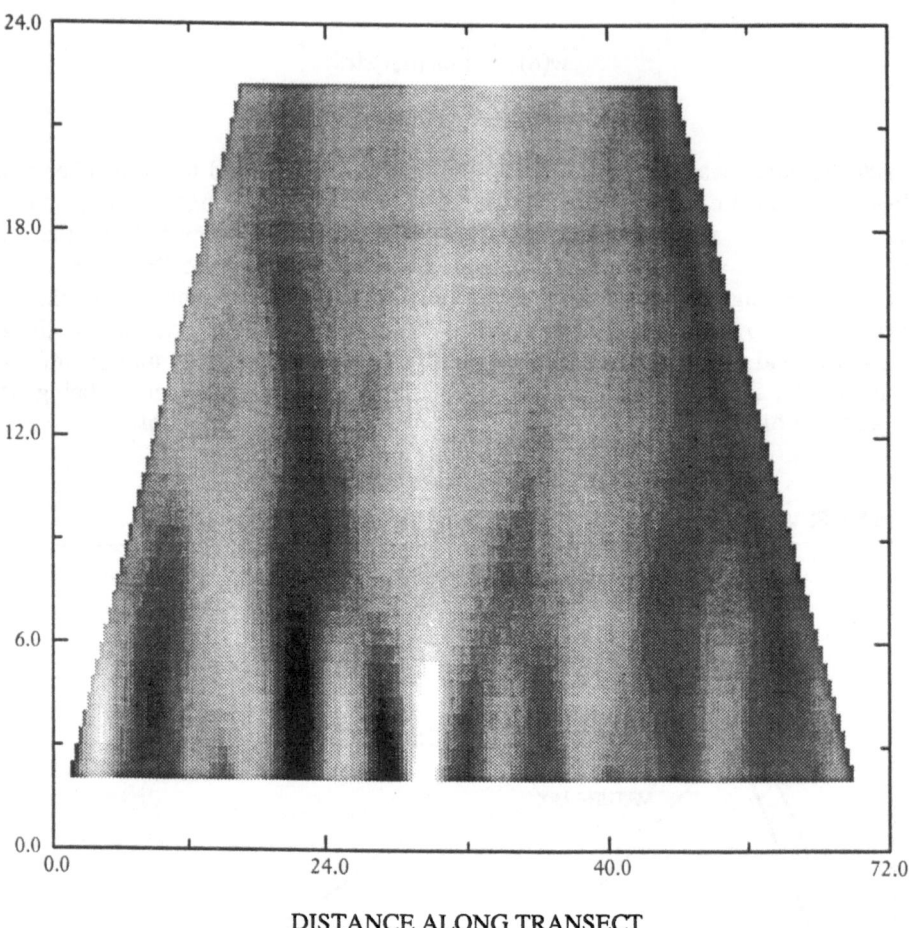

DISTANCE ALONG TRANSECT

Figure 7. Wavelet transform showing changes in the intensity of the variance in krill abundance with scale and distance traveled along acoustic transect. The relative change in intensity occurred from white (high variance) to black (low variance).

The wavelet transform and the wavelet variance are performed using univariate data. A third function, the wavelet cross-covariance, was derived from the wavelet transform to quantify the spatial (or temporal) correlation between two variables as a function of scale and lag (Bradshaw 1991). The lag is defined as the absolute distance between two points. The wavelet cross-covariance function is obtained by first calculating the wavelet transform for each variable, followed by the calculation of the cross-covariance function of the transformed data at each scale. The scale(s) at which the maximum cross-covariance occurs indicates the patch size at which the two variables are correlated. The lag at which the maximum cross-covariance occurs provides a measure of how closely coupled the two variables are in space. For example, a maximum wavelet cross-covariance at 10 m zero lag between variables A and B indicates that the two variables are coupled in space in 10 m patches. Conversely, a lack of correlation or a maximum cross-covariance at a non-zero lag at a given scale suggests that the two variables are poorly or less closely coupled. Thus, the wavelet cross-covariance function may be used to infer dominant scales of interaction. (An example is given in Deutschman et al., this volume).

Examples from Biological Oceanography. We reanalyzed the krill data presented earlier (section on spectral analysis) using the Haar transform. Wavelet transforms were calculated for four one-dimensional data sets, at 10, 15, 20, and 25 m distance from the sonar transducer. The acoustical data were integrated over a 1.5 m distance, with structure occurring at smaller scales averaged out. Due to sampling and transect limitations, the maximum scale resolved was 24 m. All four transects showed dominance of patches at ≤ 6 m, as determined by the wavelet variance function, and in agreement with the results of the spectral analysis. However, there was evidence, for the 25 m data set, that patches appear to occur at two scales, ≤ 6 m and ≥ 22 m. This may be interpreted as suggesting that at least two different mechanisms are operating to create structure at two different scales. For example, at the smaller scales, krill behavior, such as swarming, may dominate and control the observed spatial structure. At the larger scales, other physical or biological processes may dominate patch formation. Given the present data constraint this is only speculative, but it illustrates the uses of these methods.

The two-dimensional representation of the data (distance on horizontal axis and scale on the vertical axis) provide information on the intensity of the variance with scale and position along the transect (Figure 7). The wavelet transform thus suggests higher-order structure. For a clear picture of higher-order structure (≥ 24 m) we require longer acoustic transects (i.e., larger data sets). With the present data constraint, it is difficult to determine whether a nested structure is present.

Multivariate Methods

We turn now to the multivariate systems in which the data are abundances of categories (cover-types, taxa, or life-forms), with an explicit spatial reference for each subsample. Commonly, subsamples are arranged in a grid and the data can be represented as a three-dimensional matrix in which the first two dimensions are spatial coordinates and the third one, the array of categories. There are two fundamental ways to look at the variation of pattern.

How the statistical descriptors of the landscape vary. This has been popular among ecologists since Greig-Smith (1957), although several drawbacks to this particular method have been pointed out (Upton and Fingleton 1985). With the indices described below, these drawbacks can be avoided (i) if the basic subsamples used for computing are randomly located in the grid, (ii) if the maximum sample size to analyze is restricted to be less than half the whole area side, and (iii) if boundary effects are avoided. Plots of averages and/or variances, since they are equally important, calculated for each cell, versus cell size are made. However, these methods do not consider spatial scales explicitly.

*How indices vary between two grid cells separated by a distance **d** as this distance increases.* The general procedure consists of randomly placing a dipole of length **d** n times on the grid, recording values at both extremes, and computing one of the following indices.

82

The dipole length, **d,** is increased (to a maximum of 1/2 the minimum side of the grid to avoid boundary effects) and the index re-computed. Distances can be the ordinary euclidean or be arbitrarily defined.

The contingency table for length **d** is built with the co-occurrence frequencies of the species recorded at the cells intersected by the dipole's extremes. The same statistics described above can be applied to these contingency tables.

In the case of <u>multivariate dissimilarity</u>, the procedure is similar to the semivariogram, but a multivariate dissimilarity is substituted for the squared difference (Mackas 1984).

The <u>Mantel index</u> is defined as

$$\sum [X_{i,j} * Y_{i,j}] \tag{3}$$

where $X_{i,j}$ is the dissimilarity between cells i and j, and $Y_{i,j}$ is the corresponding distance. It tests the independence of both matrices. Well-documented ecological applications can be found in Upton and Fingleton (1985), Sokal (1986), Oden and Sokal (1986), and Legendre and Fortin (1989). The Mantel index can be presented in spectral form ("Mantel correlogram" in Legendre and Fortin 1989) to explore the relationship between dissimilarity and distance.

Empirical Orthogonal Functions

Empirical orthogonal functions (EOF) are also referred to as PCA (Principal Component Analysis) or Proper Orthogonal Decomposition. PCA is a tool extensively used for analysis of the spatial/temporal variability of physical fields (Preisendorfer 1988) such as winds, and of sea surface temperature (SST). An attempt to correlate physical fields (SST) and biological patchiness (pigment concentrations) will be the basis for the example presented herein. The technique is used to look at the aggregation of the variance to explain the spatial (temporal) structure. The data need only consist of real numbers.

The analysis reduces the dimensionality of the data, grouping the variance of the data to examine characteristics of induced or autonomous variables in the description of patchiness. A review by Matta and Marshall (1984) expands on the advantages of using PCA to examine phytoplankton variation.

Examples from biological oceanography: Sea surface temperature and water color (satellite). Oceanographic features such as warm core rings (WCR) are defined as distinctive patches for the purposes of this case study. WCR are clockwise rotating features, with Sargasso Sea water in the center, which have broken away from the meandering Gulf Stream.

Color data are from the Coastal Zone Color Scanner (CZCS), and sea surface temperature (SST) are from the Advanced Very High Resolution Radiometer (AVHRR). A total of 13 (single) ocean color images and 13 (single) matching SST images,[1] irregularly spaced in time but matching pairs within 24 hours, were used for the EOF analysis. These were processed using the Miami DSP image processing system for the VAX. NASA's SEAPAK-PC package and MATLAB were used for data extraction and manipulation.

---

[1]The CZCS sensor is described by Hovis et al. (1980). Pigment concentrations are estimated following details given by Gordon et al. (1980) and Gordon and Morel (1983).

The images covered the period from March 31 to June 9, 1982. The area of interest, the Mid-Atlantic Bight, includes the shelf and slope between 35° and 42° North. The WCR under study was formed early in 1982 (Celone and Price 1983). The ring measured about 200 km in diameter (Figure 8).

Data were extracted from the images in a 100 x 100 pixel array, from 4 km resolution data, whose center changed as the WCR moved in a southwesterly direction. The two-dimensional data sets lend themselves to a Principal Component Analysis (PCA). Due to the small number of images, the 100 x 100 matrix was divided into a 3 x 4 matrix (12 bins). To assure full complement of positive eigenvalues for the covariance matrix, the

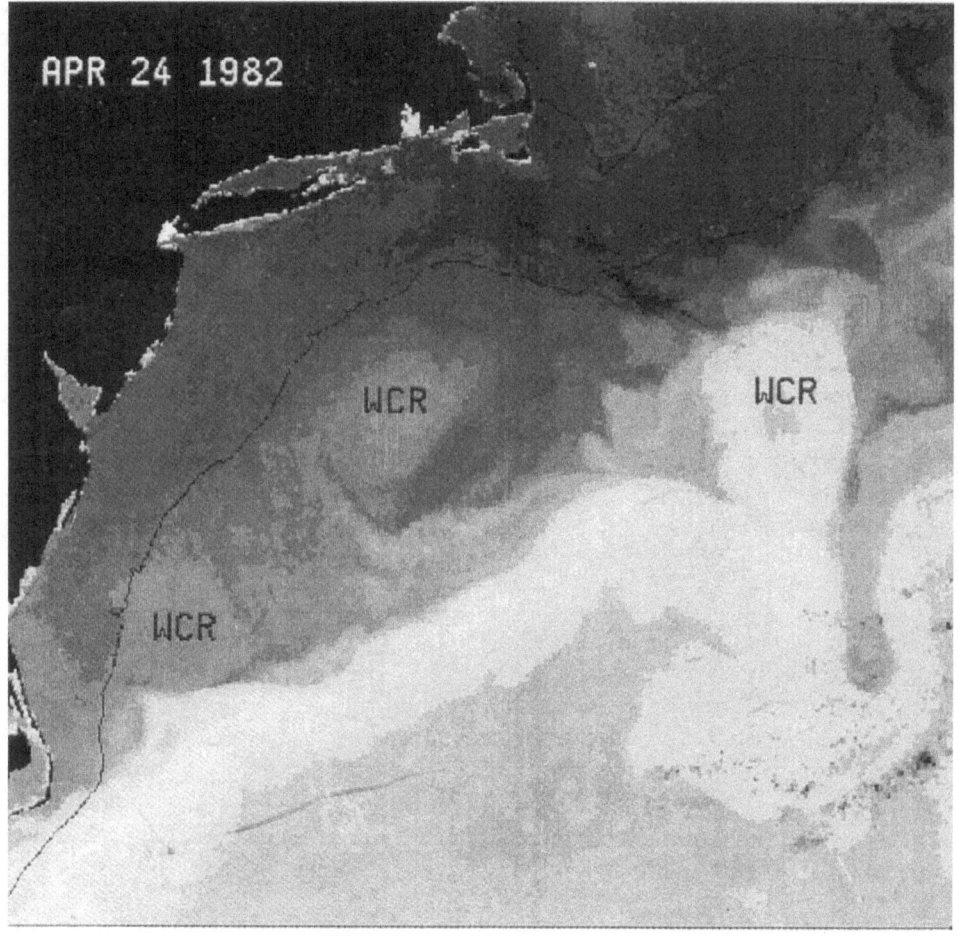

Figure 8. Satellite image of the Mid-Atlantic Bight (4 km resolution) for April 24, 1982. Three warm core rings (WCR) are shown. The WCR under use in the EOF analysis is shown in the center.

number of bins must be less than the number of images (Jassby et al. 1990). Once the n x p matrix was obtained (n = days, p = variable, mean temperature or pigment concentration per box), the covariance was calculated (yielding a square matrix) and the eigenfunction obtained from the covariance matrix. The mean is removed from all values to determine the anomalies in pigment concentrations and temperature. No rotation of eigenvectors was carried out.

The contour plot of the first eigenvector is shown in Figure 9, as an example of the patterns observed in the satellite images. The standard deviation showed greater variability in the outside of the rings.

Table 3 shows the total variance (as a percentage) explained by the PCs. About 90% of the variability is explained by the first eigenvector in the pigment data. The first eigenvector explains about 80 - 85% of the variance in the temperature field. To pull out more modes, a larger data set is needed.

The most important point of the exercise is to demonstrate ways in which moving spatial features can be tracked in the ocean. The technique reveals recognizable patterns (anomalies) in both the SST and pigment fields. In addition, it allows for the inference of mechanisms (e.g., advection) that may be responsible for the variability observed.

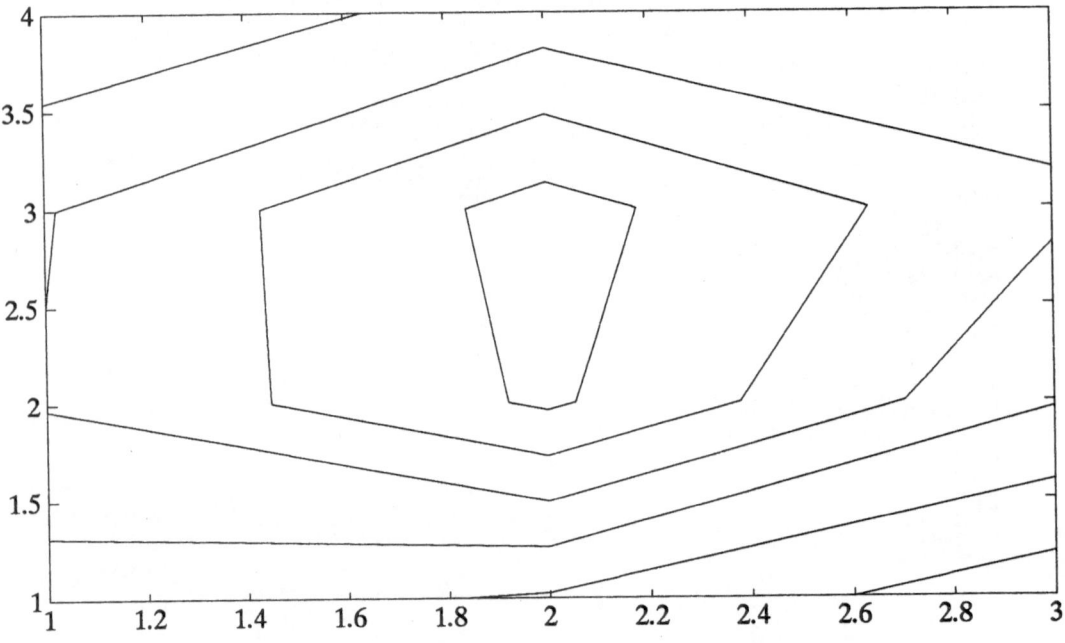

Figure 9. Contour plot of first eigenvector for sea-surface temperature (SST) showing lows in center and highs toward the bottom right-hand corner of the plot.

Table 3. Percentage total variance explained by the first most important principal components (PC) of temperature (SST) and pigment concentration (PIG) of a warm-core ring.

PC	SST	PIG
1	84.2	89.8
2	7.1	5.0
3	4.9	2.9
4	4.3	1.2

Fractal Geometry: Applications in Analysis and Description of Patch Patterns and Patch Dynamics

Mandelbrot's concept of a fractal extends our usual ideas of classical geometry beyond those of point, line, circle into the realm of the irregular, disjoint, and singular (Mandelbrot 1983) and so provides us with a new way to understand and analyze spatial phenomena. Fractals are shapes in which an identical motif repeats itself on an ever-diminishing scale. A fractal is a non-integer dimension.

For ecological applications, Krummel et al. (1987) used fractal models to examine how patch boundary varies with patch size. The relationship is applied to landscape patches by the equality between area and perimeter length,

$$A = BL^D$$

where A is patch area, L is the patch perimeter, B is a constant, and D is the fractal dimension. By using data for patch area and patch perimeter, and relationships

$$logA = logB + DlogL$$

we can estimate the fractal dimension D.

The fractal characteristics over a landscape may be important measures in ecological diversity, stability, and function. Many recent studies have included measures of the fractal geometry of landscapes and patch patterns (Burrough 1981, 1983; Krummel et al. 1987, Gardner et al. 1987, Milne 1988, O'Neill et al. 1988, DeCola 1989, Wiens and Milne 1989, Rex and Malanson 1990, Williamson and Lawton 1991).

Recently, Milne (1990) and Sugihara and May (1990) reviewed fractal applications in ecological research and landscape. Here, we shall select some interesting aspects associated with spatial patterns and patch dynamics and introduce them as follows.

Patch hierarchical scaling. Different observational scales capture different aspects of structure, and these transitions are signaled by shifts in apparent fractal dimension of the object. This suggests an interesting application of fractals as a method for distinguishing hierarchical size scales of patches in nature; for example, to determine boundaries between hierarchical levels, or the scaling rules for extrapolation within each level.

Bradbury et al. (1984) examined the possibility of hierarchical scaling in an Australian coral reef. They used the dividers method (Mandelbrot 1983) in transects across the reef to determine whether D depends on the range of length scales. They found that three ranges of scale correspond nicely with the scales of three major reef structures: 1 - 10 cm corresponds to the size of anatomical features within individual coral colonies; 20 - 200 cm corresponds to the size range of whole adult living colonies; and 5 - 10 m is the size range of major geomorphological structures. This showed that the shifts in D at different spatial scales appear to signal where the boundaries occur in the hierarchical organization of reefs.

Fractal spatial patterns and modified Brownian dynamics (diffusion). Hastings et al. (1982) and Mandelbrot (1983) have discussed how fractal exponents may be incorporated into diffusion processes, as a scaling factor for normalizing increments in space and time. They find that D may be used as an index of succession in circumstances where simple patch-extinction models are reasonable. A recent study has showed that many of nature's seemingly patchy shapes can be effectively characterized and modeled as random fractals based upon generalizations of fractional Brownian motion (Voss 1988).

DISCUSSION

It is worthwhile to recap the concepts of Kotliar and Wiens (1990) on the scales of patchiness. Recognition of spatial pattern is dependent on the scale of observation as well as on the scales of the processes under study. It is the coupling of pattern and process (i.e., patch dynamics) at different scales, be it autonomous or induced, that leads to the subjectivity involved with describing patchiness. Heterogeneity is recognized not only in terms of the processes affecting the variability of patches, but also in the changes that occur due to this variability. The description and analysis of patchiness is thus limited by the techniques for observation, sampling, and analysis. The problem is to derive methods for statistically recognizing spatio-temporal pattern. The examples used in this chapter are representative of the techniques available. The techniques allow for inferences regarding interactions between the physical and biological phenomena creating pattern at various scales. The causes and mechanisms of patch formation are discussed by Deutschman et al. (this volume) and demonstrate the complexity of these interactions.

In view of the above discussion, we might conclude that simple comparisons between terrestrial and marine systems are often artificial. Differences in scale are inherent properties of the systems and must be considered. Better comparisons of techniques across scales and ecosystems are needed. This perspective seems appropriate when we realize that all terrestrial and marine ecosystems are tightly coupled and our ecological perspective is expanding to an all-inclusive global outlook.

ACKNOWLEDGMENTS

The authors would like to thank Drs. S.A. Levin, T. Powell, and J. Steele for their kind invitation and encouragement. The authors would also like to thank all the people who were involved in the development of this chapter!

REFERENCES

Abbott, M.R. (This volume.). Phytoplankton patchiness: Ecological implications and observation methods.

Argoul, F., A. Arneodo, G. Grasseau, Y. Gagne, E.J. Hopfinger, and U. Frisch. 1989. Wavelet analysis of turbulence reveals the multi-fractal nature of the Richardson cascade. *Nature* 338: 51-53.

Barale, V. 1987. Remote observations of the marine environment: Spatial heterogeneity of the mesoscale ocean color field in CZCS imagery of California near-coastal waters. *Remote Sensing of the Environment* 22: 173-186.

Barale, V., and C.C. Trees 1987. Spatial variability of the ocean color field in CZCS imagery. *Advances in Space Research* 7(2): 95-100.

Barry, J.P., and P.K. Dayton. 1991. Physical heterogeneity and the organization of marine communities. In: J. Kolasa and S.T.A. Pickett (eds.). *Ecological Heterogeneity*. Springer-Verlag, New York.

Bradbury, R.H., R.E. Reichelt, and D.G. Green. 1984. Fractals in ecology: Methods and interpretation. *Marine Ecology Progress Series* 14: 295-296.

Bradshaw, G.A. 1991. Hierarchical pattern and process in Douglas-fir forests using wavelet analysis. Ph.D. Dissertation. Oregon State University, Corvallis, OR.

Bradshaw, G.A., and T.A. Spies. In press. Characterizing forest canopy structure using wavelet transform. *Journal of Ecology*.

Burrough, P.A. 1981. Fractal dimensions of landscapes and other environmental data. *Nature* 294: 241-243.

_____. 1983. Multiscale sources of spatial variation in soil: I. Application of fractal concepts to nested levels of soil variations. *Journal of Soil Science* 34: 577-597.

Celone, P.J., and C.A. Price. 1983. Anticyclonic warm-core Gulf Stream rings off the northeastern United States during 1982. *Annales Biologiques* 39: 19-23.

Chatfield, C. 1984. *The Analysis of Time Series: An Introduction*. Chapman and Hall, NY.

Daubechies, I. 1988. Orthogonal basis of compactly supported wavelets. *Communications of Pure and Applied Mathematics* 41: 909-996.

DeCola, L. 1989. Fractal analysis of a classified landsat scene. *Photogrammetric Engineering and Remote Sensing* 55: 601-610.

Denman, K.L. (This volume.) The ocean carbon cycle and climatic change: An analysis of interconnected scales.

Denman, K., and T. M. Powell. 1984. Effects of physical processes on planktonic ecosystems in the coastal ocean. *Oceanography and Marine Biology Annual Review* 22: 125-168.

Deutschman, D., G.A. Bradshaw, W.M. Childress, K. Daly, D. Grünbaum, M. Pascual, N. Schumaker, and J. Wu. (This volume.) Mechanisms of patch formation.

Downing, J.A.. 1991. Biological heterogeneity in aquatic ecosystems. In: J. Kolasa and S.T.A. Pickett (eds.). *Ecological heterogeneity*. Springer-Verlag, New York.

Forman, R.T.T., and M. Godron. 1986. *Landscape Ecology*. Wiley, New York.

Gamage, N.K.K. 1990. Detection of Coherent Structures in Shear Induced Turbulence Using the Wavelet Transform Methods. AMS Symposium on Turbulence and Diffusion. Roskilde, Denmark, April 1990.

Gardner, R.H., B.T. Milne, M.G. Turner, and R.V. O'Neill. 1987. Neutral models for the analysis of broad-scale landscape pattern. *Landscape Ecology* 1: 19-28.

Gordon, H.R. and A.G. Morel. 1983. *Remote Assessment Ocean Color for Interpretation of Satellite Visible Imagery: A Review*. Springer-Verlag, New York.

Gordon, H.R., D.K. Clarck, J.L. Muller, and W.A. Hovis. 1980. Phytoplankton pigments from the Nimbus-7 Coastal Zone Color Scanner: Comparisons with surface measurements. *Science* 210: 63-66.

Greene, C.H., E.A. Widder, M.J. Youngbluth, M. Tamse, and G.E. Johnson. (In press). The fine structure, migration behavior, and bioluminescent activity of krill sound-scattering layers. *Limnology and Oceanography*.

Greig-Smith, P. 1957. *Quantitative Plant Ecology*. Academic Press, New York.

Grünbaum, D. 1991. Three unrelated projects in mathematical biology. Ph.D. Dissertation. Univ. of Washington, Seattle, WA.

Hamner, W.M. 1988. Behavior of plankton and patch formation in pelagic ecosystems. *Bulletin of Marine Science* 43(3): 752-757.

Hastings, H.M., R. Pekelney, R. Monticciolo, D.V. Kannon, and D.D. Monte. 1982. Time scales, persistence and patchiness. *Biosystems* 15: 281-289.

Haury, L.R., J.A. McGoan, and P.H. Wiebe. 1978. Patterns and processes in the time-space scales of plankton distributions. In: J.H. Steele (ed.). *Spatial patterns in plankton communities.* Plenum Press, New York.

Hovis, W.A., D.K. Clark, F. Anderson, R.W. Austin, W.H. Wilson, E.T. Baker, D. Ball, H.R. Gordon, J.L. Muller, S.Z. El-Sayed, B. Strum, R.C. Wrigley, and C.S. Yentsch. Nimbus-7 Coastal Zone Color Scanner: System description and initial imagery. *Science* 210: 60-63.

Issacs, J.D. 1977. The life of the open sea. *Nature* 267: 778-780.

Jassby, A.D., and T.M. Powell. 1990. Detecting changes in ecological time series. *Ecology* 71(6): 2044-2052.

Jassby, A.D., T.M. Powell, and C.R. Goldman. 1990. Interannual fluctuations in primary production: Direct effects and the trophic cascade at Castle Lake, California. *Limnology and Oceanography* 35(5):1021-1038.

Kolasa, J. and S.T.A. Pickett (eds.). 1991. *Ecological Heterogeneity.* Springer-Verlag, New York.

Kotliar, N.B., and J.A. Wiens. 1990. Multiple scales of patchiness and patch structure: A hierarchical framework for the study of heterogeneity. *Oikos* 59: 253-260.

Krummel, J.R., R.H. Gardner, G. Sugihara, R.V. O'Neill, and P.R. Coleman. 1987. Landscape patterns in disturbed environment. *Oikos* 48: 321-324.

Legendre, L., and S. Demers. 1984. Towards dynamic biological oceanography and limnology. Canadian *Journal of Fishery Aquatic Science* 41(1): 2-19.

Legendre, L., and P. Legendre. 1983. *Numerical Ecology.* Elsevier, Amsterdam.

Legendre, P., and M.-J. Fortin. 1989. Spatial pattern and ecological analysis. *Vegetatio* 80: 107-138.

Levin, S.A. 1990. Physical and biological scales and the modelling of predator-prey interactions in large marine ecosystems. In: K. Sherman, L.M. Alexander, and B.D. Gold (eds.). *Large Marine Ecosystems: Patterns, Processes, and Yields.* AAAS, Washington, DC.

Lloyd, M. 1967. Mean crowding. *Journal of Animal Ecology* 36: 1-30.

Mackas, D.L. 1984. Spatial autocorrelation of plankton community composition in a continental shelf ecosystem. *Limnology and Oceanography* 29(3): 451-471.

Mackas, D.L., K.L. Denman, and M.R. Abbott. 1985. Plankton patchiness: Biology in the physical vernacular. *Bulletin of Marine Science* 37(2): 652-674.

Mallat, S.G. 1988. Review of Multi-frequency Channel Decompositions of Images and Wavelet Models. Robotics report no. 178, Courant Institute of Mathematical Sciences, New York University, New York.

Mandelbrot, B.B. 1983. *The Fractal Geometry of Nature.* W.H. Freeman and Company, New York.

Marquet, P.A., M.-J. Fortin, J. Pineda, D.O. Wallin, J. Clark, Y. Wu, S. Bollens, C.M. Jacobi, and R.D. Holt. (This volume). Ecological and evolutionary consequences of patchiness: A marine-terrestrial perspective.

Matta, J.F. and H.G. Marshall. 1984. A multivariate analysis of phytoplankton assemblages in the western North Atlantic. *Journal of Plankton Research* 6(4): 663-675.

Milne, B.T. 1988. Measuring the fractal geometry of landscapes. *Applied Mathematics and Computation* 27: 67-79.

_____. 1990. Lessons from applying fractal models to landscape patterns. In: M.G. Turner and R.H. Gardner (eds.). *Quantitative Methods in Landscape Ecology.* Springer-Verlag, New York.

O'Neill, R.V., B.T. Milne, M.G. Turner, and R.H. Gardner. 1988. Resource utilization scales and landscape pattern. *Landscape Ecology* 2: 63-69.

Oden, N.L., and R.R. Sokal. 1986. Directional autocorrelation: An extension of spatial correlograms to two dimensions. *Systematic Zoology* 35: 608-617.

Paine, R.T., and S.A. Levin. 1981. Intertidal landscapes: Disturbance and the dynamics of pattern. *Ecological Monographs* 5: 145-178.

Parsons, T.R. 1976. The structure of life in the sea. In: D.H. Cushing and J.J. Walsh (eds.). *The Ecology of the Seas.* Blackwell Scientific Publications, Oxford.

Pickett, S.T.A., and P.S. White. 1985. *The Ecology of Natural Disturbance and Patch Dynamics.* Academic Press, Orlando, Florida.

Platt, T., and K.L. Denman. 1975. Spectral analysis in ecology. *Annual Review of Ecology and Systematics* 6: 189-210.

Powell, T.M. 1989. Physical and biological scales of variability in lakes, estuaries, and the coastal ocean. In: J. Roughgarden, R.M. May, and S.A. Levin (eds.). *Perspectives in Ecological Theory*. Princeton University Press, Princeton, New Jersey.

Preisendorfer, R.W. 1988. *Principal Component Analysis in Meteorology and Oceanography*. Elsevier, New York.

Pringle, C.M., R.J. Naiman, G. Bretschko, J.R. Karr, M.W. Oswood, J.R. Webster, R.L. Welcomme, and M.J. Winterbourn. 1988. Patch dynamics in lotic systems: The stream as a mosaic. *Journal of the North American Benthological Society* 7: 503-524.

Rex, K.D., and G.P. Malanson. 1990. The fractal shape of riparian forest patches. *Landscape Ecology* 4: 249-258.

Sokal, R.R. 1986. Spatial data analysis and historical processes. In: E. Diday et al. (eds.). *Data Analysis and informatics*. IV Proc. 4th Intl. Sym. Data Anal. Informatics (Versailles, France, 1985). North Holland, Amsterdam pp. 29-43.

Steele, J. H. (ed.). 1978. *Spatial Pattern in Plankton Communities*. Plenum Press. New York.

Steele, J. H. 1985. A comparison of terrestrial and marine ecological systems. *Nature* 313: 355-358.

_____. 1989. The ocean landscape. *Landscape Ecology* 3(3/4): 185-192.

Stommel, H. 1963. Varieties of oceanographic experience. *Science* 139: 572-576.

Sugihara, G., and R.M. May. 1990. Applications of fractals in ecology. *TREE* 5: 79-86.

Turner, M.G., and R.H. Gardner (eds.). 1991. *Quantitative Methods in Landscape Ecology*. Springer-Verlag, New York.

Upton, G.J.G., and B. Fingleton. 1985. *Spatial Data Analysis by Example. Vol. I: Point Pattern and Quantitative Data*. John Wiley & Sons, Chichester.

van Es, H. (This volume). The spatial nature of soil variability and its implications for field studies.

Voss, R.F. 1988. Fractals in nature: From characterization to simulation. In: H. Peitgen and D. Saupe (eds.). *The Science of Fractal Images*. Springer-Verlag, New York.

Weber, L.H., S.Z. El-Sayed, and I. Hampton. 1986. The variance spectra of phytoplankton, krill and water temperature in the Antarctic Ocean South of Africa. *Deep-Sea Research* 33: 1327-1343.

Wiens, J.A. 1976. Population responses to patchy environments. *Annual Review of Ecology and Systematics* 7: 81-120.

Wiens, J.A., and B.T. Milne. 1989. Scaling of "landscapes" in landscape ecology, or, landscape ecology from a beetle's perspective. *Landscape Ecology* 3: 87-96.

Williamson, M.H., and J.H. Lawton. 1991. Fractal geometry of ecological habitats. In: S.S. Bell, E.D. McCoy, and H.R. Mushinsky (eds.). *Habitat Structure*. Chapman and Hall, London, pp. 69-86.

PART III

CONCEPTS AND MODELS: AN OVERVIEW

John H. Steele

The chapters in this section provide a rich assembly of models used in terrestrial and marine ecology. Their applications can be very general, as with Durrett's stochastic models, or fairly specific, like Hofmann's applications to coastal waters. But there is also a diversity of techniques used to portray features of ecological systems. We may ask whether there are any general correspondences between methods and environments, particularly along the land/water axis.

Caswell and Etter categorize their models according to whether they are continuous or discrete in space and in time, allowing a four-way split. There would appear to be fairly natural correspondences between certain kinds of environments and different ecological rules. Especially, we tend to view spatial changes in the ocean as continuous, whereas on land we regard patches as discrete entities with defined boundaries. This difference emerges clearly in a comparison of the papers by Leibovich and by Nisbet et al. Leibovich describes the consequences of the vertical Langmuir circulation on the spatial patterns of biological materials. The focus in this approach is on the detailed physics of the water advection and diffusion. The only significant "biological" property is that the particles have a density different from the water, and so have an added natural velocity component. Yet this is sufficient to explain the regular windrows of material observed on the sea surface under certain weather conditions.

In contrast, Nisbet et al. consider two-patch metapopulation stability. The "patches" are internally homogenous. One of the patches is regarded as the environment of the other. Movement between patches is a simple exchange process. The "biology" in each patch is Lotka-Volterra prey-predator dynamics. The stability of these simple models is very dependent on the biological rates as well as the exchange rates. The ability to obtain analytical results depends on the simplicity of the spatial exchange. In particular, by having only two boxes, there can be no explicit second-order diffusive term.

Caswell and Etter describe the progression from two-box models through patch occupancy (statistical rather than spatial) models to large gridded systems where exchange processes can simulate second-order spatial processes. Obviously, in the limit, these models converge on the numerical simulations of space in oceanographic models such as Hofmann's. Are there reasons, apart from computer requirements, to retain the discrete structure?

If we accept the fractal paradigm as a description of spatial change, we have a system that is not amenable to traditional dynamic modeling. The continuous v. discrete approaches can be regarded as two simplifications where the discontinuities (in the spatial derivative) at all scales are aggregated to one scale—very small in the continuous models and relatively large for discrete simulations. The choice is dictated in part by our human scales of perception of land and sea.

Such simplifications in scales also relate to the temporal domain. There is a similar separation into discrete and continuous time. The Leslie matrix (Caswell 1989) has been prominent in ecological theory. The applications of cellular automata provide a spatial context. Both involve discrete time sequences at which biological events (reproduction, mortality, or movement) take place. The duration of this time step is, usually, linked to "biological" time scales. Or, by setting the time step to unity, particular transition rates are assumed. Thus

$$p_{t+\Delta t} - p_t = r \cdot \Delta t \cdot p_t (1 - p_t/c) \tag{1}$$

is a numerical approximation to the continuous logistic equation. However, the discrete form

$$p_{\tau+1} - p_\tau = \rho \cdot p_\tau (1 - p_\tau/c)$$

can also be regarded as an approximation using $\tau = \Delta t \cdot t$ with a new growth rate $\rho = r \cdot \Delta t$. Thus the numerical problem of size of time step is formally equivalent to the biological problem of appropriate growth rate (Hassell et al. 1976).

This is not trivial, since the continuous logistic has only single stable solutions, whereas the discrete logistic has been shown to have a whole array of responses up to chaos, depending on the value of ρ (May 1974).

It must also be pointed out that the opposite sequence can occur. A numerical solution of differential equations that exhibit chaos will degenerate to limit cycles if too large a time step is used. (This can be demonstrated with the Lorenz equations (1984).)

This question of discrete v. continuous becomes more complicated when space and time variables are combined. In aquatic simulations, there is usually a condition relating the time and space steps (e.g., $\Delta t/\Delta x^2 < 0.5(\text{diffusion rate})^{-1}$ for forward difference schemes). Such restrictions, imposed by the physics, define time steps very much shorter than the lifetimes of the organisms. Thus the "biology" is, perforce, continuous. It is possible to mix discrete and continuous by regarding a population as a set of discrete cohorts each of which develops continuously in time (Diekmann, this volume). This approach is useful when populations are varying in space (e.g., Steele and Mullin 1977), and it is the basis of fish population dynamics (Rothschild 1986).

For systems discrete in both space and time, there are obvious problems in deciding whether the results suggest biological realizations or are numerical experiments (Hassell et al. 1991). These systems are obviously dependent on the magnitudes of Δt and Δx —in fact, this is their fascination.

One other reason for this discrete/continuous dichotomy concerns the logistics of data collection. It is much easier to make quasi-continuous transects in the ocean than on land for some biological measurements—phytoplankton chlorophyll and echo-density from zooplankton or fish. Further, these biological parameters give simple sequences of numbers for a trophic character—"plants" or "herbivores."

This apparent simplicity, combined with the density of observations, allows the use of methods such as Fourier analysis to describe the data and also to test models of the animal behavior (Deutschman et al., this volume). At the same time, the intricacies of individual cycles of growth, reproduction, and dispersion may be lost.

These various options are illustrated in the final chapter in this section (Deutschman et al.). The analysis and modeling of krill data exemplifies the applications to dense data sets using not only the traditional Fourier methods but also the newer wavelet analysis, which has previously been used for forest data.

The problem of spatial resolution, which is particularly acute in terrestrial systems, can be studied by using various grid scales combined with different degrees of clustering. This can show properties that are and are not scale dependent. In particular, this can be applied to the effects of varying dispersion rates in prey-predator systems.

Lastly, spatial "chaos" can be investigated with continuous rather than discrete models. This has the advantage of focusing on the mechanisms rather than the time and space steps, and reveals the relation between the biological behavior and the physical dispersion.

In conclusion, the initial suggestion—that the discrete/continuous axis corresponds to the terrestrial/aquatic—has some basis in modeling practice and some justification in the nature of the two environments. But there is considerable overlap in actual systems and in ideas. The terrestrial concept of ecotones, as regions with strong gradients, is similar to the marine definition of fronts as physically dynamic areas separating more static water masses. Both ecotones and fronts are recognized as regions with active ecological processes distinct from the abutting regions. The technical problem is how to simulate such abrupt features theoretically, or in numerical models. Thus the discrete and continuous models are approximations which, with greater ecological insight and numerical expertise, may converge. This can be enhanced by continued exchange of concepts and methods between practitioners in the different ecological sectors.

REFERENCES

Caswell, H. 1989. *Matrix Population Models*. Sinauer Associates, Sunderland, MA.

Deutschman, D., G.A. Bradshaw, W.M. Childress, K.L. Daly, D. Grünbaum, M. Pascual, N. Schumaker, and J.Wu. (This volume.) Mechanisms of patch formation.

Diekmann, O. (This volume.) An invitation to structured (meta)population models.

Hassell, M.P., H.M. Comins, and R.M. May. 1991. Spatial structure and chaos in insect population dynamics. *Nature* 353:255-258.

Hassell, M.P., J.H. Lawton, and R.M. May. 1976. Patterns of dynamical behaviour in single-species populations. *Journal of Animal Ecology* 45:471-486.

Lorenz, E.N. 1984. Irregularity: A fundamental property of the atmosphere. *Tellus* 36A:98-110.

May, R.M. 1974. Biological populations with nonoverlapping generations: Stable points, stable cycles and chaos. *Science* 186:645-647.

Rothschild, B.J. 1986. *Dynamics of Marine Fish Populations*. Harvard University Press, Cambridge, MA.

Steele, J.H., and M.M. Mullin. 1977. Zooplankton dynamics. In: E.D. Goldberg (ed.). *The Sea: Ideas and Observations on Progress in the Study of the Seas*, Vol. 6. John Wiley and Sons, New York, 857-890.

7

Ecological Interactions in Patchy Environments: From Patch-Occupancy Models to Cellular Automata

Hal Caswell and Ron J. Etter

Introduction

The ecological theory of species interactions rests largely on the competition and predator-prey models of Lotka, Volterra, Nicholson, and Gause (e.g., May 1973). These models neglect spatial structure in general, and patchiness in particular. In this paper we introduce *cellular automata* (CA) as a new class of models for population interactions in space. We will discuss the relations between CA models and the more familiar reaction-diffusion and patch-occupancy formulations, and compare the results of a simple CA competition model to the corresponding Markov chain patch-occupancy model. This comparison reveals some of the factors that determine when simple patch-occupancy models are successful approximations, and when spatially explicit CA models are more appropriate.

Cellular Automata

We begin with some basic information about cellular automata (von Neumann 1966). A cellular automaton, in its simplest form, is a regular array of cells,[1] each of which is characterized by a discrete state variable. The state of the entire CA is given by the configuration of cell states. At each discrete time step, the state of each cell is updated as a function of its own state and the states of the neighboring cells. In the simplest case, the state transition rules are deterministic and identical for all cells, but stochastic and heterogeneous CA are possible, and important for ecological applications.

The simplicity of CA is deceiving. Although the transition rules operate strictly locally and uniformly, they can produce dramatic global spatiotemporal patterns. Because of this, the study of CA has become a growth industry among statistical physicists interested in problems of complexity *per se* (e.g., Wolfram 1986, Toffoli and Margolus 1987).

In an exhaustive study of the behavior of simple CA, Wolfram (1983) found four types of dynamics: (1) the initial pattern dies out to homogeneity, (2) the pattern evolves

---

[1] In our discussion of ecological CA, we will use the terms *cell* and *patch* interchangeably. We will refer to the entire array of cells as a landscape; this should not be taken as a restriction to terrestrial ecosystems, or even to systems on a solid substrate.

to a fixed periodic pattern, (3) the pattern grows indefinitely at fixed speed, usually exhibiting complex fractal self-similarity, and (4) the pattern grows and contracts irregularly. Types 1, 2, and 3 correspond to fixed points, periodic attractors, and chaotic attractors of dynamical systems. The complicated spatial propagating structures in Type 4 automata are effectively unpredictable except by actually constructing the automaton (e.g., on a computer) and watching what it does. The same categories are reported by Langton (1990) in a statistical sample of a much larger space of more complex CA and by Packard and Wolfram (1985) in simple two-dimensional CA.

Thus even the *simplest* CA are capable of extremely rich and complex behavior in space and time. More complex CA have been applied to the study of fluid dynamics, parallel computation, image processing, and developmental pattern formation (see Wolfram 1986, Toffoli and Margolus 1987). Some (e.g., the simple two-dimensional, two-state CA "Life," which is popular in recreational mathematics columns) have been shown to be capable of universal computation. That is, they are in principle capable of computing *any* computable function, and thus of simulating the behavior of anything that can be simulated by any digital computer.

Much of the literature on CA is devoted to their study as metaphors for complexity, as algorithms, or as computational objects. However, they can also be used *as models*; that is, their rules can be intended to describe the dynamics of some biological or physical system. It is this aspect of CA that we investigate in this paper. To help put ecological CA models into perspective, we begin with a brief comparison with other models for spatial population interactions.

CA Models and Reaction-Diffusion Models

Approaches to modeling population interactions in space and time can be classified by whether space, time, and population density are continuous or discrete. Some of the more important categories are:

1. **Reaction-diffusion-advection models**, where space, time, and population density are all continuous. They describe the local "reaction" of individuals (i.e., local population dynamics) and movement of individuals by diffusion and advection. They are formulated as partial differential equations and date back to the work of Fisher (1937) on diffusion of mutant genotypes, Skellam (1951) on dispersal of introduced species, and Kiersted and Slobodkin (1953) on patchiness in phytoplankton (see Okubo 1980, Britton 1986, Murray 1989 for reviews). They are widely used in oceanography to study planktonic ecosystems, because they couple naturally to fluid dynamic equations describing the movement of water masses.

2. **Reaction-diffusion networks** result when space in a reaction-diffusion system is cut up into discrete patches (Rosen 1981). Turing (1952) studied such equations as a model for pattern formation in morphogenesis (see Othmer and Scriven 1971, 1974). Their application in ecology has focused on conditions under which weak dispersal can destabilize a uniform species distribution and maintain coexistence of competing species. A perturbation theorem due to Karlin and McGregor (1972) and to Levin (1974) identifies one way in which small amounts of dispersal can have major effects on community structure. If, in the absence of dispersal, each of a set of

species is capable of dominating a patch (e.g., by a priority effect), then a sufficiently small amount of dispersal produces a stable equilibrium with every species present in every patch (Levin 1974). This equilibrium is eventually destabilized by further increases in dispersal, which, in effect, turn the system into a single patch.

3. **Coupled map lattices** are the discrete-time analogs of reaction-diffusion networks. A coupled map lattice is a discrete array of discrete-time maps, each interacting with its neighbors. The study of such map lattices has been encouraged by the recent interest in chaotic dynamics of nonlinear maps; coupling the maps into a lattice permits the study of bifurcations and chaos in both space and time (see review by Crutchfield and Kaneko 1987; for ecological examples see Sole and Valls 1991, Hassell et al. 1991).

4. **Cellular automata** result from simplifying a coupled map lattice by reducing the continuous patch state variable (population density) to a discrete index (e.g., presence/absence). The dynamics, expressed as a partial differential equation in the reaction-diffusion models, as a system of ordinary differential equations in a reaction-diffusion network, and as a system of difference equations in a coupled map lattice, now appear as a transition rule for the local cell state.

   CA models are relatively unexplored in ecology (or in biology as a whole). They have been used to model the development of pigment patterns in reptiles, mammals, fish, and mollusks (Cocho et al. 1987a,b, Gunji 1990, Vincent 1986) and to simulate growth patterns of clonal plants (Inghe 1989, Korona and Bystrykh 1978). Recently Hassell et al. (1991) used a CA to model host-parasitoid interactions and found that the temporal dynamics and spatial patterns were similar to those obtained from a coupled map lattice based on the Nicholson-Bailey model.

5. **Interacting particle systems** are stochastic, continuous time analogs of cellular automata (Liggett 1985, Durrett 1989). Ecological metaphors (forest fires, epidemic spread) have been used to motivate some particle system models (e.g., Grassberger 1983, MacKay and Jan 1984, von Niessen and Blumen 1986, Durrett 1988, Griffeath 1988). Interacting particle systems are related to percolation models (Grimmett 1989), which have recently been applied in landscape ecology (e.g., Gardner and O'Neill 1991).

   In this comparison, cellular automata appear as discrete approximations to the reaction-diffusion models that are the most detailed description of spatiotemporal interactions. The justifications for their use are many. They are much easier to simulate than reaction-diffusion models (indeed, simulation of reaction-diffusion models requires some discretization of space and time), and their expression in terms of transition rules lays bare the factors determining the dynamics.

CA Models and Patch-Occupancy Models

Cellular automata are also related to another important class of ecological models for patchy environments. Consider a species living on a landscape of discrete patches. Define the state of each patch by the presence or absence of the species (hence the name, *patch-occupancy model*). Empty patches are colonized from occupied patches; occupied patches become empty when the species becomes locally extinct. Assume that the system is well mixed, so that every patch interacts equally with every other. Then empty patches are colonized at a rate proportional to the product of the frequency of empty and occupied patches, while occupied patches go extinct at a constant rate. Let p denote the fraction of occupied patches. Then

$$\frac{dp}{dt} = -ep + cp(1 - p) \tag{1}$$

where e is the extinction rate and c is a colonization parameter. Such models were introduced by Levins (1970) and Cohen (1970) and have since been used, in either continuous- or discrete-time versions, to study competition (Slatkin 1974, Hanski 1983), predation (Vandermeer 1973, Hastings 1977), predator-mediated coexistence (Caswell 1978, Hastings 1978, Crowley 1979), diversity, succession, and species-area relationships (Caswell and Cohen 1991a,b, 1992). They are of current interest as a tool for studying metapopulations (Gilpin and Hanski 1991).

Patch-occupancy models are a first step in the incorporation of spatial processes, but in comparison with the family of models descended from reaction-diffusion models they are strangely limited. Models like the Lotka-Volterra equations assume that the population is uniformly distributed in space; they thus admit only a single spatial scale. Patch occupancy models admit precisely *two* spatial scales: that of the single patch and that of the entire landscape. Because they reduce the dynamics of an array of patches to a set of equations in the *proportion* of patches in each state, these models cannot include the effects of spatial arrangement. Thus patch-occupancy models are spatially averaged or "mean-field" approximations to corresponding CA models; the CA models have local rather than global interactions.

Being spatially explicit, the CA models include more information than their mean-field approximations. In simple one-dimensional, two-state CA, the mean-field approximation yields equilibrium state distributions within 10 – 20% of the correct values in most, but not all, cases (Wolfram 1983). More complex situations, where the spatial relationships are more important, may give very different results. In a later section, we will compare a cellular automaton model of competition with the corresponding patch-occupancy model.

Competition and disturbance. Caswell and Cohen (1991a,b) introduced a general method for constructing patch-occupancy models for any ecological interaction, directly from hypotheses about the time-scales of disturbance, dispersal, and the interaction itself. This approach can also be used to develop CA models; here we develop the competition model that we will analyze in later sections.

Consider two competing species (S_1 and S_2) on a patchy landscape. A patch may be occupied by S_1, S_2, or both. S_1 always excludes S_2 when they co-occur, with a characteristic time scale (i.e., mean time to competitive exclusion) τ_c. Patches are disturbed following a Poisson process with characteristic time scale (i.e., mean time between disturbances) τ_d; a disturbance produces an empty patch available for colonization.

Colonization is a Poisson process in which the mean number of propagules arriving at an empty patch per unit time is proportional to the frequency of occurrence of the species. Let x_1, x_2, x_3, and x_4 denote the proportion of patches containing, respectively, neither species, only S_1, only S_2, and both species. Then the species frequencies f_i are given by

$$f_1 = x_2 + x_4$$
$$f_2 = x_3 + x_4$$

The patch occupancy model for this interaction can be written as a nonlinear Markov chain

$$\mathbf{x}_{t+1} = \mathbf{A}_x \mathbf{x}_t \tag{2}$$

where the transition matrix $\mathbf{A}_x$ is given by

$$\begin{pmatrix} (1 - C_1)(1 - C_2) & p_d & p_d & p_d \\ C_1(1 - C_2) & 1 - p_d & 0 & (1 - p_d)p_c \\ (1 - C_1)C_2 & 0 & (1 - C_1)(1 - p_d) & 0 \\ C_1 C_2 & 0 & C_1(1 - p_d) & (1 - p_d)(1 - p_c) \end{pmatrix} \tag{3}$$

where $p_d = 1/\tau_d$ is the disturbance rate, $p_c = 1/\tau_c$ is the competitive exclusion rate, and the C_i are colonization rates, given by

$$C_i = 1 - e^{-d_i f_i} \tag{4}$$

where d_i is a dispersal rate coefficient. This model has been extensively analyzed (Caswell and Cohen 1991a,b). When disturbance is frequent enough, the fugitive species S_2 persists regionally even though it is excluded locally. Its frequency f_2 is maximized at intermediate disturbance frequencies, as is the alpha or within-patch diversity. The disturbance frequency at which diversity is maximized varies directly with the rate of competitive exclusion; the height of the maximum varies inversely with the rate of competitive exclusion.

Notice that the model (3), like all patch-occupancy models, contains no explicit spatial configuration. Any landscape with the same cell state proportions behaves identically, regardless of the cell arrangement. CA models relax this crucial assumption.

From Ecological Interactions to CA Models

To construct an ecological CA model is to translate a set of ecological hypotheses into a set of CA transition rules. The approach to constructing patch-occupancy models in the preceding section can be applied directly to CA models by defining the colonization probabilities C_i in terms of *local* species frequencies (i.e., in the immediately neighboring cells) rather than in terms of the global occurrence of the species. This permits an exact comparison of the results of the CA model and the corresponding mean-field patch-occupancy model.

CA AND PATCH-OCCUPANCY MODELS COMPARED

We set out to compare the patch-occupancy model (3) with the corresponding CA model. Even the simplest CA models resist analytical approaches. The lengths to which able mathematicians must go to prove even the simplest theorems about them (e.g., Durrett 1989, Liggett 1985) makes these models an arena for "experimental mathematics" (Griffeath 1988). Computer simulation is the fundamental tool for their investigation.

In principle, CA simulation is easy. An array stores the values of the cell states, and each cell is updated as a function (possibly stochastic) of its own state and the states of its neighbors. However, storing large arrays of cells and updating them all repeatedly is computationally demanding. Special-purpose coprocessor boards are available for CA simulation (Toffoli and Margolus 1987), but they are too limited in the number of states per cell to be of use in ecological models. Massively parallel computers are ideally suited for CA simulation, since each cell can in principle be allocated its own processor. The results in this paper, however, were obtained on a microcomputer (Apple Macintosh IIci programmed in Lightspeed Pascal); the resulting code is slow but serviceable.

Our CA contained 65,536 cells (256 × 256), with periodic boundary conditions (i.e., the upper and lower and the left and right edges were connected to form a torus). Species frequencies for colonization calculations were evaluated using a Moore (8-cell) neighborhood. The model was initialized with equal frequencies of the four states in a random spatial distribution.

Simulations were run until there was no consistent directional change in the frequency of any of the four states. CA were considered to have reached equilibrium if the slopes of the regression lines fit to the last 100 values of each state were all less than 0.001 in absolute value.

The output of patch-occupancy models like (3) includes state frequencies, species frequencies, the mean and variance of alpha diversity, beta diversity, interspecific association patterns, and a variety of measures of turnover rate and recurrence times. This is a subset of the output available from the corresponding CA model, since the patch occupancy model produces no explicitly spatial output, whereas the CA does. Here we will focus on comparisons of species frequencies and alpha diversity. More extensive analyses will be presented elsewhere.

For these simulations we varied the competitive exclusion probability p_c (from 0.01 to 1), the disturbance probability p_d (from 10^{-3} to almost 1) and the dispersal coefficients d_1 and d_2 (all four combinations of 1 and 10).

Results

Like the patch-occupancy models, the cellular automata converge, usually rapidly, to an equilibrium cell state frequency distribution. Figure 1 shows a three-dimensional time series of the complete landscape. The top face of the figure shows a spatial cross-section at equilibrium; there is no apparent structure to the cell states, all four of which are clearly present. The sides of the figure show the temporal development of transects across the landscape; the process of disturbance, colonization, and exclusion can be seen by following individual patches upwards in the figure.

The equilibrium properties of the landscape depend on the model parameters. Figures 3 and 4 show the frequency of the losing competitor (S_2) and the mean alpha diversity as functions of the disturbance rate, the dispersal coefficients, and the competitive exclusion rate, for the same parameter values used in the patch-occupancy model of Caswell and Cohen (1991b).

The two models agree closely, except when disturbance rates or dispersal rates are low, in which case both f_2 and alpha diversity are slightly *lower* in the CA model than in the Markov chain model. These differences reflect the inter-neighborhood variance

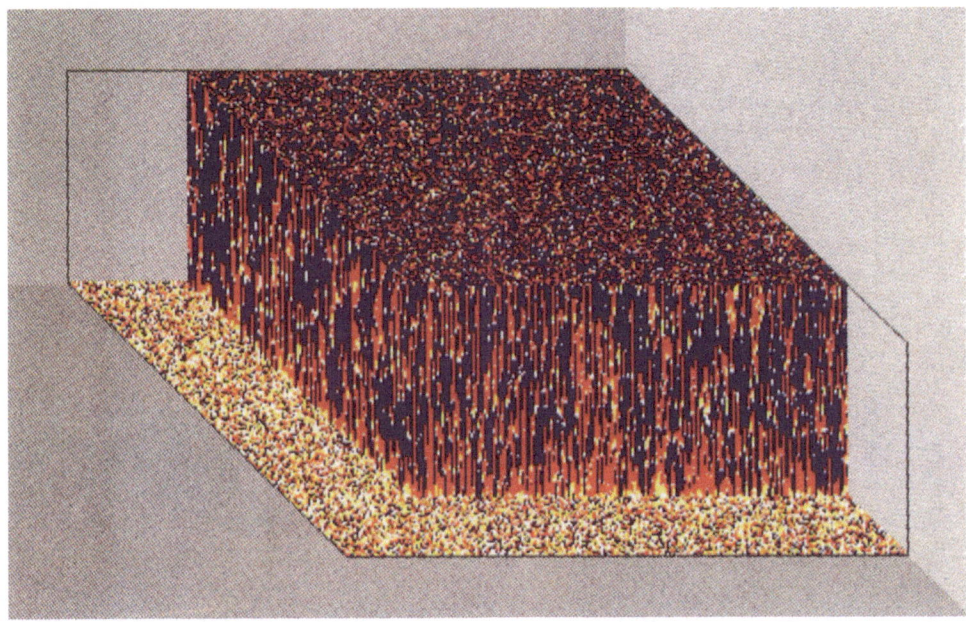

Figure 1: A three-dimensional spatiotemporal series for the CA model of competition, with parameter values $d_1 = 1$, $d_2 = 10$, $p_d = 0.098$ and $p_c = 0.1$. White $= x_1$ (empty), blue $= x_2$ (S_1 only), yellow $= x_3$ (S_2 only), red $= x_4$ (both species).

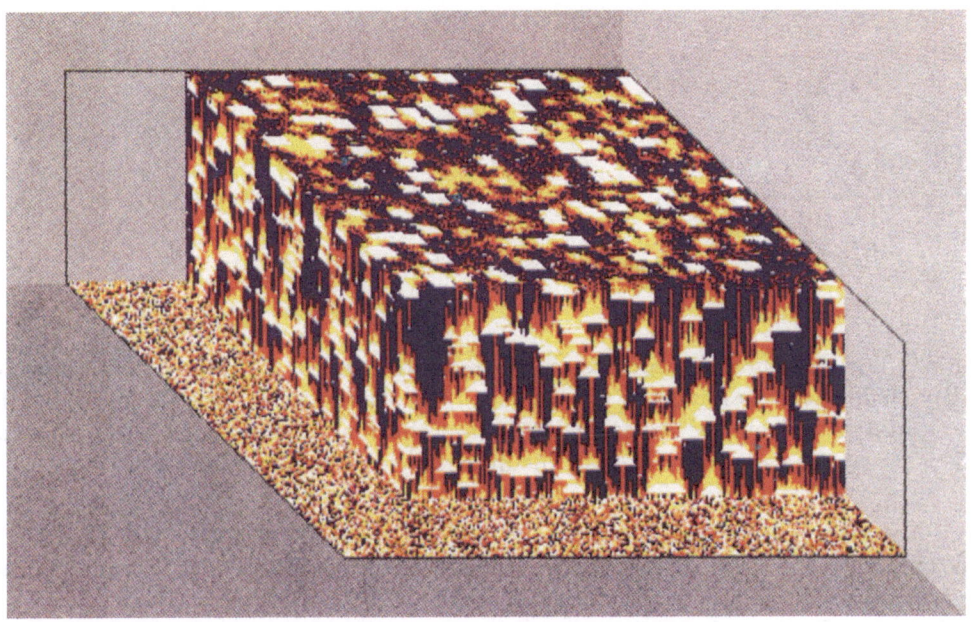

Figure 2: A 3-dimensional spatiotemporal series for the CA model of competition with variable disturbance sizes. Parameter values and scale as in Figure 1.

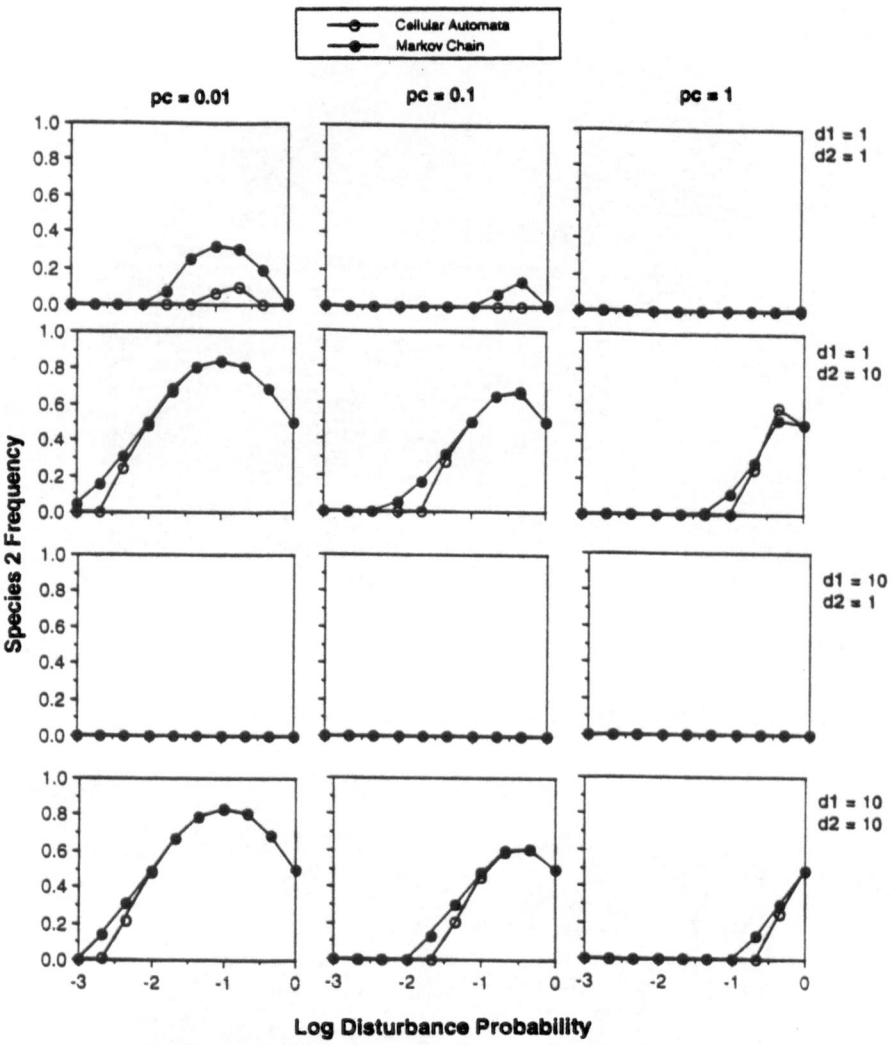

Figure 3: The frequency of occurrence of the losing competitor S_2 in the patch-occupancy Markov chain model (•) and the CA model (o) for competition. Results are shown for three values of the competitive exclusion rate p_c and four combinations of the dispersal coefficients d_1 and d_2, as a function of the disturbance probability p_d.

in cell state frequencies, which exists only in the CA model. Consider S_2, which relies on colonization of empty patches for persistence. Its colonization rate C_2 is a concave downward function of f_2 (Eq. 4); thus the effect of inter-neighborhood variance in f_2 is to reduce the effective colonization rate.

     We can turn to the patch-occupancy model to determine the consequences of this reduction. Most obviously, it reduces the equilibrium frequency of S_2 and increases the

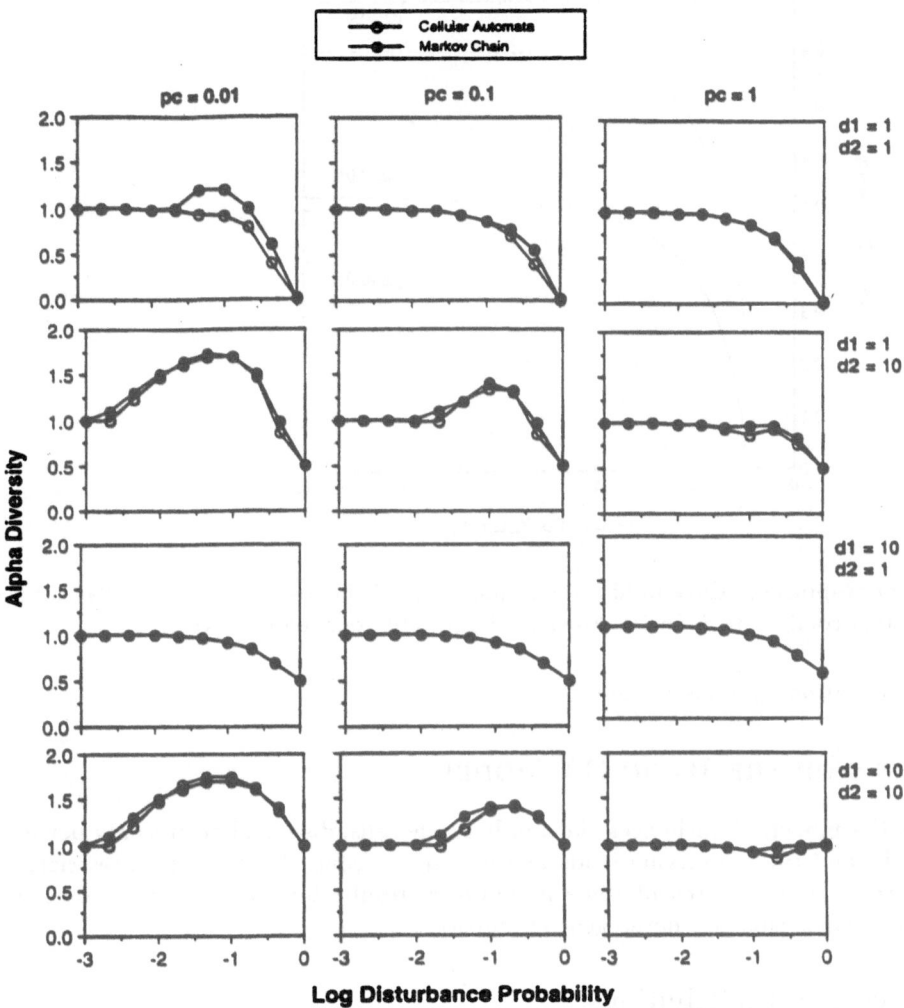

Figure 4: Mean local (alpha) diversity in the patch-occupancy Markov chain model (•) and the CA model (o) for competition. Results are shown for three values of the competitive exclusion rate p_c and four combinations of the dispersal coefficients d_1 and d_2, as a function of the disturbance probability p_d.

critical disturbance rate for S_2 to persist. Because the equilibrium frequency f_2 is itself a concave downward function of d_2 (Figure 5), it is more sensitive to reductions in d_2 when d_2 is small. This is just where the difference between the Markov chain and CA models is greatest (Figure 3). When d_2 is large enough, the Markov chain and CA models converge at high disturbance rates. This also follows from Figure 5: when p_d is large, f_2 becomes insensitive to changes in d_2 when d_2 is large. Thus the variance-induced reduction of the

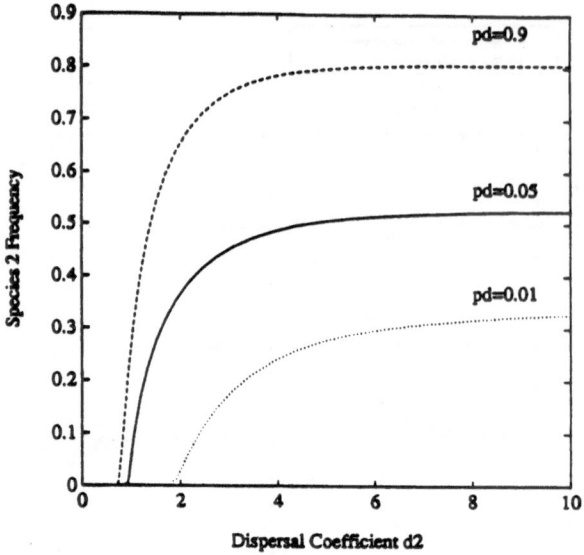

Figure 5: The response of the equilibrium frequency f_2 of the losing competitor to changes in the dispersal coefficient d_2 in the Markov chain patch-occupancy model.

effective colonization rate has no effect.

EXTENSIONS OF THE BASIC CA MODEL

In the basic CA model, disturbances affect only single cells, dispersal occurs only between adjacent cells, and the underlying landscape is homogeneous. Each of these restrictions can be relaxed. Here we present some preliminary results for two of these extensions: variable disturbance size and heterogeneous substrates.

Disturbance Size Distributions

Because patch-occupancy models assume that each patch interacts equally with every other patch, the disturbance process is completely specified by its frequency. The arrangement of the disturbances in space has no effect. By contrast, when patches interact only with their neighbors, the size of disturbed areas is important, because the center of the disturbance can be colonized only from the edges. At least some reports suggest that disturbance size distributions are skewed to the right (i.e., many small and a few large disturbances). Lognormal distributions of disturbance size have been reported by Paine and Levin (1981) in a rocky intertidal community and by Runkle (1985) in a temperate forest. Brokaw (1982) and Denslow (1987) reported skewed gap size distributions in tropical forests.

To evaluate the impact of variable disturbance sizes, we simulated the CA competition model with a simple skewed size distribution (more complete analyses will be reported elsewhere). Each cell in the landscape may be chosen, with independent probability $\hat{p}_d$, as the focal point of a disturbance event. The relation between $\hat{p}_d$ and p_d is given below. The size of the resulting disturbance is $j \times j$ cells with probability g_j; in our simulation we set

$$
\begin{aligned}
g_1 &= 0.75 \\
g_2 &= 0.24 \\
g_{10} &= 0.01
\end{aligned}
\tag{5}
$$

That is, 75% of the disturbance events affect only the focal point cell. Twenty-four percent affect a 2×2 patch of cells, and one event in 100 devastates a 10×10 patch. The probability $\hat{p}_d$ was chosen so that the per-cell probability of disturbance is fixed at p_d, for comparison with the Markov chain patch-occupancy model and the CA model with single-cell disturbances. The probability $\hat{p}_d$ must satisfy

$$
p_d = 1 - (1 - \hat{p}_d) \prod_{j=1}^{\infty} (1 - X_j)
\tag{6}
$$

where $X_j = 1 - (1 - \hat{p}_d g_j)^{j^2 - 1}$ is the probability that a randomly chosen cell falls in the 'shadow' of disturbance of size $j \times j$.

The effect of large disturbances is visible in the 3-dimensional time series (Figure 2). At equilibrium, the landscape shows a clear pattern of patchiness on a scale defined by the disturbance size distribution. Large empty (i.e., recently disturbed) patches, and patches in the process of colonization, are readily apparent. The temporal dynamics (the edges of the figure) show the creation and gradual elimination of disturbed patches, as the species colonize from the edges.

The CA model with variable disturbance size produces lower frequencies of S_2 and lower alpha diversities than the corresponding Markov chain model (Figures 6 and 7). The discrepancy is greater than in the case of the CA model with single-cell disturbances. Moreover, it does not disappear at high disturbance frequencies. This is because in the variable disturbance size model, not only is the effective colonization rate reduced by inter-neighborhood variance, but the effective disturbance rate is reduced because some of the disturbances come in 10×10 patches; the cells in the interior of such a patch are not immediately available for colonization. In terms of scale, the disturbance size distribution introduces a scale intermediate between the single cell and the complete landscape; since the Markov chain model includes only those two scales, it is no surprise that it is a poorer approximation to the CA model in this case.

Substrate Heterogeneity

The preceding simulations examine biological patterns produced on a homogeneous landscape.[2] Real ecological patterns, however, are the result of both ecological interactions and substrate heterogeneity. Here we present some preliminary results on the interaction of the two sources of heterogeneity.

The simplest model for landscape heterogeneity is a two-phase landscape with different rules for the two phases. The two phases can differ in any parameter. They might represent different micro-environmental regimes, which change the outcome of competition, or different nutrient levels, which change the rates of competitive exclusion. Figure 8

---

[2]The homogeneity of the landscape is defined in the only terms that matter in CA: the rules. If the rules are the same in every patch, the landscape is homogeneous.

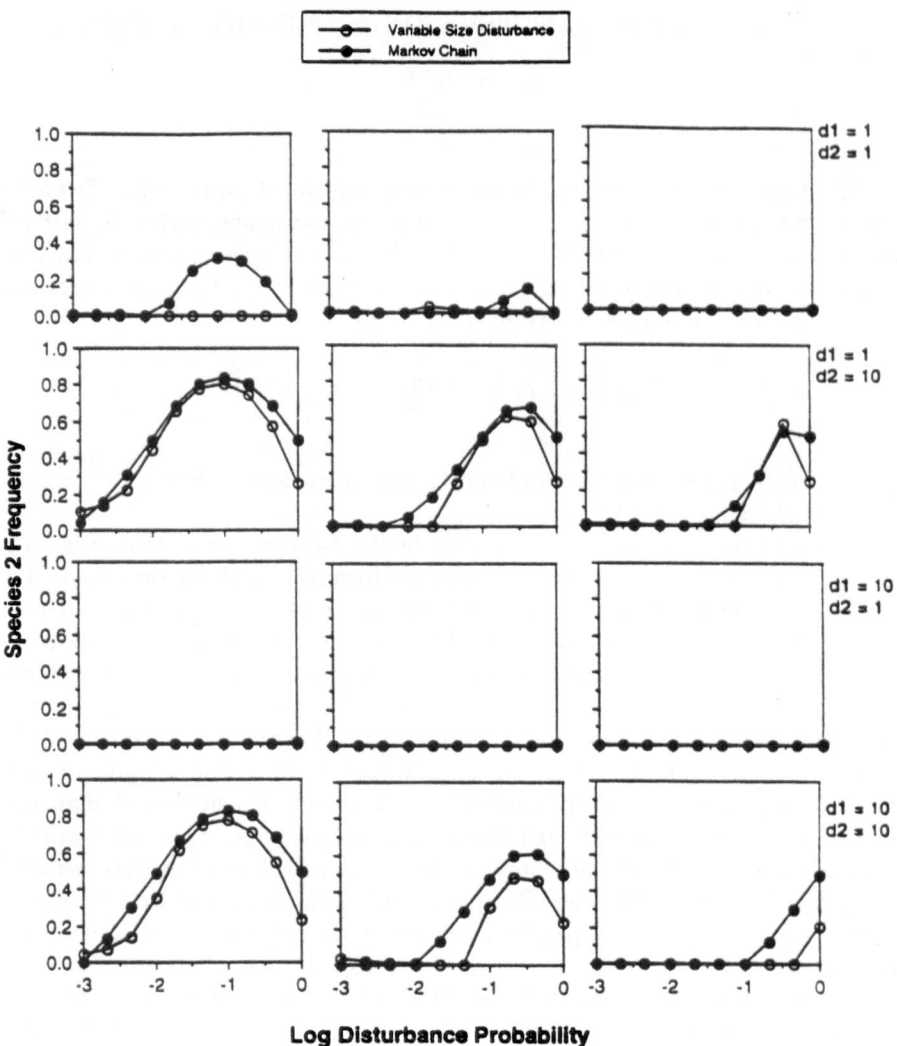

Figure 6: The frequency of occurrence of the losing competitor S_2 in the patch-occupancy Markov chain model (•) and the CA model with variable disturbance sizes (o). Results are shown for three values of the competitive exclusion rate p_c and four combinations of the dispersal coefficients d_1 and d_2, as a function of the disturbance probability p_d.

shows a fractally heterogeneous landscape; such a landscape is heterogeneous at all scales down to the level of single cells (cf. Milne 1991).

The intensity of the landscape heterogeneity can be adjusted by specifying the difference in the parameter(s) between the two phases; in the limit as this difference goes to zero, we obtain a homogeneous landscape. For example, Figure 8 shows our two-species competition model (3), with single-cell disturbances, on a fractal landscape (shown in the

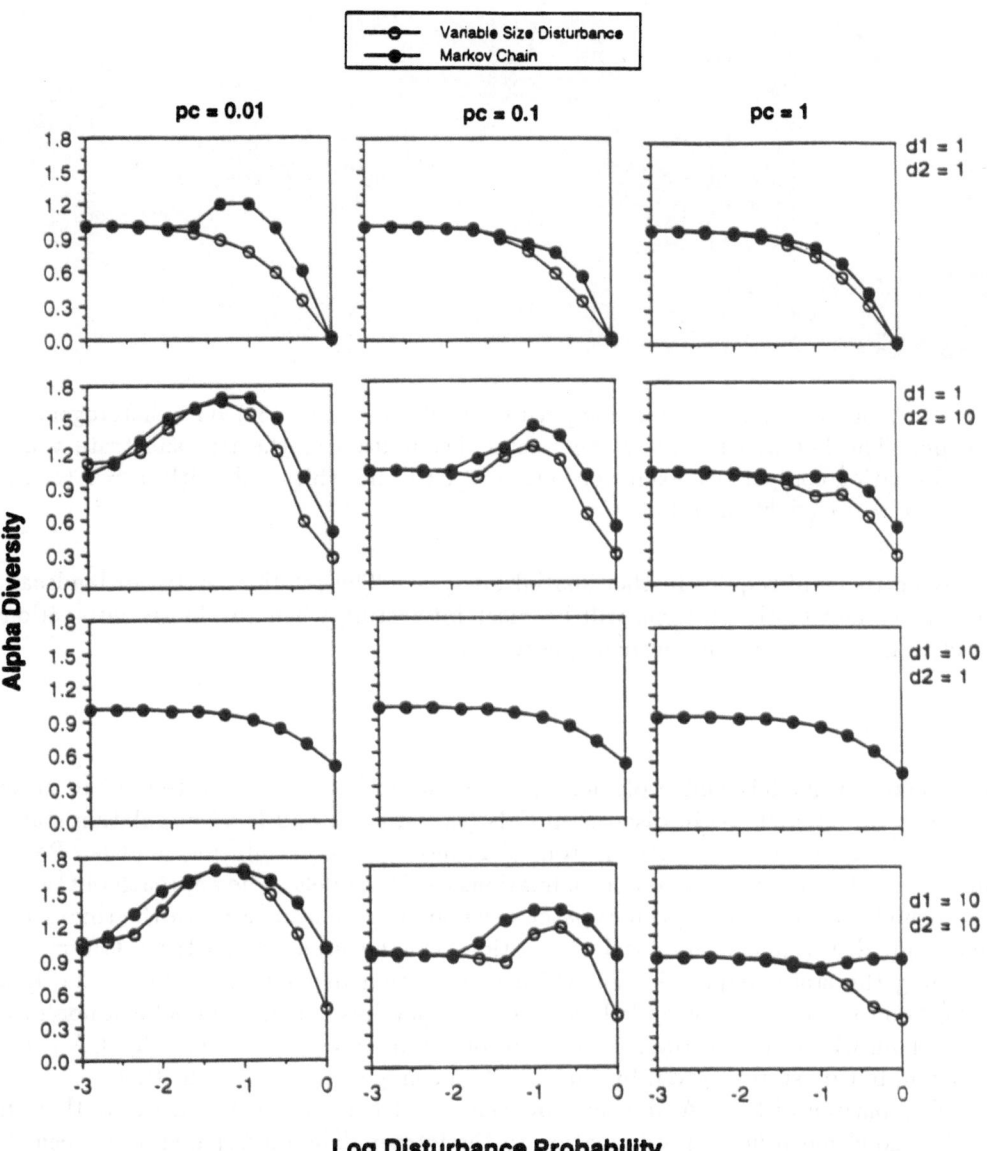

Figure 7: Mean local (alpha) diversity in the patch-occupancy Markov chain model (•) and the CA model with variable disturbance sizes (o). Results are shown for three values of the competitive exclusion rate p_c and four combinations of the dispersal coefficients d_1 and d_2, as a function of the disturbance probability p_d.

left panel) in which the phases differ in competitive exclusion rate. The center panel shows the results of a large difference ($p_c = 0.01$ vs. $p_c = 1$) between the phases. In the right panel the intensity of the landscape heterogeneity is reduced ($p_c = 0.01$ vs. $p_c = 0.1$); and it becomes more difficult to detect in the final biotic pattern. Close examination (in color rather than in the black and white reproduced here) suggests that as the intensity of the heterogeneity is reduced, the pattern disappears first at smaller spatial scales.

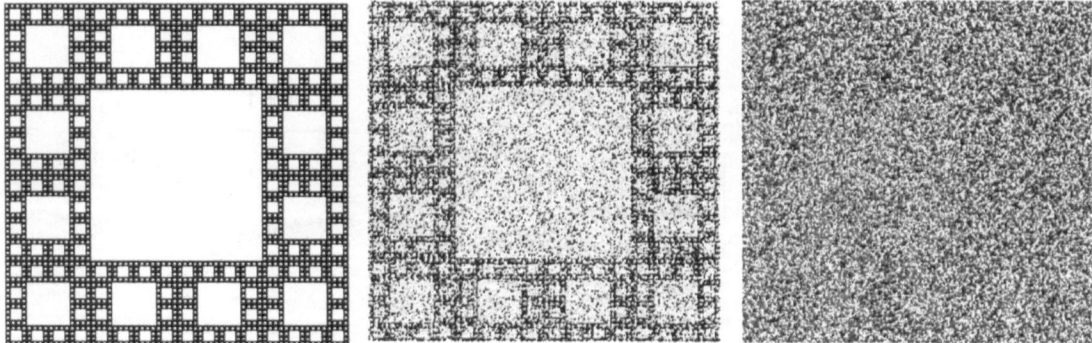

Figure 8: A patch CA model for competition (following Eq. (3)) on a heterogeneous landscape. The left panel depicts the fractal distribution of the two substrate types: black cells with a competitive exclusion rate $p_c = 0.01$ and white cells with $p_c = 1$ (center panel) or $p_c = 0.1$ (right panel).

Quantifying this will require further spatial analyses. The relations between landscape heterogeneity and biotic patterns will be most interesting when the biotic interactions themselves generate structure on multiple scales.

CONCLUSIONS

Patch-occupancy models and ecological cellular automata are both attempts to study spatiotemporal interactions in discrete models (space, time, and local population state), at the opposite end of the spectrum from continuous reaction-diffusion models. Patch-occupancy models are mean-field approximations to CA models. The approach of Caswell and Cohen (1991a,b), which specifies patch-occupancy models in terms of the time scales of dispersal, disturbance, and local interaction, can be used to construct CA models by defining the state frequencies on which colonization depends in terms of the local neighborhood rather than the global average. The patch-occupancy model is a nonlinear Markov chain whose state variable is a vector of patch state proportions. The CA is also Markovian, but its state is given by the complete configuration of patch states.

Comparison of the CA and Markov chain models for competition shows that the latter is a good mean-field approximation to the former. The discrepancies between the two can be explained in terms of inter-neighborhood variance in patch proportions; this variance reduces the effective colonization rate of the losing competitor. At sufficiently high disturbance rates and dispersal rates, the two models agree extremely closely, because the results are insensitive to changes in colonization rates under these conditions.

The patch-occupancy model is a poorer approximation to a CA model with variable disturbance sizes, because the disturbance size distribution introduces a spatial scale intermediate between the single patch and the entire landscape. Similarly, a model with substrate heterogeneity yields spatial patterns that begin to reflect the variability among patches when the pattern becomes sufficiently intense.

We conjecture that patch-occupancy models will provide good mean-field approximations to CA models as long as there are no important processes operating at scales intermediate between the single patch and the complete landscape. Of course, even in these cases, mean-field approximations produce only mean-field results (species frequencies and the like); they are blind to a wealth of spatially explicit information that can be revealed by a CA model. Cellular automata models are a powerful tool for investigating the spatiotemporal patterns resulting from ecological interactions. Because space, time, and state are all discrete, CA models are less quantitatively precise than reaction-diffusion models; but because they can be formulated directly in terms of the time scales of the processes involved they can reveal important qualitative patterns.

ACKNOWLEDGMENTS

This research was supported by grants from NSF (OCE-8900231 and DEB-9119420), DOE (DE-FG02-89ER6088W) and ONR (N00014-92-J-1527), by a Guggenheim Fellowship to HC and by Postdoctoral Scholarships at WHOI and at Rutgers to RJE. We thank Gay Bradshaw, Joel Cohen, Richard Durrett and John Steele for comments. WHOI Contribution 7944.

REFERENCES

Britton, N.F. 1986. *Reaction-Diffusion Equations and Their Applications to Biology.* Academic Press, New York.

Brokaw, N. 1982. The definition of a treefall gap and its effect on measures of forest dynamics. *Biotropica* 14:158–160.

Caswell, H. 1978. Predator-mediated coexistence: A nonequilibrium model. *American Naturalist* 112:127–154.

Caswell, H. and J. E. Cohen. 1991a. Communities in patchy environments: A model of disturbance, competition, and heterogeneity. In: J. Kolasa and S.T.A. Pickett (eds.). *Ecological Heterogeneity.* Springer-Verlag, New York, pp. 97–122.

——————— 1991b. Disturbance, interspecific interaction, and diversity in metapopulations. *Biological Journal of the Linnean Society* 42:193–218.

——————— 1992. Local and regional regulation of species-area relations: A patch-occupancy model. In R.E. Ricklefs and D. Schluter (eds.). *Community Diversity.* University of Chicago Press (in press.)

Cocho, G., R. Pérez-Pascual and J.L. Rius. 1987a. Discrete systems, cell-cell interactions and color pattern of animals. I. Conflicting dynamics and pattern formation. *Journal of Theoretical Biology* 125:419–435.

Cocho, G., R. Pérez-Pascual, J.L. Rius and F. Soto. 1987b. Discrete systems, cell-cell interactions and color pattern of animals. II. Clonal theory and cellular automata. *Journal of Theoretical Biology* 125:419–435.

Cohen, J.E. 1970. A Markov contingency-table model for replicated Lotka-Volterra systems near equilibrium. *American Naturalist* 104:547–560.

Crowley, P.H. 1979. Predator-mediated coexistence: An equilibrium interpretation. *Journal of Theoretical Biology* 80:129–144.

Crutchfield, J.P. and K. Kaneko. 1987. Phenomenology of spatio-temporal chaos. In B. Hao (ed.). *Directions in Chaos.* World Scientific, Singapore.

Denslow, J.S. 1987. Tropical rainforest gaps and tree species diversity. *Annual Review of Ecology and Systematics* 18:431–451.

Durrett, R. 1988. Crabgrass, measles, and gypsy moths: An introduction to interacting particle systems. *Mathematical Intelligencer* 10:37–47.

——————— 1989. *Lecture Notes on Particle Systems and Percolation.* Wadsworth and Brooks/Cole, Pacific Grove, California.

Fisher, R.A. 1937. The wave of advance of advantageous genes. *Annals of Eugenics* (London) 7:355–369.

Gardner, R.H. and R.V. O'Neill. 1991. Pattern, process, and predictability: The use of neutral models for landscape analysis. In: M.G. Turner and R.H. Gardner (eds.). *Quantitative Methods in Landscape Ecology.* Springer-Verlag, New York.

Gilpin, M. and I. Hanski (eds.). 1991. *Metapopulation Dynamics: Empirical and theoretical investigations.* Academic Press, London.

Grassberger, P. 1983. On the critical behavior of the general epidemic process and dynamical percolation. *Mathematical Biosciences* 63:157–172.

Griffeath, D. 1988. Cyclic random competition: A case history in experimental mathematics. *Notices of the American Mathematical Society* 35:1472–1480.

Grimmett, G. 1989. *Percolation.* Springer-Verlag, New York.

Gunji, Y. 1990. Pigment color patterns of molluscs as an autonomous process generated by asynchronous automata. *Biosystems* 23:317–334.

Hanski, I. 1983. Coexistence of competitors in patchy environment. *Ecology* 64:493–500.

Hassell, M.P., H.N. Comins, and R.M. May. 1991. Spatial structure and chaos in insect population dynamics. *Nature* 353:255–258.

Hastings, A. 1977. Spatial heterogeneity and the stability of predator prey systems. *Theoretical Population Biology* 12:37–48.

——————— 1978. Spatial heterogeneity and the stability of predator-prey systems: Predator-mediated coexistence. *Theoretical Population Biology* 14:380–395.

Inghe, O. 1989. Genet and ramet survivorship under different mortality regimes—a cellular automata model. *Journal of Theoretical Biology* 138:257–270

Karlin, S. and J. McGregor. 1972. Polymorphisms for genetic and ecological systems with weak coupling. *Theoretical Population Biology* 3:210–238.

Kierstead, H. and L.B. Slobodkin. 1953. The size of water masses containing plankton blooms. *Journal of Marine Research* 12:141–147.

Korona, V.V. and L.V. Bystrykh. 1978. Clump formation in *Festuca rubra* (Poacea) as a growth process of cellular automata. *Bot. Zh.* (Leningrad) 63:1199–1202. (In Russian).

Langton, C.G. 1990. Computationat the edge of chaos: Phase transitions and emergent computation. *Physica D* 42:12–37.

Levin, S.A. 1974. Dispersion and population interactions. *American Naturalist* 108:207–228.

Levins, R. 1970. Extinction. pp.77–107 in M. Gerstenhaber (ed.). *Lectures on Mathematics in the Life Sciences* , vol. 2. American Mathematical Society.

Liggett, T.M. 1985. *Interacting Particle Systems.* Springer-Verlag, New York.

MacKay, G. and N. Jan. 1984. Forest fires as critical phenomena. *Journal of Physics A: Mathematical and General* 17:L757–L760.

Markus, M. and B. Hess. 1990. Isotropic cellular automaton for modelling excitable media. *Nature* 347:56–58.

May, R.M. 1973. *Stability and Complexity in Model Ecosystems.* Princeton University Press, Princeton.

Milne, B. T. 1991. Lessons from applying fractal models to landscape patterns. In: M.G. Turner and R.H. Gardner (eds.). *Quantitative Methods in Landscape Ecology.* Springer-Verlag, New York, pp. 199–235.

Murray, J. 1989. *Mathematical Biology.* Springer-Verlag, New York.

Okubo, A. 1980. *Diffusion and Ecological Problems: Mathematical Problems.* Springer-Verlag, New York.

Othmer, H.G. and L.E. Scriven. 1971. Instability and dynamic pattern in cellular networks. *Journal of Theoretical Biology* 32:507–537.

——————— 1974. Non-linear aspects of dynamic pattern in cellular networks. *Journal of Theoretical Biology* 43:83–112.

Packard, N. and S. Wolfram. 1985. Two-dimensional cellular automata. *Journal of Statistical Physics* 38:901.

Paine, R. T. and S. A. Levin. 1981. Intertidal landscapes: Disturbance and the dynamics of pattern. *Ecological Monographs* 51:145–178.

Rosen, R. 1981. Pattern generation in networks. *Progress in Theoretical Biology* 6:161–209, Academic Press, New York.

Runkle, J.R. 1985. Disturbance regimes in temperate forests. pp. 17–33 in S.T.A. Pickett and P.S. White (eds.). *The Ecology of Natural Disturbance and Patch Dynamics.* Academic Press, New York.

Skellam, J.G. 1951. Random dispersal in theoretical populations. *Biometrika* 38:196–218.

Slatkin, M. 1974. Competition and regional coexistence. *Ecology* 55:128–134.

Sole, R.V. and J. Valls. 1991. Order and chaos in a 2D Lotka-Volterra coupled map lattice. *Physics Letters A* 153:330–336.

Toffoli, T. and N. Margolus. 1987. *Cellular Automata Machines: A New Environment for Modeling.* MIT Press, Cambridge.

Turing, A. M. 1952. The chemical basis of morphogenesis. *Philosophical Transactions of the Royal Society* B237:37–72.

Vandermeer, J.M. 1973. On the regional stabilization of locally unstable predator-prey relationships. *Journal of Theoretical Biology* 41:161–170.

Vincent, J.F.V. 1986. Cellular automata: A model for the formation of color patterns in molluscs. *Journal of Molluscan Studies* 52:97–105.

von Niessen, W. and A. Blumen. 1986. Dynamics of forest fires as a directed percolation model. *Journal of Physics A: Mathematical and General* 19:L289–L293.

Von Neumann, J. 1966. *Theory of Self-Reproducing Automata.* Univ. of Illinois Press.

Wolfram, S. 1983. Statistical mechanics of cellular automata. *Reviews of Modern Physics* 55:601.

——————— 1986. *Theory and Application of Cellular Automata.* World Scientific, Singapore.

8
SPATIAL AGGREGATION ARISING FROM CONVECTIVE PROCESSES

Sidney Leibovich

INTRODUCTION

Irving Langmuir (1938) carried out the first scientific investigation of the phenomenon now called Langmuir circulation. This physical process became known to him by the spatial aggregation of biological material that it caused, which he observed as rows of *Sargassum* on the ocean surface during a cross-Atlantic voyage by ship. Had the patchiness in the *Sargassum* distribution been less apparent, it is very likely that the discovery of the physical phenomenon would have taken a much longer time.

The rows were essentially parallel to the direction of the surface wind, and are called windrows. Earlier published references to rows aligned with the wind on natural bodies of water can be found (see the review in Leibovich 1983), but the underlying causes had not been considered. Langmuir, in a subsequent series of experiments on Lake George in New York State, showed that the collection of surface material into rows was due to convective motions, and inferred that the phenomenon owed its existence to the action of the wind by some unknown mechanism.

The potential biological significance of Langmuir circulations has been commonly appreciated (see, for example, the reviews by Fasham 1978 and by Denman & Powell 1984) since first pointed out by Woodcock (1944). In an important short paper, Stommel (1948) recognized that Langmuir circulations, as well as other convective motions, can lead to a suspension of particulate matter well removed from the surface at which, by virtue of their buoyancy, they would normally be imagined to congregate. This observation enlarges the scope of spatial aggregation from a phenomenon situated at locations of surface water convergence to a three-dimensional question. The vertical transport in convective motion apparently is of greater biological import than the horizontal sweeping that creates the most readily observed patterns.

The objective of this chapter is to see how patterning comes about by convective processes. Convective processes are mainly up-down water motion set up by vertical gradients. They tend to equalize the vertical properties in the water column responsible for the convection. This means that they are inherently self–limiting, since they act to erase the agency that gives rise to them.

Convective processes such as Langmuir circulation are not the only physical phenomena leading to the formation of patterned arrangements of marine organisms in the upper layers of the ocean. These arrangements are commonly in the form of rows, and reports of such patterns may be traced back hundreds of years (Bainbridge 1957), mostly without recognition that the material was living. Other physical mechanisms which may produce such aggregative results include thermal convection (see Owen 1966), which occurs when there is cooling of the water at the air-sea interface; thermohaline convection, which is due to very disparate rates of the molecular diffusivities of salt and heat; internal wave activity; and fronts separating water masses with different properties. There presumably also are many other possible causes for patterning.

In the next section, the physical mechanisms responsible for thermal convection, the prototypical convective activity, will be discussed in a nonmathematical fashion. The relevant scales for these processes are extracted heuristically, with the aim of developing an understanding of the prospects for interaction with marine biota. Then the emphasis will shift to Langmuir circulations, with a brief discussion of the available experimental information and current theoretical understanding. The relevant scales in this case will be extracted from experiment and theory following, more or less, the method used for thermal convection. Finally, aggregation caused by convective processes is discussed, taking as an example, either a non-motile organism with positive buoyancy (such as *Sargassum* or certain pelagic fish eggs), or an organism that exhibits negative geotaxis or positive phototaxis, swimming vertically upwards at a mostly constant speed.

THERMAL CONVECTION: MECHANISM AND SCALES

The standard idealized model for thermal convection, upon which more realistic problems are based, contemplates a horizontal layer of fluid (liquid or gas) of finite depth, say d. Its top surface (coinciding with the plane $z = d$) is maintained at one fixed temperature, T_{top}, and the bottom surface (at the plane $z = 0$) is held at a another fixed temperature, $T_{bottom} > T_{top}$. Under most circumstances, fluids expand when heated and thereby become less dense, or lighter per unit volume; and contract and become more dense, hence heavier per unit volume, when cooled. Heavier fluid lying above lighter fluid will tend to sink. "Tend" is the operative word, because if the fluid temperature and hence its density were stratified in perfectly horizontal layers, the temperature and density varying only in the vertical direction, z, the fluid would be in equilibrium and could maintain this (unnatural) state of affairs forever if left unperturbed. The equilibrium is one with no motion, and a temperature variation

$$T = T_{bottom} - \Delta T \frac{z}{d}, \qquad \Delta T \equiv T_{bottom} - T_{top}.$$

It is impossible to prevent very small disturbances to this (Platonic) situation, however, and so one must suppose that any equilibrium will be disturbed so that there are (at least very small) variations of the density in the horizontal as well as the vertical direction, and the equilibrium force balance will be disrupted, causing motion. The motion that results could further disrupt the equilibrium, leading to an increase in the disturbance; or it could decrease, returning the system to its original condition. The first possibility is the hallmark of an *unstable* equilibrium, and the second of a stable one.

It is an empirical fact that disturbances do tend to damp out, implying a stable situation, when the temperature difference across the layer is smaller than a clearly defined amount that depends on the depth of the layer, d, and the properties of the fluid involved.

Conversely, disturbances grow, implying instability, when the temperature difference exceeds this threshold value. Why doesn't the heavy fluid always continue to sink and the light fluid always rise, even when the temperature difference is small? The answer is that the cold fluid gains heat from the warmer surroundings it falls through, and the rising hot fluid cools down as it loses heat to the colder surroundings it traverses during its rise. If the temperature anomaly of a falling or rising blob starts out being too small, it can be dissipated before the blob has had a chance to move any significant distance. This can be seen from the following simplified analysis (based loosely on Normand et al. 1977).

Suppose our static equilibrium state is slightly disturbed so as to produce a small blob of fluid hotter than the surrounding fluid, with a temperature excess δT. Suppose the blob has volume V, so its characteristic length scale is $a = V^{1/3}$. Because the blob is hotter, it is lighter than the surroundings, with a density difference $\delta\rho = -\rho\beta(\delta T)$, where β is the coefficient of thermal expansion of the fluid. Since it is lighter, the blob experiences an Archimedian buoyancy force, $\mathcal{B} = -g\delta\rho V$, where g is the acceleration due to gravity. The buoyancy force causes the blob to rise (slowly, because $|\delta\rho|$ is small). If the upward speed of rise is U, the frictional drag, $\mathcal{F}$, with which the surrounding fluid resists the motion of the blob is proportional to U, to the coefficient of viscosity, μ, and to the length scale of the blob, a, or

$$\mathcal{F} = C_1 a \mu U,$$

where C_1 is a dimensionless number that depends only on the shape of the blob. For example, if the blob were a rigid sphere, $C_1 = 6\pi$.

The blob loses heat to the colder surroundings. If this were not the case, the higher the blob would rise, the greater its temperature excess relative to the fluid it encountered, causing the buoyancy force to become larger, and its upward speed larger, and so on. The heat transfer between the blob and its surroundings takes place on a time scale τ proportional to the surface area, a^2, and inversely proportional to the thermal diffusivity, κ, of the fluid, so

$$\tau = C_2 a^2 / \kappa,$$

for some dimensionless constant C_2. In the absence of heat transfer, a fluid blob displaced vertically upward a distance ℓ will keep the temperature of the background fluid at the location from which it started. Its temperature excess relative to its new location is therefore $(\Delta T/d)\ell$. But heat transfer does occur, and the temperature tends to equalize in a time interval of order τ. For the purpose of estimating a *sustainable* temperature excess of a blob moving vertically upward, we therefore may suppose that ℓ is of order of

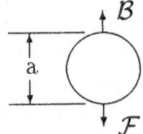

Figure 1: Sketch of forces on a blob of light fluid surrounded by heavy fluid.

the distance the blob may move in a time of order τ. This estimate indicates a temperature excess of

$$\delta T = \frac{\Delta T}{d} U\tau,$$

and a corresponding buoyancy force of

$$\mathcal{B} = \frac{\beta g \Delta T}{\kappa d} \rho V U C_2 a^2.$$

Newton's second law applied to the vertical direction gives

$$\begin{aligned}
\rho V \frac{dU}{dt} &= \mathcal{B} - \mathcal{F} \\
&= \rho V U \{C_2 \frac{\beta g \Delta T a^2}{\kappa d} - C_1 a^{-2}\}.
\end{aligned}$$

The blob will accelerate upwards when the term in {} > 0, causing the departure from equilibrium to be accentuated, or the system to be unstable. If this quantity is negative, then the force on the blob is a restoring force, so the system is stable. Now the term in {} is greatest when a is as large as it possibly can be, or $a = C_3 d$ (where C_3 is a number no larger than 1/2, which would correspond to a spherical blob), and this will be connected with the smallest ΔT for which the bracket is positive, so instability can occur. This smallest ΔT may be found from the relation

$$\frac{\beta g \Delta T d^3}{\kappa \nu} > \frac{C_1}{C_2 C_3^4},$$

where $\nu \equiv \mu/\rho$ is the so-called kinematic viscosity. The left hand side of this expression is a dimensionless quantity, called the "Rayleigh number," Ra, and the right hand side is the "critical value" of the Rayleigh number, denoted Ra_c. Our estimate of Ra_c depends on the constants of proportionality that appeared in our parameterization of the fundamental physical processes that are involved. In a more complete mathematical treatment, Ra_c is found to depend on the boundary conditions imposed on thermal and mechanical processes. For example, if the top and bottom boundaries support no shear stresses and are held at constant temperature (i.e., isothermal boundaries), $Ra_c = 657.5$. The important point to make here is that stability depends on the parameter

$$Ra = \frac{\text{Archimedean buoyancy force}}{\text{Diffusive "resistances"}}$$

exceeding a threshold value, Ra_c.

When instability sets in, it does so in the form of cells having a characteristic horizontal length scale. The only geometrical length scale involved in the setting up of this stability issue is the depth of the layer, d. It is not surprising then, that according to linear stability theory, the preferred horizontal length scale is proportional to d. The constant of proportionality depends on the boundary conditions. When the boundaries are isothermal, for example, the cells are nearly square, with roughly equal horizontal and vertical extents, the cell shape depending only weakly on the mechanical boundary conditions. With this in mind, the natural choice for the prevailing horizontal length scale is d.

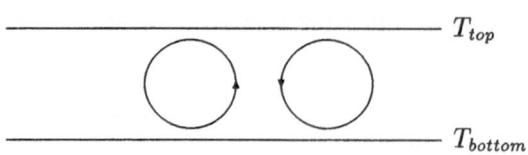

Figure 2: Sketch of a counter-rotating pair of convection cells with isothermal boundaries.

To find the expected velocity scale for the vertical motions that take place in an unstable layer, we note that the driving force for convection is buoyancy, and so the maximum work per unit mass of fluid that can be obtained is the maximum buoyancy force times the maximum distance over which the buoyancy force can act, or $\beta g \Delta T \times d$. The kinetic energy/mass must be smaller than this maximum work, so the greatest possible kinetic energy/mass derivable from the temperature difference across the layer is $\beta g d \Delta T$, implying the maximum possible speed of the convecting fluid layer is $U_{max} = \sqrt{2\beta g d \Delta T}$, or

$$U_{max} = \frac{\kappa}{d}\sqrt{\sigma Ra}, \text{ where } \sigma \equiv \frac{\nu}{\kappa} \text{ is the "Prandtl number".}$$

This is a suitable scale for the convection, and in fact is an upper bound. It ignores dissipation entirely. For example, we know that for $Ra < Ra_c$, there is no motion induced by instability. With this in mind, an improved estimate would be

$$U_{max} \sim \frac{\kappa}{d}\sqrt{\sigma(Ra - Ra_c)}.$$

Langmuir Circulation: Mechanism and Scales

The density of the water in the ocean varies due to variability in temperature and in salt content. Like all natural water bodies, the ocean is for the most part stably stratified, with the density increasing with depth. The upper layer of the ocean, being in contact with and driven by the wind, is typically more turbulent than the main body of the ocean. The mixing action of the turbulence in this region tends to even out variations in density, as well as vertical variations of other quantities. This region is called the "mixed layer" and has a depth that varies depending on location and time of year. Its lower boundary is marked by much higher temperature and density gradients than appear in the main body of the ocean, and of course, by a very much larger density gradient than in the mixed layer itself. This bounding layer of high gradients is called the "thermocline zone."

Langmuir circulations, which will be abbreviated as "LC," are wind-driven convective motions in the mixed layer. While they are known because of the windrows of floating material or compressed organic film they produce on the surface, presumably they occur also in circumstances where no suitable visible marker is present.

While visual observations of LC in the ocean are numerous, detailed measurements are very difficult. The measurements made by Weller et al. (1985), by Smith et al. (1987), and by Weller & Price (1988) in a series of three cruises (the third as part the MILDEX

– Mixed Layer Dynamics Experiment) in 1982 and 1983, are unique and extremely valuable. The measurements were made from the Research Platform *FLIP* of the Scripps Institution of Oceanography as it drifted from 200 to 850 km off the coast of California. The experiment is unique, because it provides accurate three-dimensional velocity and temperature data from the surface to the thermocline, together with good records of wind speed and direction, sea surface temperature, surface visualizations of windrows that could be correlated with velocity features, and finally, a rough indication of sea state.

This experiment confirmed the following characteristics of LC, which had previously emerged as a composite picture of many earlier and more fragmentary studies.

- LC cause surface convergences, or windrows, approximately parallel to the surface wind direction.

- The penetration depth is a significant fraction of the mixed layer depth, which during the experiments was on the order of 50 m. Strong effects of LC were found down to about half (20 – 35 m) the mixed layer depth. Reports from other studies indicate penetration throughout the mixed layer.

- Convergences were found in a hierarchy of spacings, ranging from several meters up to 100 m. The hierarchy has been observed before and is predicted in computations (see Leibovich 1983) made from the theory (referred to as CL) of Craik and Leibovich (1976).

- Vertical downwelling speeds below surface convergences in excess of 25 cm/s were measured, amounting to as much as 2% of the wind speed. This is as much as twice the maximum vertical speeds, as a fraction of the wind speed, that had previously been reported. Downwelling zones are narrow compared to a broad and diffuse upwelling that occurs between convergences. This narrow "jet" has been reported by virtually all observers, and is predicted by the CL theory.

- Associated with the downwelling jet is an increase of speed in the downwind direction of a comparable amount. Thus, the jet is oriented roughly 45° to the horizontal. Furthermore, this downwind speed anomaly is greatest a few meters *below* the surface. That this might happen was predicted earlier in calculations from the CL theory (see Leibovich 1977, Leibovich and Paolucci 1980).

Figure 3 is from the Weller and Price paper. This figure graphically presents their data and concisely captures the general structure of Langmuir circulations.

The CL theory predicts the qualitatively observed features of Langmuir circulations. Small-scale, incoherent turbulence is parameterized in the theory by an eddy viscosity, a quantity not directly measurable. Quantitative agreement can be arrived at by appropriate choice of this parameter, which introduces a semi-empirical aspect to the model. The important point, however, is that 'reasonable' choices of the parameter, consistent with other oceanographic invocations of eddy viscosity in the upper layers, give time scales for formation of LC, depth of penetration, windrow spacings, intensity of downwelling, and so forth that all are consistent with the data. This gives confidence that the physical mechanisms incorporated in the theory are appropriate, and that the theory can be used to predict the three-dimensional velocity fields in these circulation systems.

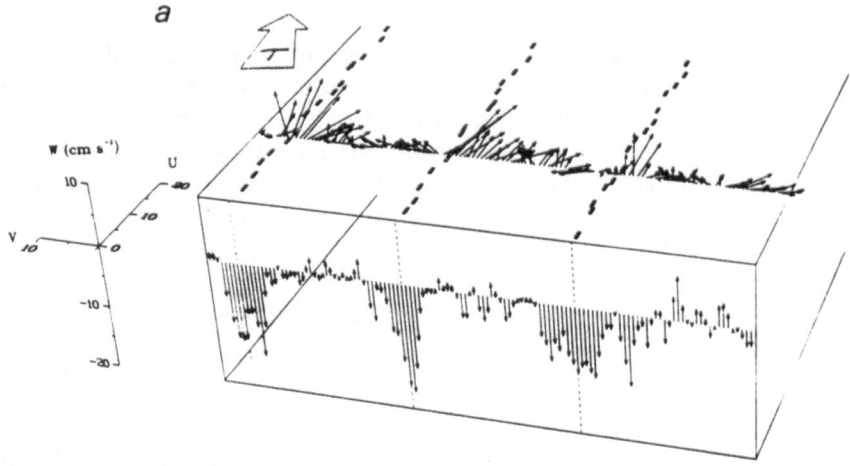

Figure 3: Measured data in Langmuir circulations, from Weller and Price (1988).

The development of the CL theory may be found in literature reviewed in Leibovich 1983). The central result is a mathematical model for the velocity and temperature field due to the circulations, with the detailed variations due to fluctuations caused by surface gravity waves filtered out. The form of the model is precisely the same as the equations of motion would be prior to filtering, but with an extra force representing the rectified effects of the surface waves. This apparent force has been called the "vortex force" and is given (per unit volume) by

$$\mathbf{F}_v \equiv \rho \, \mathbf{u}_s \times \boldsymbol{\omega}$$

where $\mathbf{u}$ is the velocity vector and $\boldsymbol{\omega} \equiv \nabla \times \mathbf{u}$ is the vorticity vector of the rectified motion. In this definition, $\mathbf{u}_s$ is the "Stokes drift" due to the surface gravity waves. The classical Stokes drift (Stokes 1847) is the Lagrangian velocity (motion of identifiable mass elements of fluid) caused by the propagation of waves on the fluid, and decays with depth on a length scale comparable to the wavelength (λ) of the most energetic waves. Given the characteristics of the surface waves, $\mathbf{u}_s$ can be calculated provided the wave slopes are not large. For example, for a monochromatic deep water wave train (one with a single wavelength and frequency) and a waveheight H (twice the wave amplitude), the Stokes drift speed is

$$|u_s| = \left(\pi \frac{H}{\lambda}\right)^2 \sqrt{\frac{g\lambda}{2\pi}} \exp[4\pi(z/\lambda)],$$

where z is measured upwards from the mean free water surface. For example, if the wave height were 1.5 m and and the wave length were 24 m, this formula would give 24 cm/s for the surface value of the Stokes drift.

In the CL theory, $\mathbf{u}_s$ is regarded as given. Typically, $\mathbf{u}_s$ is a decreasing function of depth only, and if the waves are generated by the local wind, it is reasonable to suppose that $\mathbf{u}_s$ is parallel to the wind direction.

It is possible to have "equilibrium flows" with $\mathbf{u}_s$ and $\mathbf{u}$ parallel to each other and to the wind direction. In such cases, the vortex force tends to destabilize this equilibrium, much like a destabilizing temperature distribution. When this tendency is sufficiently large to cause instability, vertical convective motions ensue, disturbing the "structureless" equilibrium.

Since the driving force behind Langmuir circulations is the vortex force (at least according to the CL theory), we can estimate the intensity of convection due to this phenomenon. The maximum possible vortex force occurs when $\mathbf{u}_s$ and $\boldsymbol{\omega}$ are at right angles. Now consider a layer of depth d in which Langmuir circulation takes place. Let $\max|\boldsymbol{\omega}| = \Omega_c$ be the maximum of the vorticity vector in the current. Then the work done by the vortex force over a distance equal to the depth of the layer does not exceed

$$\rho \Omega_c \int_{-d}^{0} u_s dz < \rho \Omega_c \int_{-\infty}^{0} u_s dz = \Omega_c |\mathcal{M}|, \tag{1}$$

where $|\mathcal{M}|$ is the total momentum carried by the surface waves in the layer, per unit area of the water surface. The maximum convective kinetic energy obtainable by work done by the vortex force is therefore $\Omega_c|\mathcal{M}|$, so an upper bound for the vertical speed in the Langmuir circulation is estimated to be

$$U_{LC} = \sqrt{2\frac{|\mathcal{M}|}{\rho}\Omega_c}. \tag{2}$$

Again, taking the case of a monochromatic train of deep water waves for illustration, (2) leads to

$$U_{LC} = \sqrt{2\mathcal{U}_s \frac{\lambda}{4\pi}\Omega_c},$$

where $\mathcal{U}_s$ is the surface value of the Stokes drift. If we have a direct means of estimating the current shear, then we can substitute it and $\mathcal{U}_s$ into this expression. In the absence of an observed value for the shear, we can suppose that the vorticity in the current is derived only from the local applied wind stress. In this case, the maximum vorticity tends to occur at the surface. The surface value of $|\boldsymbol{\omega}| = \tau_{wind}/\mu$, where τ_{wind} is the stress applied by the wind to the water surface, and μ is the viscosity (in natural circumstances, the water is more often than not turbulent, in which case μ must be interpreted as an eddy viscosity). This leads to

$$U_{LC} = \sqrt{2\mathcal{U}_s \frac{\lambda}{4\pi}\frac{\tau_{wind}}{\mu}} = u_* \sqrt{2\mathcal{U}_s \frac{\lambda}{4\pi\nu}}. \tag{3}$$

The last replacement is a bow to the convention of defining a "friction velocity," u_*, by the relation

$$u_*^2 \equiv \frac{\tau_{wind}}{\rho},$$

where ρ, as before, is the water density.

We can give an estimate of U_{LC}, based, for example, on (3). Suppose the wind speed is 10m/s, then we can estimate (very crudely) that $u_* \approx 1.5$cm/s. The wave characteristics used above to illustrate rough magnitudes for Stokes drift are reasonable for the so-called "significant waves" generated by a wind speed of this level, so we can take $\mathcal{U}_s$ to be 24cm/s and $\lambda \sim 24$ m. The price we pay for using the wind stress as our method of estimating the current shear is the introduction of a semi-empirical eddy viscosity, to which it is difficult to assign a value. From standard ways to guess at an appropriate value for this parameter when the wind speed is 10m/s, the value $\nu \approx 23$cm^2/s is plausible (see Leibovich and Radhakrishnan 1977, pg. 504), which leads to $U_{LC} = 29$cm/s as an estimate for the upper bound. By way of comparison, the maximum circulation speeds reported by Weller and Price (1988) are about 2% of the wind speed, which for this example is 20cm/sec.

SPATIAL AGGREGATION

The vertical velocity in a convection cell vanishes at the surface, and grows in magnitude with depth below cell boundaries. In this section, the origin of the convection is unimportant, but for purposes of illustration, I will assume that it is caused by Langmuir circulation. Suppose an organism has the tendency to move vertically upwards with respect to the surrounding water with a speed V_s, which may be, for example, its vertical swimming velocity, or, if the organism lacks a means of locomotion, its buoyant speed relative to stationary water. Examples of the latter circumstance are certain pelagic fish eggs (see Sundby 1983 for a survey of some relevant data). If the organism is situated at a point below the convergence line where downwelling speeds exceed V_s, the particle will be carried down. On the other hand, the particle will eventually be swept laterally away from its initial position into a region where water downwelling is less than V_s, and the particle will tend to rise back towards the surface. A fascinating result of Stommel's (1948), in which an artificial velocity field was used to show that convective motion can prevent heavy particles from sinking, is that a buoyant particle in some sense can be trapped "forever" in a zone some distance below the surface if it is injected into that zone and if turbulence is ignored. A subsurface zone in which particles can become trapped will be called a "Stommel retention zone," or SRZ, here.

Suppose a two-dimensional Langmuir circulation system has developed, with the z coordinate measured vertically upward, and the horizontal coordinate y. Then the horizontal and vertical velocity components, v and w, may be given in terms of a streamfunction ψ, as

$$v = \frac{\partial \psi}{\partial z}, \quad w = -\frac{\partial \psi}{\partial y}.$$

If we assume, as Stommel did, that the velocity of a particle relative to the surrounding water is always in the vertical direction with *constant* speed V_s, then the trajectory of a particle $y = Y(t)$, $z = Z(t)$ is given by

$$\frac{dY}{dt} = v = \frac{\partial \psi}{\partial z} \tag{4}$$

$$\frac{dZ}{dt} = w + V_s = -\frac{\partial}{\partial y}(\psi - V_s y). \tag{5}$$

If the fluid motion is steady in time, then $\partial \psi / \partial t = 0$, and trajectories of all particles in the cross-plane determined by (4) coincide with level curves of the function

$$\Psi(y, z) = \psi(y, z) - V_s y. \tag{6}$$

It is not hard to see that, if $w + V_s < 0$ for given V_s anywhere on a downwelling plane, then a Stommel retention zone exists for all particles with terminal velocity of rise less than or equal to V_s. These zones are nested regions for which the contours of Ψ are closed. Each of these nested closed contours (if any exist) is enclosed within its largest member, which forms the boundary of the SRZ. The disposition and extent of an SRZ depends upon V_s,

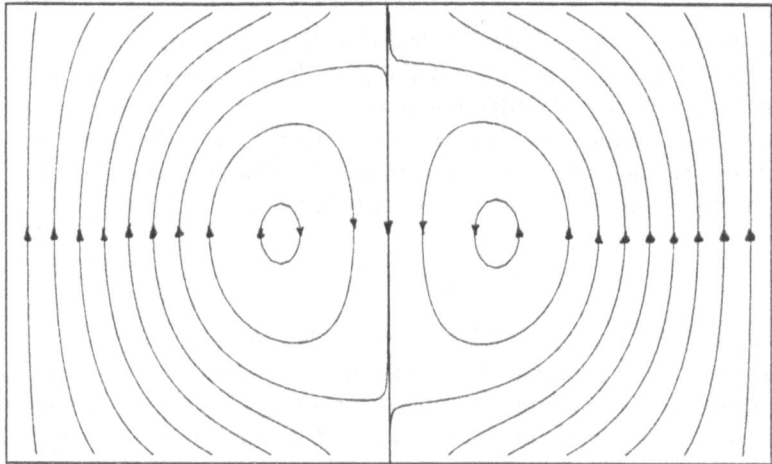

Figure 4: Particle trajectories in a convection cell containing a SRZ. Trapping occurs within the closed contours. This example is similar to one considered by Stommel.

the convective velocity and length scales ℓ. If $\mathcal{C}$ is the boundary contour of an SRZ for a given value of V_s in a given convective field, then any particle with this value of V_s or smaller placed inside $\mathcal{C}$ will, according to this model, remain trapped within $\mathcal{C}$ forever. If V_s is reduced, the SRZ is enlarged; if increased the SRZ is reduced. An example of a retention zone is given in Figure , which is constructed from an artificial convective field, described later in equation (16), with a ratio of V_s to the maximum fluid downwelling speed of 0.2.

Now, retention zones are not permanent in the sense that a particle once in them stays in them, as would be suggested by the idealized model presented by Stommel. Instead, due to neglected effects of turbulence, waves, and other sources of fluctuations, particles can cross the SRZ boundary in either direction. Nevertheless, the SRZ serves to imprint the water column with a region in which particles have a tendency, if not an obligation, to congregate. To have a more complete picture, it is necessary to examine how particles get into, and out of, the trapping zones.

A Diffusion Model for Transport Across a Trapping Zone Boundary

I will describe an analysis due to Leibovich and Lumley 1981 for the concentration of particles as a function of position and time in a convective zone. This is equivalent (Tennekes and Lumley 1972, Monin and Yaglom 1971) to asking for the probability that a single particle may be found in the subvolume centered on the position of interest. In the present context, it is more natural to cast the discussion in terms of the number density (number/volume) of organisms being considered. Restricting attention to two space dimensions, the number density, $n(\mathbf{x}, \mathbf{t})$, is taken to be described by a diffusion equation

$$\frac{\partial n}{\partial t} + v\frac{\partial n}{\partial y} + (w + V_s)\frac{\partial n}{\partial z} = \frac{\partial}{\partial y}\left(D_1\frac{\partial n}{\partial y}\right) + \frac{\partial}{\partial z}\left(D_2\frac{\partial n}{\partial z}\right), \qquad (7)$$

where the parameters $D_{1,2}$ are diffusivities resulting from turbulent buffeting that may be different for the horizontal and the vertical directions, and are permitted to depend on depth. It is the diffusivities that permit the particles, now represented only in terms of their concentration, to cross the SRZ boundary.

Suppose that the total number of organisms in the region of the ocean considered is fixed; that is, there are no births or deaths in the time interval considered. The total number of organisms in the volume of the ocean being considered is

$$N = \int_V n\,dx\,dy\,dz. \qquad (8)$$

Its rate of change with time is zero, by hypothesis. The convection cells may be idealized as periodic in the cross-wind horizontal direction y and are independent of x. Consider a volume V of unit distance in the x-direction, and consisting of one period of the convection system in the spanwise (y) direction, and of depth d. Since the spanwise distribution of particles is determined strictly by the convective activity, n will have the same periodicity. Integrating (7) over this volume, the invariance of N implies that the net flux of organisms across the bounding planes, $z = 0, -d$ vanishes, or, on assuming that V_s is a constant,

$$\int_L [V_s n(y, 0, t) - D_2(y, 0, t)\frac{\partial n}{\partial z}(y, 0, t)]dy \;=$$
$$\int_L [V_s n(y, -d, t) - D_2(y, -d, t)\frac{\partial n}{\partial z}(y, -d, t)]dy, \qquad (9)$$

where L is the period (two counter-rotating cells per period).

Thus the numbers are conserved if we set the integrands, the flux density of organisms, to be zero at each point of the boundary, so we take

$$V_s n(y, 0, t) - D_2(y, 0, t) \;=\; V_s n(y, -d, t) - D_2(y, -d, t)\frac{\partial n}{\partial z}(y, -d, t) = 0$$
$$n(y + L, z, t) \;=\; n(y, z, t), \qquad (10)$$

where the last condition imposes periodicity in the horizontal.

Of course, the swimming velocity of an organism, V_s, which we have for simplicity taken to be constant, should really vanish at the horizontal boundaries, in which case the vanishing of the normal derivative of n would be the appropriate condition, but we can get away here with the simple assumption of constant swimming speed providing we use (10) to make the appropriate global adjustment. The boundary conditions (10) hold for unsteady as well as steady conditions, although we will concern ourselves only with the time-independent case.

The solution of the problem posed by (7) and (10) will be sketched momentarily. Two observations are in order at this stage. First, the level of n is not determined; if we find a solution, any multiple of it is also a solution. The magnitude of n is determined by the additional condition (8), which is just an initial condition for the unsteady problem whose steady limit we seek. Second, the role of the Stommel retention zones can now be further clarified. If we make (7) dimensionless using the Langmuir circulation velocity scale U_{LC}, and the length scale d, then the dimensionless measures of the diffusivities are

$$\epsilon_1 = \frac{D_1}{U_{LC}d}, \text{ and } \epsilon_1 = \frac{D_1}{U_{LC}d}. \qquad (11)$$

These parameters typically are very small. Consequently, it is clear that the reduced equation found by neglecting diffusion altogether will play an important role. Returning now to dimensional variables, the reduced equation is

$$v\frac{\partial n}{\partial y} + (w + V_s/U_{LC})\frac{\partial n}{\partial z} = J_{y,z}(n, \Psi) = 0, \tag{12}$$

where J is the Jacobian, and Ψ is, as before, the "streamfunction" for the SRZ contours. Thus, as usual in problems of boundary layer character, the leading approximation satisfies an equation of reduced order and we are not able to enforce all boundary conditions.

The general solution to (12) is

$$n = n(\Psi). \tag{13}$$

Thus, n is constant on lines of constant Ψ. These lines, in turn, are the level curves found by Stommel; a closed Ψ = constant line defines a SRZ. The restoration of the diffusivities and the enforcement of the boundary conditions must next be dealt with.

The Effects of Small Diffusivity

When the diffusivities are small compared with the advective effects, as they are usually expected to be when strong organized convection is present, the structure of the trapping region can be analyzed using boundary layer techniques. The details will take us too far afield, and the interested reader can find them explained in Leibovich and Lumley (1981). A descriptive survey that reveals the essential elements will be given here.

Consider the region outlined by the diffusion-free SRZ. When diffusive effects are small but nonzero, we expect that organisms that enter this region, although not trapped forever, will remain inside for a relatively long time. One effect of nonzero diffusivity is to smear the concentration out so that it is nearly uniform inside the SRZ. We therefore expect that n is essentially constant in this region.

If an organism is anywhere outside the SRZ, then with diffusion neglected, it will rise to the surface, and then be swept laterally into a convergence zone that lies above the SRZ. In this simplified view, the entire water volume outside the SRZ except for a thin region near the surface is eventually swept free of organisms. The preliminary picture, then, is one in which organisms are either in the SRZ with a uniform number density, or collected into windrows on the surface. This is a singular distribution, with discontinuities in the number density. Where spatial gradients are large, the diffusion terms are not insignificant, even though the diffusivities are small. Therefore, those surfaces at which formally infinite gradients are found are in reality thin layers where diffusion is not negligible.

Consequently, we expect the surface defining the SRZ to be a thin diffusion layer, and that organisms in this region can and will slowly escape by diffusive transport across the SRZ boundary. They too will rise to the surface and be collected into windrows. If there were no other effects, leakage across the boundary will eventually empty the retention zone.

The diffusion layer that wraps around the SRZ has a thickness δ, where

$$\delta \sim \sqrt{\frac{D\ell}{U_{LC}}}. \tag{14}$$

Here ℓ is a length of the order of the height of the SRZ, and D is either of the diffusivities, since both are of the same order for phenomena on this scale. If the top point of the SRZ, say z_+, is not within a distance of order δ, then the windrow at the surface cannot communicate with the SRZ. In this case, we expect that the effect of diffusion is to cause the SRZ slowly to clear. If, on the other hand,

$$|z_+| = O(\delta) \tag{15}$$

or less, then the windrow at the surface lies within the diffusion layer, and material collected in the windrow can be reinjected into the SRZ by diffusive transport. In this case, the SRZ acts as a semi-permanent zone of large residence time, and one then expects to see substantial subsurface concentrations of organisms.

It is useful to consider numerical examples, to see the prospect of subsurface trapping zones, and their extent when they are anticipated based upon the qualitative arguments give here. In the spirit of the estimates given so far (and following in Stommel's footsteps), let me take a modelled convective field appropriate to the example given in §3, and suppose that the thermocline delineating the mixed layer has a depth $d = 24$m, and that the thermocline acts as an effective slippery bottom that the convective motion cannot penetrate. Then a simple model for a fluid velocity field with square convection cells is

$$\psi = \frac{U_{LC}d}{\pi} \sin\left(\frac{\pi y}{d}\right) \sin\left(\frac{\pi z}{d}\right), \tag{16}$$

and $\Psi(y,z)$, the function whose contours describe trajectories of buoyant or swimming organisms is given by (6) with (16) substituted. The illustrative example (16) fails to imitate LC because the motion given by (16) is symmetric about the mid-layer plane and, more important, has upwelling and downwelling of equal intensity. A faithful surrogate for LC, by contrast, would have an intense and narrow downwelling "jet," and a diffuse and more gentle upwelling away from the jet. The velocity along the trajectory of an organism is

$$\begin{aligned}
v_{org} &= U_{LC} \sin\left(\frac{\pi y}{d}\right) \cos\left(\frac{\pi z}{d}\right), \\
w_{org} &= V_s - U_{LC} \cos\left(\frac{\pi y}{d}\right) \sin\left(\frac{\pi z}{d}\right).
\end{aligned}$$

The plane $y = 0$ defines a cell boundary on which the fluid is moving vertically downwards. If there is a SRZ, it will be centered on this downwelling line, which lies below a line of surface convergence. For a SRZ to exist, we must have w_{org} vanish at some depth $z = z_+$; if it does so, it will also vanish at a second depth, say $z = z_- < z_+$. This pair of points are roots of the equation

$$\sin\left(\frac{\pi z_\pm}{d}\right) = \frac{V_S}{U_{LC}},$$

and so the SRZ exists if $V_s < U_{LC}$. For a convection cell that is symmetric about the mid-layer level (like our example), $z_- = -d + z_+$. The length ℓ appearing in (14) is then (roughly speaking) $d - 2|z_+|$. Provided $|z_+| \ll d$, the condition (15) is

$$\frac{|z_+|}{d} < \frac{\sqrt{Dd}}{U_{LC}}\{1 + \frac{\sqrt{Dd}}{U_{LC}}\}^{-1}. \tag{17}$$

In the example, I will take $U_{LC} = 20$cm/s, which is slightly less than the estimated · maximum possible convective speed of Langmuir circulations at a wind speed of 10m/s. The vertical swimming speed of *Oikopleura longicauda* is estimated by Owen (1966) to be about 2mm/s, which is very close to the terminal velocity of rise of mackerel eggs, estimated by Sundby (1983) to be about 1.8mm/s. In taking $V_s = 2$mm/s, the distribution of either type of organism may be estimated. The important ratio, V_s/U_{LC}, has the value 0.01, so $|z_+| \approx 7.6$cm, and $|z_-| = d - z_+$. Thus, the SRZ extends almost the entire depth of the convecting layer in this case.

Is the SRZ in the example "connected" to the windrow at the surface? If so, we expect the SRZ to be a region of enhanced concentration of organisms. A reasonable guess at the diffusivity D in (14) is to take it equal to the eddy viscosity, ν, which in the example is 23cm^2/s. This implies that $\delta \approx 49$cm. Since this exceeds $|z_+|$, the SRZ is indeed connected to the windrow. We therefore would expect a nearly uniform concentration of organisms suspended throughout most of the mixed layer in this case.

CONCLUSIONS

The biological consequences of spatial aggregation and vertical transport by physical processes in the ocean have long been a matter of interest and concern. Among those processes, the convective activity associated with Langmuir circulations is almost always mentioned when near-surface aggregation is discussed.

In this chapter, I have focussed on Stommel's concept of spatial collection of material, both in the horizontal direction and in depth, with special attention paid to forming estimates of the role of Langmuir circulations. It should be understood, of course, that order-of-magnitude estimates like those done here are but a first step. A detailed analysis of the physics appropriate to a given oceanic condition (sea state, wind forcing, preexisting current, and mixed layer structure) may prove that the kind of aggregation anticipated here may not be realized. Furthermore, it goes without saying that other physical processes are taking place contemporaneously with the convective activity and may act to complicate the issue considerably.

ACKNOWLEDGMENTS

This work was supported by the National Science Foundation under Grant OCE–9017882. Figure 3 is reprinted from *Deep Sea Research* and appears by permission of the Pergamon Press.

REFERENCES

Bainbridge, R. 1957. The size, shape, and density of marine phytoplankton concentrations. *Camb. Phil. Soc. Biol. Rev.* 32: 91–115.

Craik, A. D. D., and Leibovich, S. 1976. A rational model for Langmuir circulations. *J. Fluid Mech.* 73: 401–426.

Denman, K.L., and Powell, T.M. 1984. Effects of physical processes on planktonic ecosystems in the coastal ocean. *Oceanogr. Mar. Biol. Ann. Rev.* 22: 125–168.

Fasham, M.J.R. 1978. The statistical and mathematical analysis of plankton patchiness. *Oceanogr. Mar. Biol. Ann. Rev.* 16: 43–79.

Langmuir, I. 1938. Surface motion of water induced by wind. *Science* 87: 119–123.

Leibovich, S. 1977. On the evolution of the system of wind drift currents and Langmuir circulations in the ocean. Part I. Theory and the averaged current. *J. Fluid Mech.* 79: 715–743.

Leibovich, S. 1983. The form and dynamics of Langmuir circulations. *Ann. Rev. Fluid Mech.* 15: 391–427.

Leibovich, S., and Lumley, J.L. 1981. A Theoretical Appraisal of the Joint Effects of Turbulence and of Langmuir Circulations on the Dispersion of Oil Spilled in the Sea. U.S. Coast Guard Report No. CG-D-26-82, 119pp.

Leibovich, S., and Paolucci, S. 1980. The Langmuir circulation instability as a mixing mechanism in the upper ocean. *J. Physical Oceanography* 10: 186–207.

Leibovich, S., and Radhakrishnan, K. 1977. On the evolution of the system of wind drift currents and Langmuir circulations in the ocean. Part 2. Structure of the Langmuir vortices. *J. Fluid Mech.* 80: 481–507.

Monin, A.S., and Yaglom, A.M. 1971. *Statistical Fluid Mechanics*, Vol. 1, MIT Press, Cambridge, MA.

Normand, C., Pomeau, Y., and Velarde, M.G. 1977. Convective instability: A physicist's approach. *Rev. Mod. Phys.* 49: 581–623.

Owen, R. W. Jr. 1966. Small scale, horizontal vortices in the surface layer of the sea. *J. Marine Res.* 24: 56–65.

Smith, J. Pinkel, R., and Weller, R. A. 1987. Velocity structure in the mixed layer during MILDEX. *J. Phys. Ocean.* 17: 425–439. Stokes, G. G. 1847. On the theory of oscillatory waves. *Trans. Camb. Phil. Soc.* 8: 441-455.

Stommel, H. 1948. Trajectories of small bodies sinking slowly through convection cells. *J. Marine Res.* 8: 24–29.

Sundby, S. 1983. A one-dimensional model for the vertical distribution of pelagic fish eggs in the mixed layer. *Deep Sea Res.* 30: 645–661.

Tennekes, H., and Lumley, J.L. 1972. *A First Course in Turbulence*, MIT Press, Cambridge, MA.

Weller, R.A., Dean, J.P., Marra, J., Price, J.F., Francis, E.A., and Boardman, D.C. 1985. Three-dimensional flow in the upper ocean. *Science* 227: 1552–1556.

Weller, R. A., and Price, J.F. 1988. Langmuir circulation within the oceanic mixed layer. *Deep Sea Res.* 35: 711-747.

Woodcock, A. H. 1944. A theory of surface water motion deduced from the wind-induced motion of the *Physalia. J. Mar. Res.* 5: 196–205.

9
TWO-PATCH METAPOPULATION DYNAMICS

Roger M. Nisbet, Cheryl J. Briggs, William S. C. Gurney, William W. Murdoch, Allan Stewart-Oaten

INTRODUCTION

The concept of *metapopulation* is widely used by modelers exploring the effects of spatial heterogeneity on population dynamics. Yet *prima facie*, it is a remarkably restrictive idealization implying that a population is distributed over a number of patches, each sufficiently well defined to permit local definition of vital rates, with migration between patches occurring over time scales comparable in magnitude to that of population change. For a small number of systems, this abstraction may be defensible, but metapopulation models commonly are poor approximations to real systems. For example, it is often unclear what constitutes a patch, with a suitable definition for considering vital rates being inappropriate for modeling migration. Therefore, metapopulation models are mainly useful for developing intuitive understanding on broader questions concerning the relationship between spatial heterogeneity and population density and stability. For this reason, it is not only important to derive mathematical results on particular models, it is also essential to understand intuitively the various stabilizing and destabilizing mechanisms, as this intuition is likely to be of more general applicability than the models from which it was obtained.

Taylor (1988) recognizes two types of metapopulation model. In the first family, the primary processes are extinction and colonization on individual patches, and interest normally centers on the fluctuations in the number of occupied patches. For both deterministic and stochastic models, extensive analysis is possible provided population fluctuations on different patches are assumed to be uncorrelated (e.g., Hastings 1991, Gurney and Nisbet 1978, Nisbet and Gurney 1982, Diekmann, et al. 1988).

This chapter is concerned with a second type of metapopulation model in which abundances on individual patches are modeled explicitly. Interest focuses on the extent to which the dynamical behavior of the metapopulation differs from that of the constituent subpopulations treated in isolation. There is an extensive biomathematical literature on single-species models of this type, but rather little is known about prey-predator models (Maynard Smith 1974, Murdoch and Oaten 1975, Chewning 1975, Crowley 1981, Godfray and Pacala in press, Ives in press, Murdoch et al. in press). Here we offer some tentative ideas on stabilizing and destabilizing mechanisms in prey-predator metapopulation models, and illustrate our ideas with simulations on a two-patch system.

A SINGLE PATCH AND ITS ENVIRONMENT

One intuitively appealing approach to the study of a metapopulation on n patches is to concentrate on one of the patches and to regard all the other patches as its "environment." This idealization may also be of interest in its own right as a representation of real populations - for example, when considering a central refuge with peripheral patches (Harrison 1991). Let H and P denote respectively prey and predator populations on the patch of interest, and denote by x and y the corresponding quantities in the environment. In general x and y will be time dependent, and the equations describing their dynamics will be coupled to those for H and P. However, we might expect to obtain some insight into the population dynamics on our chosen patch by assuming a constant environment, i.e., letting x and y be constants. With this assumption, the dynamics on a patch may be influenced by the remainder of the universe (its "environment"), but there is no feedback from patch to environment.

We explore this possibility with a particular model very similar to one used by Godfray and Pacala (in press) and by Murdoch et al. (in press) to model two-patch systems. We assume prey leave all patches at a density-independent per capita rate e_h and that a fraction z of the migrating prey settle on our special patch with no mortality en route. We make similar assumptions for the predators with parameters e_p and q. We assume Lotka-Volterra dynamics on the patch. Then

$$dH/dt = aH - bHP - e_hH + ze_h(H + x),$$

$$dP/dt = fbPH - dP - e_pP + qe_p(P + y).$$

Here a is the per capita growth rate of the prey in the absence of predation or migration, b is the predator's attack rate, d is the predator death rate, and f is the conversion efficiency of prey to predators. Without migration, this model would have a single, neutrally stable, non-trivial equilibrium; we regard this as a reference point, and characterize additional features of the model (e.g., migration, aggregation) as "stabilizing" (or "destabilizing") if they move the equilibrium from this knife-edge to local stability (or instability). This approach, discussed further by Murdoch and Oaten (1975), is assailable on many grounds, of which the most compelling is the unrealistic representation of predation in the Lotka-Volterra model. Nevertheless, it remains a simple and convenient approach. In particular, close to neutral stability, it is not unreasonable to regard the negative of the real part of the dominant eigenvalue of the stability matrix as a measure of stability (May 1973).

We restrict our analysis to situations where the probabilities q and z are either constant or depend only on the prey densities H and x. Constant q and z imply density-independent (diffusive) movement between patch and environment. Density dependence of z implies that prey prefer (or avoid) regions of high prey density, while dependence of q on *prey* density covers the case where predators aggregate preferentially in regions of high prey density. Routine local stability analysis (Appendix 1) then points to a number of conclusions.

(i)   *Density-independent movement of prey or predator leads to stability.*

This can be understood intuitively by noting that in our model, the total rate of immigration to a patch is the sum of two terms, one proportional to patch density, the other constant. Thus the per capita immigration rate of (say) prey decreases with increasing prey density. This density dependence is the source of the stability. A similar argument holds for predator immigration.

(ii)   *Preferential prey movement to the area with more prey (prey aggregation) may lead to instability if the equilibrium prey population on the patch is higher than that of the environment, and has little effect on stability otherwise.*

(iii) *Preferential movement of prey to areas with fewer prey is stabilizing if the equilibrium prey population on the patch is lower than that of the environment and otherwise has little effect.*

These results may also be interpreted intuitively by considering the per capita immigration rate of prey to the patch of interest. For example, the instability in case (ii) occurs because, over some range of populations, this per capita rate increases with population size.

No immediate generalization is possible regarding predator aggregation to regions of high prey density, as the slope of a plot of q versus H does not enter the stability condition explicitly (Appendix 1). This observation is of particular interest in view of current controversy regarding the relationship between predator aggregation and stability in continuous time models (Murdoch and Stewart-Oaten 1989, Godfray and Pacala in press, Ives in press, Murdoch et al. in press). Aggregation and stability are connected, but the relationship is complex and involves the equilibrium prey density on the patch.

An example illustrates the way this works. Suppose the prey are immobile, and that the probability q that a mobile predator selects the patch depends strongly on the relative prey populations on patch and environment. A function with this property is

$$q = H^u/(H^u + x^u)$$

with u large. If the equilibrium prey density H^* is less than x, then the predator immigration term on the patch of interest is small, and the equilibrium is close to neutrally stable. If, on the other hand, $H^* > $ x, then the immigration rate remains finite even with large u, and the system remains strongly damped. Thus we can add a fourth conclusion:

(iv)   *Strong predator aggregation to regions of high prey density is stabilizing if the patch has a higher prey density than the environment (because it then attracts immigrants), and has little effect if the patch has lower prey density than the environment.*

TWO-PATCH MODELS

A Simple Two-Patch Model

Two-patch models, as well as, obviously, modeling systems in which there really are two patches, may (tenuously) be regarded as an improved approximation to the case of a central area with peripheral patches, but with the environment treated as a second patch. Feedback in both directions can now be included in the models. The two-patch analog of our previous model is:

$$dH_1/dt = a_1H_1 - b_1H_1P_1 - e_hH_1 + z_1e_h(H_1 + H_2),$$

$$dP_1/dt = fb_1P_1H_1 - d_1P_1 - e_pP_1 + q_1e_p(P_1 + P_2).$$

$$dH_2/dt = a_2H_2 - b_2H_2P_2 - e_hH_2 + z_2e_h(H_1 + H_2),$$

$$dP_2/dt = fb_2H_2P_2 - d_2P_2 - e_pP_2 + q_2e_p(P_1 + P_2).$$

As before, we assume that there is no mortality associated with inter-patch movement; thus $z_1 + z_2 = 1$ and $q_1 + q_2 = 1$.

Density-Independent Migration Rates

For the general model, determination of the conditions for existence and stability of equilibria is an exercise with many complications. However, a few results can be established by straightforward means for the particular case of *purely diffusive* interpatch movement, and an equilibrium in which all populations are strictly positive.

(i)   *If the patches are identical ($a_1 = a_2$; $b_1 = b_2$; $d_1 = d_2$), and movement is symmetrical ($z_1 = z_2 = 1/2$; $q_1 = q_2 = 1/2$), the equilibrium is locally neutrally stable.*

This amounts to saying that near equilibrium, the two identical, symmetrically linked patches behave like one big patch. However, this result is fragile against relaxation of either assumption; indeed it is straightforward to prove:

(ii)   *If only one species is mobile ($e_h = 0$ or $e_p = 0$), and either the patches are not identical or the movement is nonsymmetric, then the equilibrium is locally stable.*

This second result is understandable on the basis of the results for a single patch coupled to a constant environment. Consider (say) prey movement. Provided the patches are non-identical, the two patches will in general make unequal contributions to the pool of migrants. Thus near equilibrium, there is net movement from one patch to the other at a rate that decreases with increasing prey density on the recipient patch. Thus from the perspective of that patch, the movement is stabilizing.

However, there are cases where it is not possible to anticipate the dynamics on the basis of our understanding of a single patch plus environment. For example, by local stability analysis (Stewart-Oaten et al., in prep.), or by numerical solution of the equations with appropriate choice of parameters (Figure 1), we can establish:

(iii)   *Oscillatory instability is possible if both species are mobile.*

This result is inconsistent with intuition based on extrapolation from the single-patch dynamics. We are still investigating the mechanisms that cause this instability; however, in many (but certainly not all) cases, the parameter combinations that allow instability have opposite directions for net flow of prey and predator .

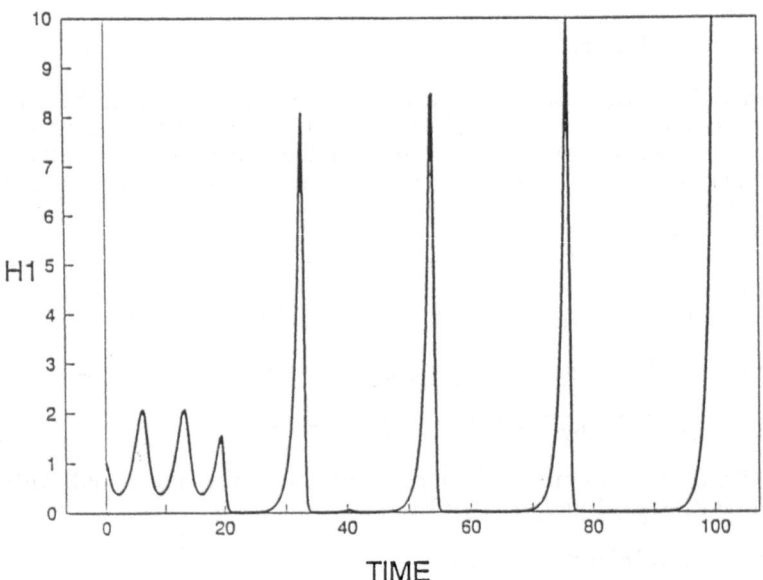

TIME

Figure 1.   An example of unstable dynamics in a two-patch model with density-independent migration rates for both prey and predator.

Density-Dependent Migration Rates

With a judicious selection of numerical investigations (some of which are detailed in Murdoch et al. in press), it is possible to reach some tentative conclusions in cases with prey or predator aggregation. Again, the analysis of a single patch can give us some insight on what to expect in such circumstances. For example, consider the effect of strong predator aggregation. Here conclusion (iv) implies that one patch (that with lower prey population) will be effectively decoupled from the other. Its populations will thus continue to cycle much as they would if the patch were totally isolated. On the other hand, the natural cycles on the patch with the higher prey equilibrium will be damped; however, this patch is receiving immigrants at a rate that itself cycles due to the oscillations on the first patch. In short, the two patches have a "master/slave" relationship, with the consequence that, as the strength of predator aggregation increases, one effectively dominates the other and neutral stability ensues. Similar reasoning can be applied to the case of prey aggregation.

These expectations are consistent with numerical results (see Murdoch et al. in press). However, there are situations where the master/slave argument appears to fail, notably those where net flows of prey and predator in opposite directions make impossible any unambiguous identification of the master or the slave patch.

CORRELATIONS BETWEEN FLUCTUATIONS ON TWO PATCHES

A common argument in the ecological literature relates metapopulation stability to *asynchrony* in fluctuations in individual subpopulations (e.g., Murdoch and Oaten 1975, Crowley 1981, Reeve 1988). This is plausible. For example, symmetric diffusion will tend to equalize populations, and hence synchronize fluctuations, on neighboring patches (Maynard Smith 1974). Spatial density-dependent population movement, while potentially stabilizing, may also have the effect of synchronizing (say) prey fluctuations on the two patches, thereby causing a metapopulation to behave more like a single (neutrally stable) patch.

We are currently engaged in a program, one of whose aims is to better understand the relationship between stability (defined previously) and asynchrony in patch models. It is possible to prove, at least in the limiting situation where the oscillations on the two patches are lightly damped, that the asymptotic phase difference (ϑ) between the prey oscillations on the two patches is independent of initial conditions, and is calculable from an eigenvector of the stability matrix (Appendix 2). If r is the correlation coefficient between prey fluctuations on the two patches, we define asynchrony to be $(1 - r)/2$. With light damping, this is approximately $\sin^2(\vartheta/2)$.

Figure 2 shows the variation of stability and asynchrony for a model with symmetric, density-independent migration. The two patches differ only in prey growth rate, and the figure illustrates the loss in stability as patches oscillate closer in phase.

Figures 3 and 4 show similar behavior for models with, respectively, density-dependent prey movement and predator aggregation. Taken together, these examples support the view that in two-patch models, the same factors that promote stability frequently desynchronize fluctuations. However, the converse is not true; instability need not involve synchronous cycles on the two patches (Figure 5).

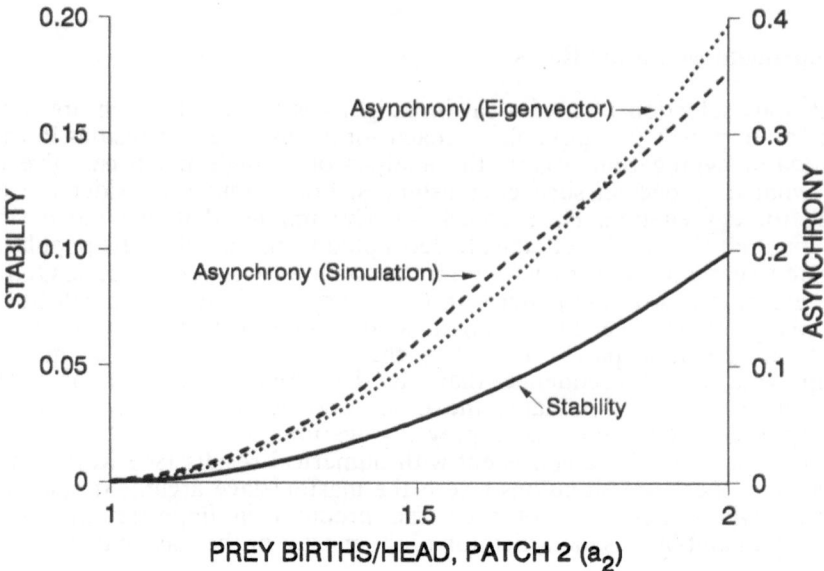

Figure 2. Changes in the degree of stability and asynchrony in response to an increase in the difference between the prey birth rates in the two patches (a_1 = 1), for a model with symmetric, density-independent prey movement (z_1 = z_2 = 0.5, e_h = 1). Stability is measured as the negative of the real part of the dominant eigenvalue of the stability matrix. Asynchrony (simulation) was calculated directly from simulations of the model as $(1 - r)/2$, where r is the correlation coefficient between H_1 and H_2. Asynchrony (eigenvector) was computed using the procedure described in Appendix 2.

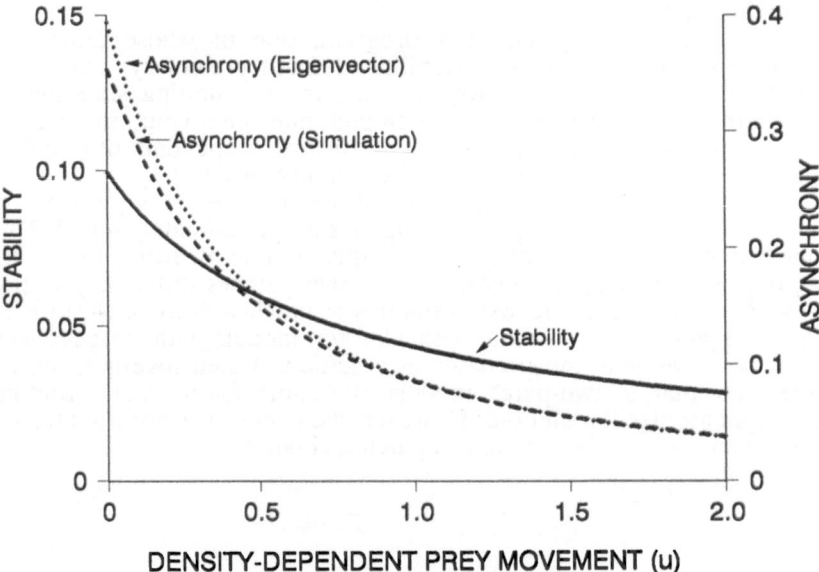

Figure 3. Changes in the degree of stability and asynchrony as the degree of density dependence in prey movement is increased by setting $z_1 = H_2^u/(H_1^u + H_2^u)$ and $z_2 = H_1^u/(H_1^u + H_2^u)$, then varying u. There is no predator movement (e_p = 0). Other parameter values are e_h = 1, a_1 = 1, a_2 = 2, b_1 = b_2 = 2, f = 1, d = 1.

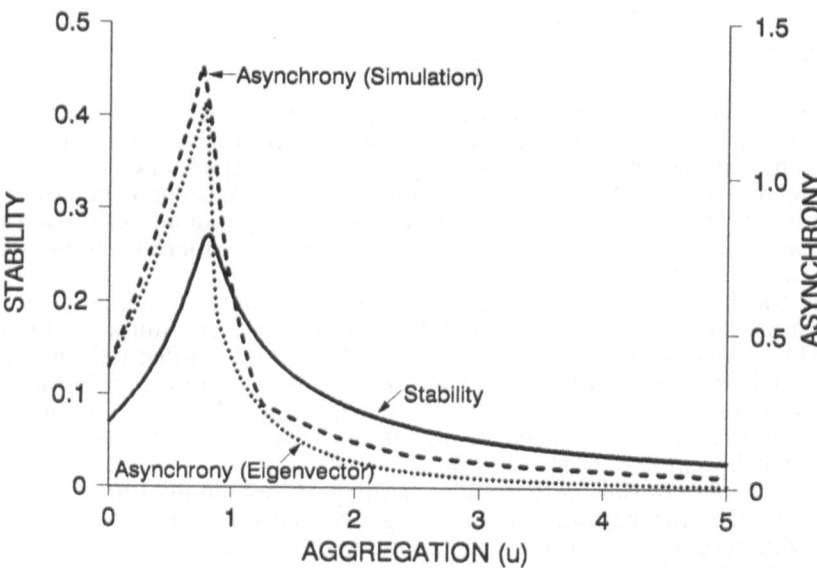

Figure 4. Changes in the degree of stability and asynchrony as the degree of predator aggregation is increased by setting $q_1 = H_1{}^u/H_1{}^u + H_2{}^u)$ and $q_2 = H_2{}^u/H_1{}^u + H_2{}^u)$, then varying u. There is no prey movement $(e_h = 0)$. Other parameter values are $e_p = 1$, $a_1 = 1, a_2 = 2, b_1 = b_2 = 2, f = 1, d = 1$.

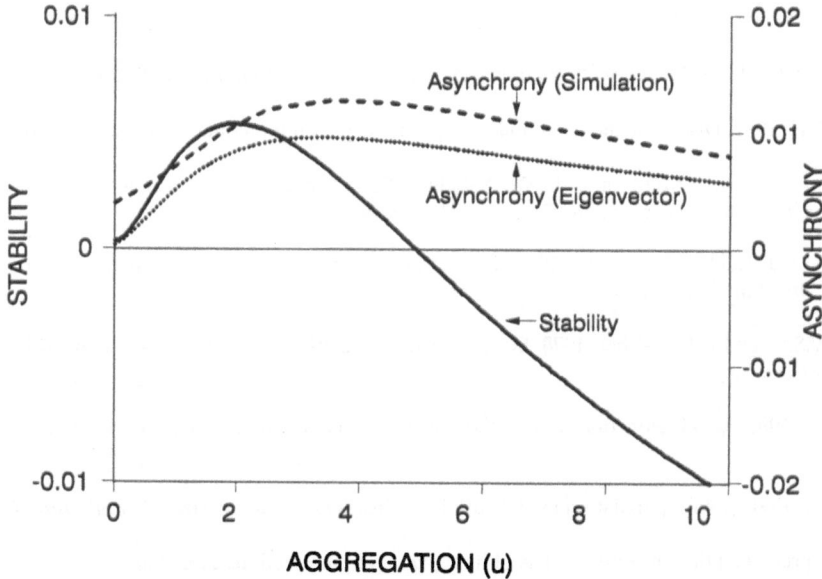

Figure 5. Changes in the degree of stability and asynchrony as the degree of predator aggregation is increased $(e_p = 0.85)$, and the prey preferentially move to the poorer patch $(e_h = 1.7, z_1 = 0.3)$. $a_1 = 1, a_2 = 0.7, b_1 = 1, b_2 = 0.7, f = 1, d = 0.6$.

DISCUSSION

The primary take-home message from our work to date is that when only one species (prey or predator) is mobile, considerable insight into two-patch metapopulation stability can be obtained by considering temporal density dependences on a single patch and treating the rest of the world as its environment. A second, apparently robust conclusion is that in many situations, the same density-dependent factors that influence stability may entrain oscillations on the two patches, thereby influencing both asynchrony and stability.

This conclusion should be tempered with the qualification that it is based on analysis of a model that assumes Lotka-Volterra population dynamics on the individual patches. The model is both biologically unrealistic and mathematically fragile. In particular, small levels of self-regulation of the prey and of saturation in the predator's functional response lead to limit cycles or a stable equilibrium. The former case is particularly fascinating, as the mechanisms for synchronizing or desynchronizing stable limit cycles may be subtle (Crowley 1981).

However, our conclusion that understanding of metapopulation stability involves understanding the mechanisms determining the correlation between patches is consistent with recent work on (non-metapopulation) models using cellular automata (de Roos et al. 1991). Possibly we are seeing a rival paradigm to the popular "reaction-diffusion" models, with the strength and range of inter-patch correlations the determinant of the appropriate abstraction in any specific context.

ACKNOWLEDGMENTS

This research was supported by grants from the US National Science Foundation and the Department of Energy (both to WWM) and from the UK Science and Engineering Research Council (to RMN and WSCG).

REFERENCES

Chewning, W.C. 1975. Migratory effects in predator-prey systems. *Math. Biosci.* 23: 253-262.

Crowley, P.H. 1981. Dispersal and the stability of predator-prey interactions. *Am. Nat.* 127: 696-715.

Diekmann, O., J.A.J. Metz, and M.W. Sabelis. 1988. Mathematical models of predator-prey-plant interactions in a patchy environment. *Exp. Appl. Acarology* 5: 319-342.

Godfray, H.C.J., and S.W. Pacala. 1992. Aggregation and the population dynamics of parasitoids and predators. *Am. Nat.*, in press.

Gurney, W.S.C. and R.M. Nisbet. 1978. Single-species population fluctuations in patchy environments. *Am. Nat.* 112: 1075-1090.

Harrison, S. 1991. Local extinction in a metapopulation context: An empirical evaluation. *Biol. J. Linnean Soc.* 42: 73-88.

Hastings, A. 1991. A metapopulation model with local disasters of varying sizes. *J. Math. Biol.* (submitted).

Ives, A. In press. Continuous time models of host-parasitoid interactions. *Am. Nat.*

May, R.M. 1973. Stability in randomly fluctuating versus deterministic environments. *Am. Nat.* 107: 621-650.

Maynard Smith, J. 1974. *Models in Ecology*. Cambridge University Press, Cambridge, UK.

Murdoch, W.W., C.J. Briggs, R.M. Nisbet, W.S.C. Gurney, and A. Stewart-Oaten. In press. Aggregation and stability in metapopulation models. *Am. Nat.*

Murdoch, W.W. and A. Oaten. 1975. Predation and population stability. *Adv. Ecol. Res.* 9: 1-131.

Murdoch, W.W. and A. Stewart-Oaten. 1989. Aggregation by parasitoids and predators: Effects on equilibrium and stability. *Am. Nat.* 134: 288-310.

Nisbet, R.M., and Gurney, W.S.C. 1982. *Modeling Fluctuating Populations*. John Wiley and Sons, Chichester, UK.

Reeve, J.D. 1988. Environmental variability, migration and persistence in host-parasitoid systems. *Am. Nat.* 32: 810-836.

de Roos, A.M., E. McCauley, and W. Wilson. 1991. Mobility versus density limited predator-prey dynamics on different spatial scales. *Proc. Roy. Soc. London* 246: 117-122.

Taylor, A.D. 1988. Large-scale spatial structure and population dynamics in arthropod predator-prey systems. *Ann. Zool. Fenn.* 25: 63-74.

APPENDIX 1: STABILITY ANALYSIS FOR A ONE-PATCH MODEL

The single-patch model is described by the pair of ordinary differential equations

$$dH/dt = aH - bHP - e_h H + ze_h(H + x),$$
$$dP/dt = fbPH - dP - e_p P + qe_p(P + y)$$

where z and q may be functions of H and x. Assume that these functional forms and the parameter values allow the existence of an equilibrium (H^*, P^*). Stability of this equilibrium requires negative real parts for the eigenvalues of the 2 x 2 matrix **A** with elements

$$A_{11} = a - e_h - bP^* + z^* e_h + e_h(H^* + x)(\partial z/\partial H)^*$$
$$A_{12} = -bH^*$$
$$A_{21} = fbP^* + e_p(P^* + y)(\partial q/\partial H)^*$$
$$A_{22} = fbH^* - (d + e_p) + e_p q^*$$

in which * denotes a quantity evaluated with the populations assigned their equilibrium values. For the equilibria to remain unique and non-zero, it is necessary that the determinant of **A** be non-zero. Consequently, the condition for local stability is

Trace **A** < 0

which with a little algebra can be shown equivalent to

$$z^* e_h x/H^* - e_h(H^* + x)(\partial z/\partial H)^* + e_p y q^*/P^* > 0.$$

Result (i) in section 2 is now immediate since density-independent movement implies that z is independent of H. Results (ii) and (iii) are a little more subtle. Suppose there is no predator movement, so that $e_P = 0$. The first part of each result simply reflects possible different signs for the second term in the stability condition. Both results, however, also discuss conditions under which the prey movement has little effect on stability. To understand these, note that with strongly density-dependent movement of prey, the plot of z versus H will be sigmoidal in form with the inflexion close to $H = x$. If H^* is significantly greater or less than x then the partial derivative $(\partial z/\partial H)$ will be close to zero. Stability is then determined by the first term in the stability condition, i.e., by the magnitude of z^*. Results (ii) and (iii) express circumstances that will make z^* small, and hence give approximately neutral stability.

APPENDIX 2: INTER-PATCH PHASE DIFFERENCES

For the analyses in this paper (and also in Murdoch et al. in press), we use a measure of the *asynchrony* of two patches based on the linear correlation coefficient, r, of the asymptotic prey or predator fluctuations on the two patches. The asynchrony is $(1 - r)/2$. With this definition, if the populations on the two patches are varying sinusoidally with the same frequency, but differ in phase by an angle ϑ, then the asynchrony is $\sin^2(\vartheta/2)$. With light damping, this is likely to remain a good approximation.

However, before we are justified in using a measure of asynchrony based on asymptotic phase difference between damped sinusoids, we must investigate:

(a) whether the phase difference between cycles on the two patches is
asymptotically independent of initial conditions, and
(b) how to compute this asymptotic phase difference.

Both investigations involve the eigenvectors of the standard stability matrix (see e.g., Nisbet and Gurney 1982, Chapter 4). Suppose that after linearization the dynamics of the two-patch system are described by linear equations of the form

$$dx/dt = \mathbf{A}x \qquad \text{where } \mathbf{x} = \begin{bmatrix} h_1 \\ p_1 \\ h_2 \\ p_2 \end{bmatrix}$$

and $\mathbf{A}$ is the stability matrix.

Except in the special case (which is of no biological importance) where the eigenvalues of $\mathbf{A}$ are not distinct, the general solution of the linear equation takes the form

$$\mathbf{x}(t) = \Sigma_i \, C_i \exp\{\lambda_i t\} u_i$$

where $\{C_i\}$ are constants whose values are determined by initial conditions.

In all the cases we considered in our investigations of asynchrony, the long-term behavior of the linearized system involved convergent or divergent oscillations. We write

$$\lambda_1 = \mu + i\omega; \quad \lambda_2 = \mu - i\omega$$

and assume that the real parts of the other eigenvalues are less than μ, thereby defining λ_1 and λ_2 as the "dominant" pair of eigenvalues. It is possible to arrange the normalization of eigenvectors so that the eigenvectors $\mathbf{u}_1$ and $\mathbf{u}_2$ are complex conjugates of each other and have first component (u_{11}) equal to unity. Then as $t \rightarrow \infty$ the solution approaches

$$\mathbf{x}(t) = \text{Real part of } \{C_1 \exp [\mu t + i\omega t]\} \, \mathbf{u}_1$$

The initial conditions enter this asymptotic solution only through the term C_1, which multiplies all components of the solution vector. Thus we conclude that the asymptotic phase difference between two components is independent of initial conditions and is equal to the *argument* of the relevant eigenvector. For example, the phase difference between the prey fluctuations on the two patches is equal to the argument of u_{13}, the third component of the first eigenvector.

In practice, there are limitations to be considered. In particular, in a stable system, if one or more eigenvalues have real parts close to that of the leading eigenvalue, then the population fluctuations may be infinitesimally small before the asymptotic form is valid. In such cases the phase relationship between population fluctuations on different patches depends on initial conditions.

10
COUPLING OF CIRCULATION AND MARINE ECOSYSTEM MODELS

Eileen E. Hofmann

INTRODUCTION

Historical Perspective

The history of modeling biological processes associated with marine plankton populations is relatively short, having its origins in the mid-1930s to mid-1940s (see Mills 1989 for an overview). The first models were developed to explain and quantify the processes that resulted in seasonal plankton cycles (e.g., Riley 1946). More recent models have retained the basic structure of the early models, but have extended their realism by incorporating advances in measurements and understanding of biological oceanographic processes.

While the first time-dependent models and subsequent models have done much to advance understanding of biological processes, they did not explicitly include the effects of advection and diffusion on marine plankton populations. Environmental effects were included through exchange coefficients that increased or decreased nutrient or biomass concentrations. However, this paramterization of the physical environment does not capture the spatial and temporal variability that is so important in structuring marine plankton populations.

It was not until the development of large multidisciplinary oceanographic programs in the late 1960s and early 1970s that measurements of sufficient breadth and detail were available to begin the development of models that included circulation as well as biological effects on the lower trophic levels. At the same time an awareness was developing that a systems approach to studying marine ecosystems (Walsh 1972) was needed. Also, computers capable of handling the storage, speed, and memory requirements of models that combined circulation and biological processes (physical-biological models) were becoming available.

The development and implementation of physical-biological models has a history that is considerably shorter than that of modeling marine plankton populations. The first of these models (Walsh 1975, Wroblewski 1977) resulted from the Coastal Upwelling Ecosystem Analysis (CUEA) program, which was a multidisciplinary oceanographic program in the 1970s. These modeling efforts combined results from biological, physical, and chemical measurements and represented an interdisciplinary approach to understanding the lower trophic levels of marine ecosystems. Subsequent physical-biological models have also, for the most part, resulted from multidisciplinary programs. This is not surprising given the enormous amount of data that is needed for initialization, calibration, and verification of these models. An overview of many of the marine interdisciplinary models is given in Wroblewski and Hofmann (1989).

Physical-biological models have been in use for about fifteen years. We are gaining experience with these models and are beginning to appreciate their limitations and strengths. New oceanographic programs and initiatives (e.g., Joint Global Ocean Flux Study [JGOFS], Global Ocean Ecosystems Dynamics [GLOBEC]) proposed for the next decade will, out of necessity, make extensive use of interdisciplinary physical-biological models for the design of field programs and for integrating and analyzing data that come from the field studies.

This chapter presents a discussion of combining circulation and biological processes in models of marine planktonic systems. First, models that primarily investigate circulation effects (Lagrangian models) on marine plankton populations are discussed. Second, models that provide time and space varying distributions (Eulerian models) of biological properties that result from biological as well as circulation effects are discussed. Finally, approaches that use ocean color data to improve physical-biological models are presented. A common feature of all physical-biological models is the need for a circulation field. Hence, a brief discussion of how this may be obtained is given first.

Construction of Circulation Fields

There are two general approaches that can be used to specify the circulation: theoretical and observational. The available theoretical approaches range from simple relationships (e.g., Ekman 1906) to fully three-dimensional primitive equation circulation models (e.g., Semtner 1974, 1986; Bryan and Sarmiento 1985; Hedstrom 1990; Haidvogel et al. 1991b) or quasi-geostrophic circulation models (Holland 1985). The approach used is dependent upon the questions being addressed. For example, if across-shelf transport of fish eggs that are found in the upper part of the water column is of interest, then surface water velocities computed from wind speed may be all that is required. However, if quantifying the flux of organic carbon from a continental shelf is of interest, then a fully three-dimensional circulation model may be required.

For some systems, a theoretical circulation model may not be available or the circulation dynamics are not understood well enough to construct a circulation model that correctly simulates the observed flow. If observations of the currents are available (e.g., current meter moorings), then it is possible to construct a circulation field from the observations by using optimal interpolation or objective analysis techniques (Bretherton et al. 1976). However, for this approach to be valid, the current measurements must be taken with spatial resolution that is within the correlation length scales of the flow. Using data to construct flow fields has the advantages over a theoretically derived circulation field that the dynamics of the flow are included in the measurement and do not need to be explicitly identified, and that comparisons with observations are straightforward. However, the application of optimally interpolated flow fields is limited to specific times and locations and does not allow for process or parameter studies. Theoretical circulation models are not limited to specific times or regions and allow parameter and process studies.

LAGRANGIAN MODELS

Model Equations

A Lagrangian model tracks individual, identifiable particles. Examples of particles may be a copepod, a phytoplankter, or a cm^3 of water that is always composed of the same particles. Lagrangian models are constructed beginning with the definition of velocity, which is

$$\frac{d\vec{x}}{dt} = \vec{U}(t) \tag{1}$$

where $\vec{x}$ is the location of the particle in the coordinate system and $\vec{U}$ is the velocity of the particle at some time, t. The location of the particle over time is obtained by integrating equation (1). However, since a continuum has a large number of particles, it is necessary to identify the specific particles of interest. The most common approach for doing this is to identify particles by their location at the initial time, t_0. Therefore, the solution to equation (1) becomes

$$\vec{x}(t;\vec{x}_0) = \int_{t_0}^{1} \vec{U}(t;\vec{x}_0)\,dt \tag{2}$$

Equation (2) states that the location of a particle at time t, that started at location $\vec{x}_0$, is obtained by accumulating all of the displacements of the particle that occurred from the initial to the current time.

Usually velocities are known only at certain discrete times. Therefore, the continuous form of the Lagrangian model (equation 2) is usually written in discrete form. For example, the discrete representation of a Lagrangian model that describes the horizontal displacement of a particle is of the general form

$$X_n = X_0 + u\Delta t \tag{3}$$

$$Y_n = Y_0 + v\Delta t \tag{4}$$

where X_n and Y_n represent the new location of a particle that results from the displacement of the particle from its previous position (X_0, Y_0) by the horizontal velocity, u and v, in a time interval, Δt. The effect of vertical advective velocities, w, on the location of a particle is given by:

$$Z_n = Z_0 + w\Delta t + w_b\Delta t \tag{5}$$

Since many plankton species can actively alter their vertical location, a biological vertical velocity is included through $w_b \Delta t$.

Interpolation. In Lagrangian calculations the velocities must be known at the location of the particle. Typically, the velocity field is specified from either an Eulerian circulation model in which the velocity has been calculated on a grid with a particular spatial resolution, or from observations that have been interpolated onto a regular spatial grid. In either case, there is no reason to expect that the particle position will coincide with the spatial grid point locations on which the velocity is known. Hence, the velocity must be interpolated to the location of the particle.

Interpolations are approximations and, therefore, can introduce errors. For some situations, accumulated interpolation errors can result in particle trajectories that depart from the true trajectories. Several techniques are available for interpolation of the velocity field, and the particular technique that is used depends on the characteristics and resolution of the circulation fields. For smooth, simple circulation distributions, linear interpolation may be adequate for producing velocities for Lagrangian calculations. However, for velocity fields with considerable structure, more sophisticated interpolation techniques, such as a fourth-order Runge-Kutta technique, may be needed (see Hofmann et al. 1991 for an example). No matter what interpolation technique is used, the space and time resolution of the circulation fields will determine the accuracy of the interpolated velocity. If the original circulation fields have undersampled the variability of the flow, then no interpolation method will restore this variability.

Passive Particles

The simplest type of Lagrangian calculation consists of releasing a series of passive particles (no behavior) into a flow field and tracking their location over time. These calculations are useful for providing first-order estimates of general transport patterns and residence times in specific regions. Two examples are given from the southeastern U.S. continental shelf in which passive particles are used to understand general transport and residence times and to analyze phytoplankton pigment fields obtained from satellite observations. A third example, which used passive particles to understand transport patterns in the Coastal Transition Zone off California, is also described.

Southern U.S. continental shelf: Gulf Stream upwelling effects. The Gulf Stream, an intense western boundary current, is located along the shelf break region of the southeastern U. S continental shelf. This current has a seasonal pattern in its proximity to the shelf break, flowing along the 75 m isobath during the winter and early spring and being further offshore during the summer. Maximum surface velocities associated with the Gulf Stream are in excess of 100 cm^{-1}. Thus, it can have a substantial effect on residence time of populations on the outer shelf. Additionally, the southeastern U. S. continental shelf is affected by two types of Gulf Stream–related upwelling; frontal eddies and bottom intrusions (Figure 1). The space and time scales associated with these two types of upwelling are different. Consequently, their effect on the shelf is different (see Atkinson et al. 1985 for an overview). Thus, this region has a seasonal signal in the Gulf Stream position which is superimposed on a nutrient supply that changes in intensity and duration throughout the year.

140

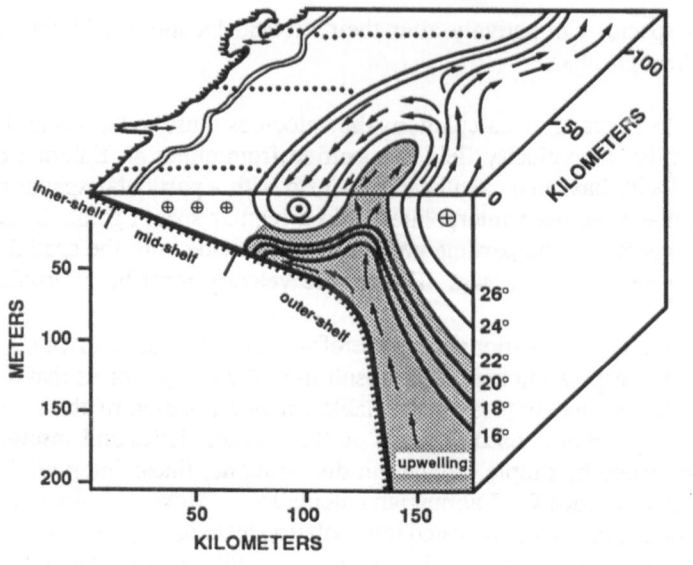

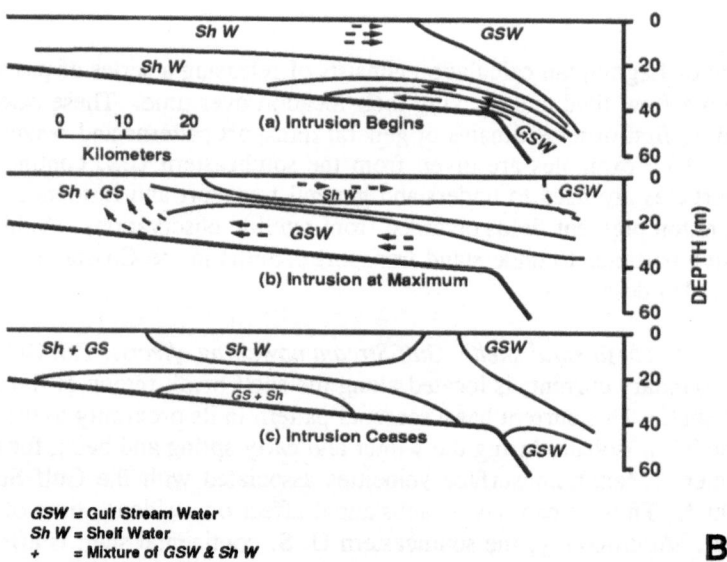

Figure 1. (A) Schematic of a cold-core Gulf Stream frontal eddy and warm-core filament on the Georgia shelf (from Lee et al. 1981). Stippled area indicates the upwelling associated with the cold-core eddy. Arrows indicate the direction of flow. (B) Schematic showing the time development of a bottom intrusion of Gulf Stream water (from Blanton 1971). Solid lines represent isopycnals. Figure is from Ishizaka and Hofmann (1988).

Ishizaka and Hofmann (1988) applied a Lagrangian model to a limited region (Figure 2) of the southeastern U. S. continental shelf. The circulation fields were constructed from an optimal interpolation of current meter measurements (from 37 m) made on the southeastern U. S. continental shelf. Current meter measurements were available for spring, which is the time during which frontal eddies are prevalent, and for summer, which is when bottom intrusions dominate. These measurements also captured the seasonal signal in the location of the Gulf stream relative to the shelf break.

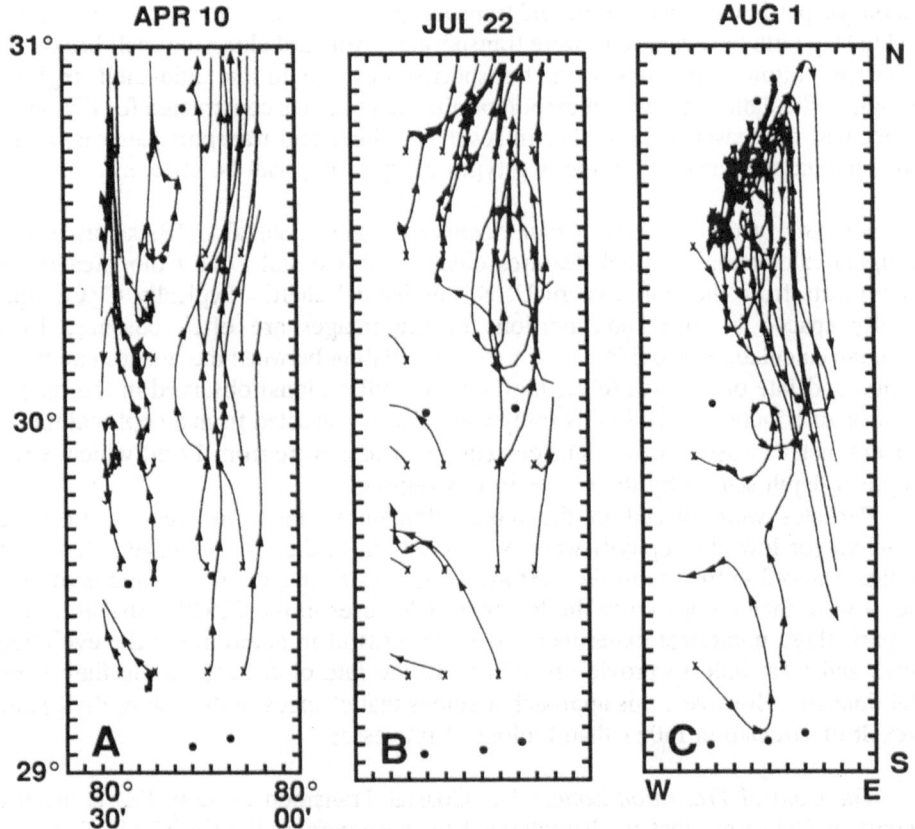

Figure 2. Trajectories followed by particles started on (A) April 10, 1980; (B) July 22, 1981; and (C) August 1, 1981. Arrows on the trajectories indicate particle positions in days and particle flow directions. Figure is from Ishizaka and Hofmann (1988).

A series of Lagrangian calculations were done using the optimally interpolated flow fields to address the effects of 1) Gulf Stream position and 2) frontal eddy and bottom intrusions upon particle residence time and transport patterns. Numerous particles were released in the flow field and tracked until the simulation ended or until they exited the model region. The first set of experiments considered the seasonal patterns in Gulf Stream position (Figure 2). Particles released in the spring, when the Gulf Stream is near the shelf break, were primarily transported to the north with the Gulf Stream. Little across-shelf transport of particles was observed in the particle trajectories and particle residence times were short (i.e., a few days). In contrast, in the summer, when the Gulf Stream is offshore, particles are retained in the model region and residence times can be as much as thirty to forty days. These results show that for the region of the southeastern U. S. continental shelf included in the model, there is a seasonal signal in transport patterns and residence times of passive particles.

Additional Lagrangian experiments considered the effect of frontal eddies and bottom intrusions on particle residence time and transport patterns. Particles released in frontal eddies tended to stay with the events and were transported northward along the shelf break. Particles released in bottom intrusions were transported onshore to the mid-shelf region. These Lagrangian calculations show that carbon or nitrogen budgets constructed for this shelf region must account for seasonal variability in residence times and transport patterns, and that the fate of material associated with the two types of upwelling can be different.

Analysis of Coastal Zone Color Scanner images. Ishizaka (1990a) used a series of Lagrangian calculations to track features observed in Coastal Zone Color Scanner (CZCS) measurements from the southeastern U. S. continental shelf. Typically CZCS images are unequally spaced in time, and portions of the images are often obscured by clouds. Lagrangian calculations provide a means of interpolating between the images and tracking the evolution and fate of features (e.g., high chlorophyll regions) observed in the images. The circulation distributions used in this study were also constructed from an optimal interpolation of current meter measurements. The current meter data were from 17 m, which is within the one optical depth sensed by the CZCS in this region.

Particles were placed in the model domain in the same regions in which high chlorophyll (or low chlorophyll) water was observed in the CZCS images. These particles were then tracked in time until the next available CZCS image. The locations of the various particles were then compared to the location of features in the CZCS distributions. For the most part, these numerical experiments were successful in determining the evolution of the features and were able to provide insight as to the fate of features while they were in the model domain. However, this approach assumes that changes in the chlorophyll patterns are the result of circulation rather than biological processes.

The Coastal Transition Zone. The Coastal Transition Zone (CTZ) is the region off the coast of California that is characterized by meanders of the California Current and by prominent, across-shelf jets and filaments. The offshore-flowing jets represent a potentially important mechanism for transporting properties such as nutrients and organic carbon from nearshore to offshore regions. Recent studies (Bucklin et al. 1989, Mackas et al. 1991) have suggested that the jets and filaments of the CTZ are important in determining the transport and structure of zooplankton populations off the coast of California. As part of a multidisciplinary study of the CTZ, a three-dimensional primitive equation circulation model has been developed to study the physical processes responsible for the formation of the filaments and jets (Haidvogel et al. 1991a). A description and analysis of the simulated circulation fields produced for the CTZ with this circulation model is given in Haidvogel et al. (1991a).

The simulated circulation fields provide three-dimensional time-dependent realizations of the velocity at intervals of a few minutes. These velocity fields were used for a series of Lagrangian calculations designed to demonstrate the role of the offshore-flowing filaments in determining plankton residence times and general transport patterns in the CTZ region (Hofmann et al. 1991). In particular, the drifter simulations were undertaken to explain zooplankton distributions that were observed during the 1988 CTZ sampling study. Thus, particle release points were chosen to provide coverage, relative to an emerging filament, comparable to that of an actual drifter released as part of the extensive CTZ field program in 1988.

Unlike the particle trajectories calculated for the southeastern U. S. continental shelf, those calculated for the CTZ are three-dimensional. The particles were considered passive because the dominant taxa observed in the CTZ did not appear to undergo vertical migration. The general particle trajectories (Figure 3) show that there is a narrow inshore region where particles started inshore of this area are transported southward with the California Current. Particles started offshore of this region are transported offshore in the filament. Further, particles released deep (i.e., 60 to 90 m) undergo little net displacement. An interesting aspect of the particle trajectories is that those transported in the filaments downwell as they move offshore. Such a pattern was also observed in zooplankton distributions in the CTZ (Mackas et al. 1991).

Particles With Behavior

Many marine plankton species exhibit sinking, swimming, and migration behaviors that, when combined with the circulation field, can affect their horizontal or vertical location. Plankton that sink or migrate vertically through a horizontally sheared velocity field can substantially alter their horizontal displacement. Migration strategies can cover a range of behaviors, from simple light-induced diel vertical migration to a behavioral response that has evolved to take advantage of particular aspects of the environment. The examples discussed below illustrate how these biological effects are incorporated into Lagrangian models.

Phytoplankton sinking. In coastal upwelling systems, surface water is moved offshore, usually by an alongshore wind stress. To replace the surface water, cold, nutrient-rich water is transported vertically (upwelled) from depth. The nutrients associated with the upwelled water stimulate primary production in the euphotic zone. In an across-shelf vertical plane representation, the circulation that characterizes upwelling systems is composed of horizontal offshore/onshore flows with a vertical flow near the coast. The strength and duration of the upwelling circulation is variable in space and time. It has been suggested (Malone 1975) that phytoplankton community composition is strongly controlled by the structure of the circulation associated with upwelling systems. In essence, netplankton (> 20 μm) dominate coastal upwelling systems because, as they are transported offshore in surface waters, their sinking rates are high enough for them to reach the onshore flowing water at a depth that returns them inshore to the surface (Figure 4). These phytoplankton provide the seed population for subsequent blooms. In contrast, nanoplankton (2 - 20 m) are selectively removed from the

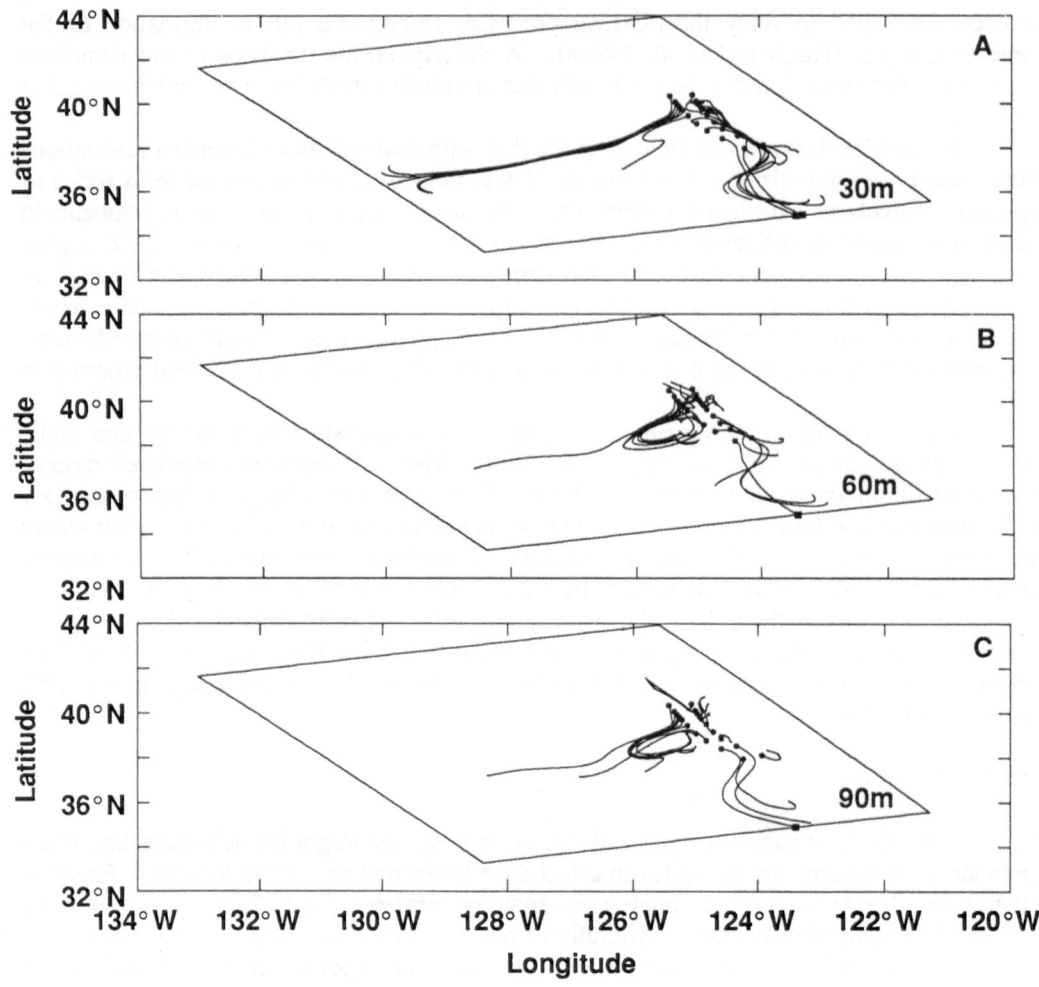

Figure 3. Simulated trajectories followed by drifters released along lines that bracket the southward flowing California Current and the region where jets and filaments form from this current. Drifters were released at initial depths of (A) 30 m, (B) 60 m, and (C) 90m. Figure is from Hofmann et al. (1991).

upwelling region because their lower sinking rates are not sufficient to transport them to the onshore flowing layer at depth.

To test the hypothesis of physical control on biological communities in upwelling systems, Smith et al. (1983) developed a Lagrangian model. The flow field was contructed from current meter measurements made in the Peru upwelling region using a variational calculus objective analysis method. This analysis provided estimates of the horizontal (onshore-offshore) and vertical velocity field at six-hour intervals for a 52-day period that started in early March 1977. This approach for constructing a flow field captures the space and time variability of the flow, but limits the application of the model to a specific time and location.

Numerous particles, with varying sinking rates, were released in the flow field and tracked until they exited the model region. Analysis of the particle trajectories showed that, for all sinking rates, for the entire 52-day period there was not a single occurrence of a particle moving offshore, sinking from the surface layer, entering the onshore flow, and being upwelled to the surface near the coast. Thus, the Lagrangian trajectory analyses, which contradicted the reseeding hypothesis (Figure 4), led Smith et al. (1983) to conclude that the reseeding mechanism of the Peru upwelling system had been disrupted in early 1977. This conclusion was consistent with observations of anomalously low phytoplankton biomass and productivity during 1977, even though nutrient concentrations and growth rates were high over the Peru shelf. However, the processes resulting in disruption of the reseeding mechanism were not apparent.

Zooplankton with vertical migration. It has been suggested (Peterson et al. 1979) that the distribution of the neritic copepod *Calanus marshallae* in the Oregon upwelling region is determined by the interaction of the upwelling circulation with the animal's vertical migration behavior. Essentially, ontogenetic migration by adult females brings the reproducing animals nearshore in newly upwelled water. The females release their eggs in the food-rich nearshore region. The developing eggs, nauplii, and early stage copepodites are transported offshore in the surface waters of the upwelling circulation. Diel migration by the late-stage copepodites reduces their transport out of the upwelling system. To investigate this sequence of events, Wroblewski (1982) developed a model that consisted of *C. marshallae* population dynamics and a time-dependent velocity field obtained from a theoretical circulation model. The older copepodite stages and adults were allowed to migrate vertically, whereas the younger copepodite stages, nauplii, and eggs were strictly planktonic.

The circulation model (Thompson 1974) used to obtain the flow field simulates the response of the coastal ocean to a wind stress in a vertical plane normal to the coast. The circulation model reproduced the seasonal pycnocline and regions of upwelling, convergence, and divergence. The population model used by Wroblewski (1982) represented the life history of *Calanus marshallae* by five developmental stages: eggs, nauplii, copepodites (CI-CIII), CIV-V, and adults. Parameter values for the population model were specified using laboratory and field observations. Vertical migration behavior for the two older developmental stages (CIV-V and adults) was specified as

$$w_n = w_s \sin(2\pi t) \tag{6}$$

where w_s is the maximum vertical migration speed of the *nth* developmental stage. This formulation goes in place of w_b in equation (5) and allows for a vertical migration with a period of twelve hours, e.g., diurnal vertical migration.

The effect of vertical migration on copepod location is illustrated by the trajectories shown in Figure 5. The nonmigrating copepod is transported about 9 km from the release point by the end of the simulation. However, migrating copepods show essentially no net displacement over the time of the simulation. Hence, by periodically migrating into the onshore flow at depth, these animals are able to maintain their position in the productive inshore waters.

The results from this model support the life cycle of *Calanus marshallae* proposed by Peterson et al. (1979) from field observations. Under upwelling-favorable conditions, it is possible for high concentrations of older copepodites and adults to be maintained in nearshore regions by diurnal vertical migrations.

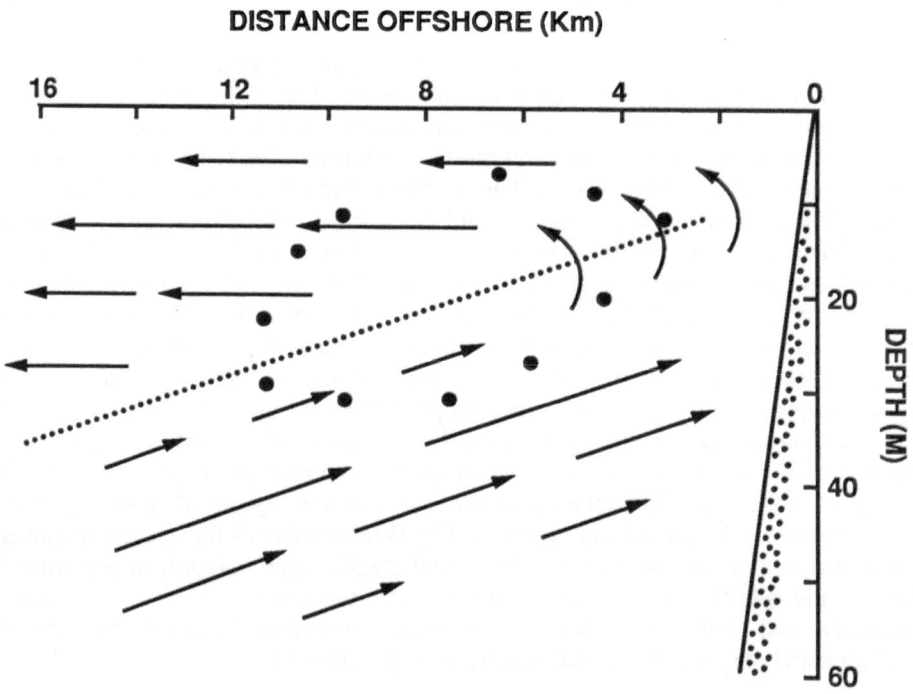

Figure 4. Schematic of the across-shelf reseeding mechanism. Phytoplankton (circles) are advected offshore in the surface layer but sink into the onshore flow at depth. The cells then move onshore and are upwelled near the coast. The dotted line represents the depth of the zero onshore-offshore flow. Figure is from Smith et al. (1983).

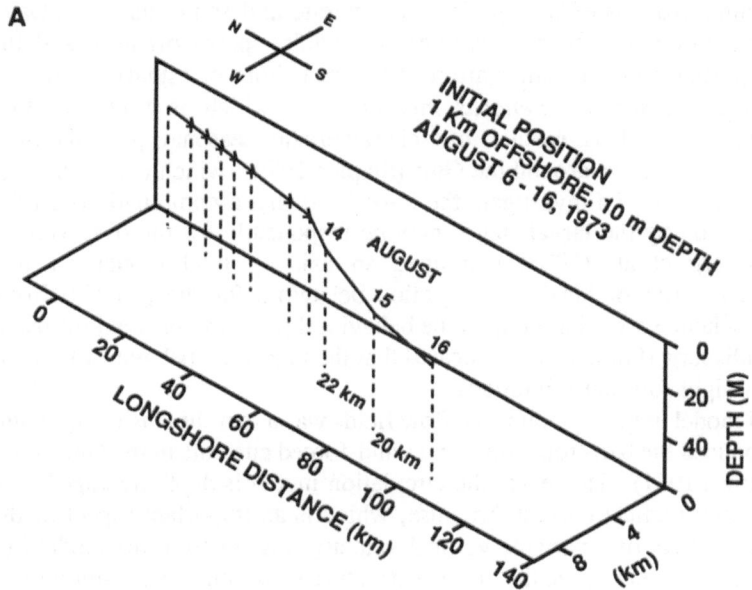

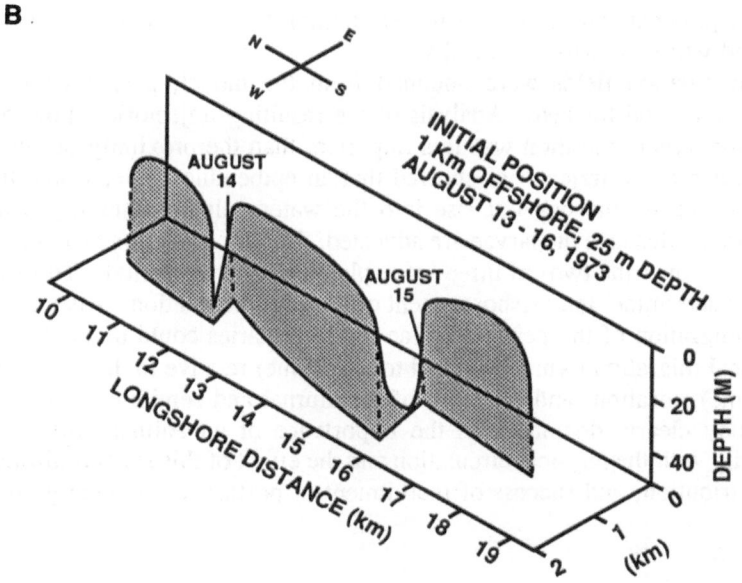

Figure 5. (A) Three-dimensional trajectory for a hypothetical nonmigrating copepod initially located 1 km offshore and at 10 m depth, and (B) at 25 m depth. The copepod migrates between 3 and 44 m depth each day. Figure is from Wroblewski (1982).

Varying vertical migration strategies. Many marine species have a life history in which the spawning grounds of the species are separate and some distance away from the nursery grounds. For many of these species, the adults spawn offshore and the nursery grounds for the postlarvae or juveniles are found in nearshore or estuarine areas. One such species is the penaeid shrimp *(Penaeus merguiensis)*, which is found in the Gulf of Carpentaria, Australia. This species has characteristic seasonal patterns of postlarval immigration into the estuaries around the Gulf (Staples 1979), which cannot be explained by simple random diffusion. To investigate the hypothesis that a combination of tidal currents and vertical migration by the larval stages may be responsible for the observed recruitment patterns, Rothlisberg et al. (1983) combined an existing tidal model for the Gulf of Carpentaria with a series of idealized migration behaviors for the peneaid shrimp larvae. These migration schemes were based upon the behavioral patterns observed for paeneid larvae in the field (Rothlisberg 1982), which indicated that the migration behavior of all larval stages was mediated by light (diurnal migration).

The tidal model used to produce the flow fields was a two-dimensional, depth-averaged model that reproduced the barotropic tidal and wind-forced currents in the Gulf of Carpentaria (Church and Forbes 1981). However, the circulation model is depth-averaged and, as such, does not contain any vertical current structure, which is an important aspect of the physical environment. The effect of a variable vertical velocity was added to the model by including an Ekman dynamics model to give the vertical structure of the currents resulting from surface and bottom stresses and from pressure gradients. This approach assumes that frictional effects are the primary mechanism producing vertical shear, and, for many shallow tidal systems, this is an adequate approximation for the vertical current structure. However, for systems where the internal density gradients are important in determining the vertical shear, this approach is inappropriate and will give erronous results.

Once the circulation fields were obtained from the model, particles with various behaviors were released and tracked. Analysis of the resulting trajectories showed that the vertical extent of the larval migration was less important than the proximity of the larvae to the bottom. The Lagrangian trajectories showed that an epibenthic mode, where the larvae settle to the bottom during the day but rise into the water column at night, significantly increased the horizontal distance the larvae are advected. In fact, advection distances of about 165 km were possible over the two- to three-week planktonic larval period. This distance is related to, and may determine, the offshore extent of the adult population. Also, the seasonal patterns in the immigration of the penaeid larvae to the estuaries could be explained by the timing of the vertical migration (which is keyed to solar time) relative to the phase of the tide (keyed to lunar time), location, and structure of the diurnal and semidiurnal tidal currents. This modeling study clearly demonstrates the importance of an animal synchronizing its reproductive activity with the physical circulation and the effect of this synchronization on the timing, spatial distribution, and success of recruitment of postlarvae to nursery grounds.

EULERIAN MODELS

Model Equations

The role of physical-biological interactions in determining observed biological distributions can be described by the equation for the time-dependent advection and diffusion of a nonconservative substance, which is of the form

$$\frac{\partial B}{\partial t} + \frac{\partial (uB)}{\partial x} + \frac{\partial (vB)}{\partial y} + \frac{\partial (wB)}{\partial z} = \frac{\partial}{\partial x}(K_x \frac{\partial B}{\partial x}) + \frac{\partial}{\partial y}(K_y \frac{\partial B}{\partial y}) + \frac{\partial}{\partial z}(K_z \frac{\partial B}{\partial z}) \quad (7)$$

+ Biological Source/Sink Terms

where the terms in equation (7) represent the time-dependent changes in B, horizontal and vertical advective effects on B, horizontal and vertical diffusive effects on B, and biological effects on B, respectively. The model given by equation (7) allows the circulation to affect the biological distributions. However, the biological processes have no effect on the circulation, so the model is not truly coupled. Most of the physical-biological models in the marine literature consider only circulation effects on the biology. Some recent physical-biological models have allowed for a feedback between the biology and physics. For example, heating of the water column due to differential absorption of solar energy by phytoplankton is sufficient to alter the mixing characteristics of the upper ocean for some systems (Lewis et al. 1983).

Reduced Two-Dimensional Models

The model described by equation (7) includes three space dimensions and time. While this is a realistic representation of the world, the understanding and measurements to implement such a model are frequently lacking. Hence, truncated versions of equation (7) are frequently used to model the space and time distributions of marine planktonic populations. In spite of neglect of one or more dimensions, these models are still instructive. Described below are examples of physical-biological models in which one space dimension has been neglected. These represent a limited sample of the available models in the marine literature.

Vertical plane models. The first physical-biological models were developed for coastal upwelling systems. At the time these models were being developed, the prevailing theories of coastal upwelling assumed that processes occurred primarily in the across-shelf direction; alongshore processes were considered to have little effect. Consequently, the geometry used for the first physical-biological models was a vertical plane normal to the coast. One example of this type of upwelling model is that developed by Wroblewski (1977) for the Oregon upwelling system. This model was designed to investigate the processes controlling primary production in coastal upwelling systems. Therefore, the ecosystem components and biological processes chosen for inclusion in the model were those that directly impact the growth of phytoplankton.

The model ecosystem consisted of five components: phytoplankton, zooplankton, ammonium, nitrate, detritus. Within each component, processes that affect phytoplankton growth, such as nutrient removal and zooplankton grazing, were specified using laboratory or field measurements. All ecosystem components were expressed in terms of nitrogen, which was assumed to be the limiting nutrient. The five-component ecosystem model was combined with circulation distributions obtained from the theoretical circulation model discussed above for the *Calanus marshallae* model. The final model then consisted of a set of five coupled partial differential equations, similar to equation (7), with the dependent variable representing each ecosystem component. The advective velocities (u and w) were specified from the circulation model output as variable parameters and the biological processes defined the source and sink terms on the right side of each equation. This set of equations was used to simulate the effect of constant and time-varying wind stress on primary production.

An example of the results obtained with the Oregon upwelling model is shown in Figure 6. The simulated flow field shows the characteristic structure expected of an upwelling circulation. The highest primary production is found in the inshore region where the upwelling circulation is providing nutrients to the euphotic zone. Phytoplankton biomass, represented as chlorophyll, is highest in the inshore region, and extends offshore and downward as a coherent plume. Note that the structure of the phytoplankton plume corresponds well to the structure of the circulation field. One important result that was obtained from this modeling study is that the highest phytoplankton concentration occurs after the upwelling winds relax. When upwelling winds are strong, the phytoplankton are supplied with nutrients, but the residence time of the cells in the euphotic zone is short. The cells are transported offshore and down to depths of decreased light, which limits growth. However, when the winds relax, the strength of the upwelling circulation decreases and cells are retained in the euphotic zone longer and can make use of the upwelled nutrients. This model result has important implications for designing field programs for upwelling systems.

In spite of the lack of alongshore variability, considerable insight as to the interactions of physical and biological processes in coastal upwelling systems was gained from this model. Also, the ecosystem model used for this study, while not including all aspects of the lower trophic levels, was adequate to represent to first order the important biological processes. However, the neglect of one spatial dimension and portions of the ecosystem limits model realism and the space and time scales over which the model solutions are valid.

Horizontal plane models. During the early 1980s a large multidisciplinary program was undertaken to study Gulf Stream–induced upwelling effects on biological production of the southeastern U. S. continental shelf (Atkinson et al. 1985). Using these data, a physical-biological model was developed to investigate the transport of nutrients (e.g., nitrate) and organic carbon across this shelf (Hofmann 1988). Circulation and temperature fields constructed from an optimal interpolation of current meter data provided the physical portion of the model. The biological model consisted of a ten-component ecosystem that was designed to describe the lower trophic levels on the southeastern U. S. continental shelf (Hofmann and Ambler 1988). This was considered to be the minimum ecosystem required to characterize biological effects on nitrate and organic carbon transport on this shelf. The biological processes associated with each ecosystem component (the terms on the right side of equation [7]) were specified using laboratory and field measurements.

The model geometry was a horizontal plane at a nominal depth of 37 m, which corresponds to the depth of the current meter measurements used for the optimal interpolation. A horizontal plane does not allow for the vertical processes that are important in upwelling regions. However, most of the model region is dominated by the Gulf Stream, and therefore, alongshore transport is a major pathway for nutrients and organic carbon.

Upwelling of nutrients was included by using a relationship between nutrients and temperature that has been established for newly upwelled waters on the southeastern U. S. continental shelf. The relationship was applied along the outer and southern boundaries of the model domain, and the temperature values along these boundaries obtained from the optimal interpolation were used to specify the nitrate value. The velocity field then transported the estimated nitrate into or out of the model domain, depending on direction.

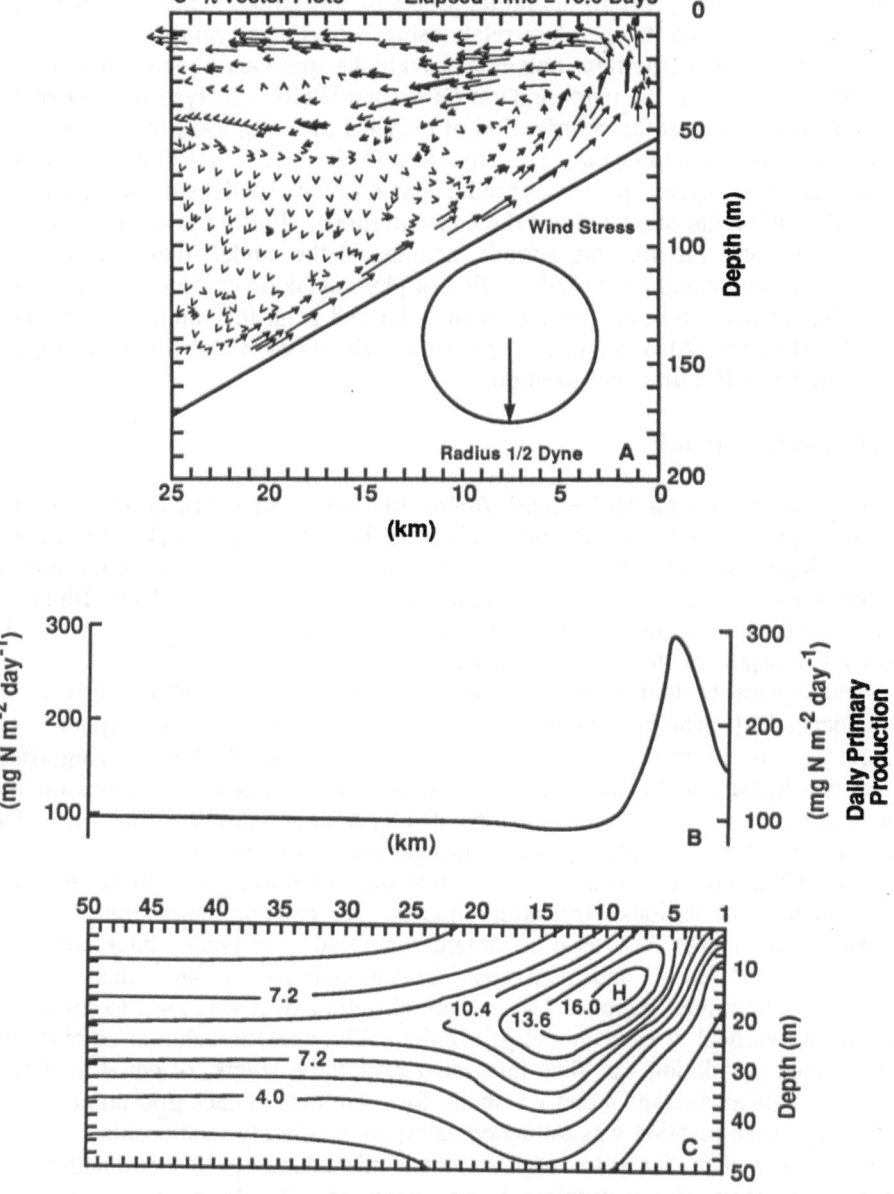

Figure 6. Simulated distributions for the Oregon upwelling after 10 days elapsed time. (A) Circulation in a transverse plane normal to the coast, the bottom topography, and the wind stress. (B) Daily gross primary production of the water column. (C) Distribution of phytoplankton. Contour intervals are 1.6 μg at N l^{-1}. Figure is from Wroblewski (1977).

This approach assumes a conservative relationship between temperature and nitrate. Therefore, it works well only in the initial stages of upwelling. Once biological processes become active, the nitrate-temperature relationship is no longer valid.

The simulated distributions obtained provide the space and time evolution of nutrient, chlorophyll, and zooplankton fields at 37 m (Hofmann 1988). Analysis of these distributions allowed conclusions about the contribution of physical or biological proceses in determining the observed spatial structures on the southeastern U. S. continental shelf. For this shelf, it was determined that physical processes produce the observed spatial structure while biological processes determine the magnitude of the concentrations. Further analysis of the simulated fields indicated that, for the time periods included in the model, there is, at 37 m, a net onshore flux of nitrate and a net offshore flux of phytoplankton carbon. However, since this model lacked vertical processes, flux estimates derived from the simulated distributions are likely underestimates. Also, the use of optimally derived flow fields limits the application of the model to a specific time and location.

Three-Dimensional Models

   Phytoplankton in the Mid-Atlantic Bight. Recently there has been considerable interest in the fate of phytoplankton carbon on continental shelves. Walsh et al. (1988) developed a physical-biological model to investigate the seasonal production, consumption, and transport of the spring bloom that occurs on the continental shelf of the Mid-Atlantic Bight. For this system, the effect of resuspension of phytoplankton due to wind events and subsequent across-shelf transport of the resuspended material was of interest.

The physical-biological model presented by Walsh et al. (1988) is three-dimensional and time dependent. The model consists of a relatively simple biological system. Only two dependent variables, nitrate and chlorophyll, were used. Zooplankton grazing effects were included as a linear loss to the phytoplankton; no explicit zooplankton equation was used. The phytoplankton were assumed to be regulated by light and nitrate availability. The effect of regenerated nutrients on phytoplankton growth was not included.

The horizontal and vertical velocities that were used to advect nitrate and chlorophyll were obtained from various circulation models. An existing circulation model for the Mid-Atlantic Bight was first used to obtain the steady, vertically integrated circulation resulting from wind forcing and river input. This model provided an estimate of the surface elevation over the model region. The surface elevation field was used to force an Ekman balance in the vertical at each model grid point. The analytic solution obtained from the Ekman balance calculation was then integrated over three layers, of equal thickness in the vertical, to obtain an average u and v velocity for each layer at each grid point. The vertical velocity, w, at each location was estimated using the continuity relationship. This approach is cumbersome and is likely to result in some mismatches of the circulation dynamics. However, the nature of the problem being addressed with the physical-biological model required that the vertical currents resulting from wind forcing be known, and the existing barotropic circulation for the Mid-Atlantic Bight lacked this information.

For the Mid-Atlantic Bight simulations, phytoplankton sinking rate, zooplankton grazing, and nitrate inputs were varied. For all cases, the model results suggested that about 25% of the total shelf carbon fixation was lost as export. The model domain covered a large portion of the Mid-Atlantic Bight; therefore, simulated distributions of nitrate and chlorophyll were obtained at frequent intervals over this entire region. Thus, the model results can be analyzed in the same way as distributions obtained from a ship or from other types of measurements. Also, simulated distributions can be produced for different parameter ranges and the results compared to determine the important processes governing phytoplankton distributions.

OCEAN COLOR MEASUREMENTS

All the models discussed above are limited in realism by the choices that are made to parameterize poorly understood processes and by the values used to specify poorly known coefficients. For some cases, errors that are introduced into the model solutions by choices of parameter values or processes can result in simulated distributions that have little relation to reality. Obviously, improved measurements and understanding of parameters and processes will minimize this problem. However, some biological processes (i.e., *in situ* mortality) are difficult or impossible to measure in the field.

Data assimilation (Haidvogel and Robinson 1989) represents one approach that is used to improve model realism. Physical-biological models are now becoming advanced enough to attempt data assimilation. This requires biological data sets with good space and time resolution. The phytoplankton pigment fields obtained from satellites such as the Coastal Zone Color Scanner (CZCS) are perhaps the most obvious choice for assimilation into physical-biological models. These data are available with fine spatial resolution, are synoptic over large areas, and, for some regions, relatively frequent in time. Examples are given of how CZCS data can be used with physical-biological models to adjust little-known parameters, verify simulated distributions, and adjust model solutions.

Parameter Studies

Physical-biological models contain parameters, usually in the closure terms, such as mortality for biological populations or diffusion for the circulation dynamics, that are poorly known, and frequently model solutions are sensitive to the choice made for these parameters. Ishizaka (1990b) presents an approach that uses CZCS-derived chlorophyll distributions to define ranges of values for these parameters in physical-biological models.

The biological model used by Ishizaka (1990b) consisted of a four-component ecosystem (nitrogen, phytoplankton, zooplankton, and detritus). The circulation field was obtained from an optimal interpolation of current meter measurements from 17 m. The physical-biological model, with a standard set of parameters, was used to produce chlorophyll distributions for April 1980. Correlations were then computed between simulated and CZCS-derived distributions for the same time. The value of a model parameter(s) was then changed, the model rerun, and correlations recomputed. The correlation (COR) is a measure of how well the patterns in the simulated and observed field agree. The root mean square error (RMSE) is a measure of how well the magnitude of the two fields match.

As an example, correlations and standard deviations between simulated and observed chlorophyll fields for a range of values in the horizontal diffusion coefficients (Figure 7) were used to bracket the value for this parameter. The value of the COR increased little with increases in the value of the horizontal diffusion coefficients. However, the RMSE of the mean fields decreased rapidly as the diffusion coefficient increased. The COR and RMSE of the standard deviation fields also increased and decreased with increases in the value of the coefficient. These results indicate that the simulated phytoplankton distributions contained

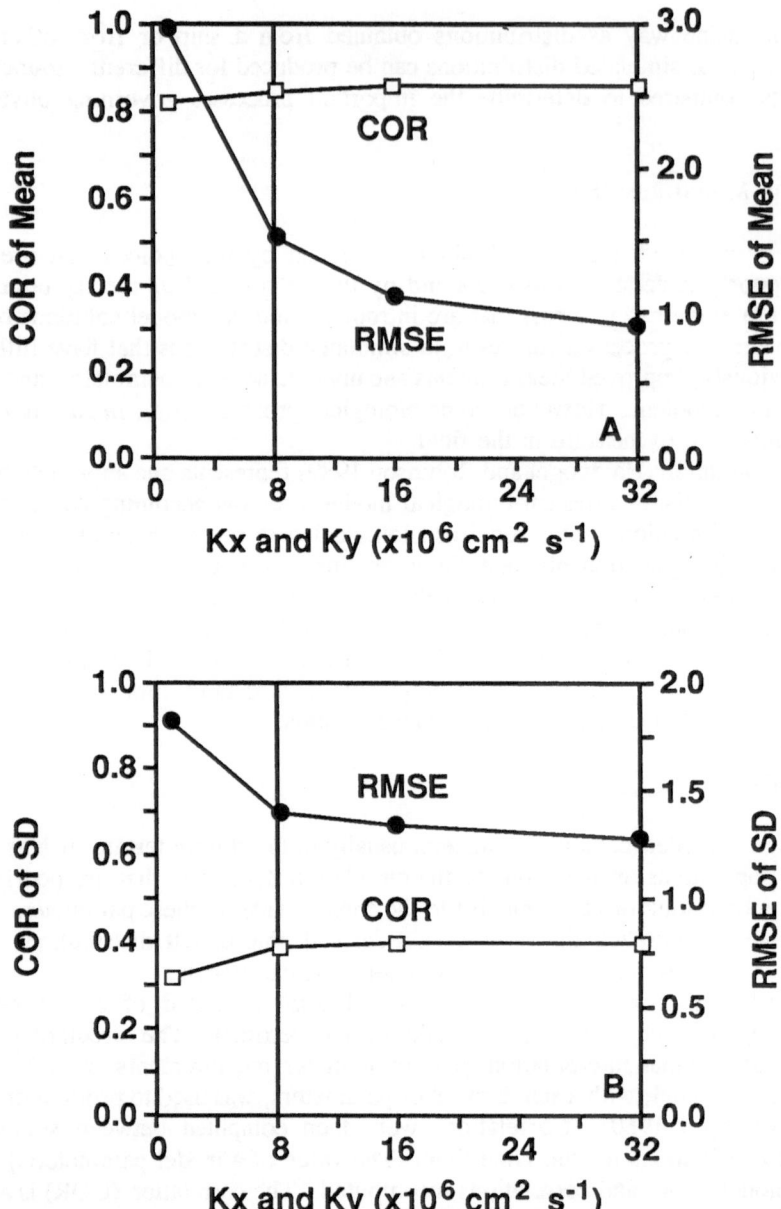

Figure 7. Changes in the COR and RMSE for changes in the horizontal eddy diffusion coefficients, K_x and K_y (A) for the time mean, and (B) for the standard deviation. Solid line indicates the standard value chosen for the physical-biological model. Figure is from Ishizaka (1990b).

large local errors and that increasing the diffusion coefficient improved the comparisons with data by diffusing away the error. Similar comparisons were done to bracket ranges of values for phytoplankton and zooplankton loss rates (Ishizaka 1990b).

The results shown in Figure 7 illustrate how CZCS data can be used to adjust parameters in physical-biological models. This approach has promise for obtaining ranges of values for some of the poorly known biological and physical parameters. However, as indicated by Ishizaka (1990b), the problem with using CZCS data to adjust model parameters is that there is no feedback between the physical and biological regimes. For example, if the flow field contains errors, then adjustment of the biological parameters with CZCS measurements may result in good statistical correspondence, but also in biological parameters that are unrealistic. Hence, for this method to be generally usable, it is likely that the physical fields will also need adjustment with a different data set, such as sea surface temperature.

Model Verification

One test of a physical-biological model is the extent to which it can reproduce observed patterns. Related to this issue is the degree to which physical or biological processes contribute to the observed patterns. One approach for verifying the distributions obtained from models (and indirectly verifying the processes included in the model) is to compare simulated chlorophyll patterns to those observed in CZCS images. At a minimum, the model should reproduce quantities such as the time average and standard deviation of the CZCS-derived chlorophyll distributions.

Ishizaka (1990b) computed the time mean and standard deviation distributions from nine CZCS images (Figure 8a) from the southeastern U. S. continental shelf from April 1980. He then used a physical-biological model to produce chlorophyll distributions that resulted from the inclusion or exclusion of various physical or biological processes. The time mean and standard deviation distributions were computed from the simulated fields and compared to those obtained from the CZCS distributions.

The first model simulation did not include biological or upwelling processes (Figure 8b). Thus, this was a test of the effects of horizontal advection on phytoplankton distributions. The average chlorophyll and standard deviation distributions from the model were similar to those obtained from the CZCS. However, the chlorophyll gradient was lower than the observed gradient. Including biological effects, but not upwelling terms, resulted in the distributions shown in Figure 8c. The average and standard deviation distributions for this case were small, and in general the spatial distributions are different from those in the CZCS distributions. Allowing upwelling processes to be active along with biological and advective processes resulted in the distributions shown in Figure 8d. These simulated fields reproduced the strong gradient, from high chlorophyll at the inner edge of the model region to low chlorophyll at the offshore edge, and showed maximum variance along the outer part of the model domain.

The results shown in Figure 8b indicate that the basic patterns and variability observed in the chlorophyll distributions are controlled by horizontal advection. Biological processes along with upwelling processes control the magnitude of the chlorophyll distribution. Biological and horizontal advective processes alone are not sufficient (Figure 8c). Thus, this study implies that any physical-biological model developed for the southeastern U. S.

156

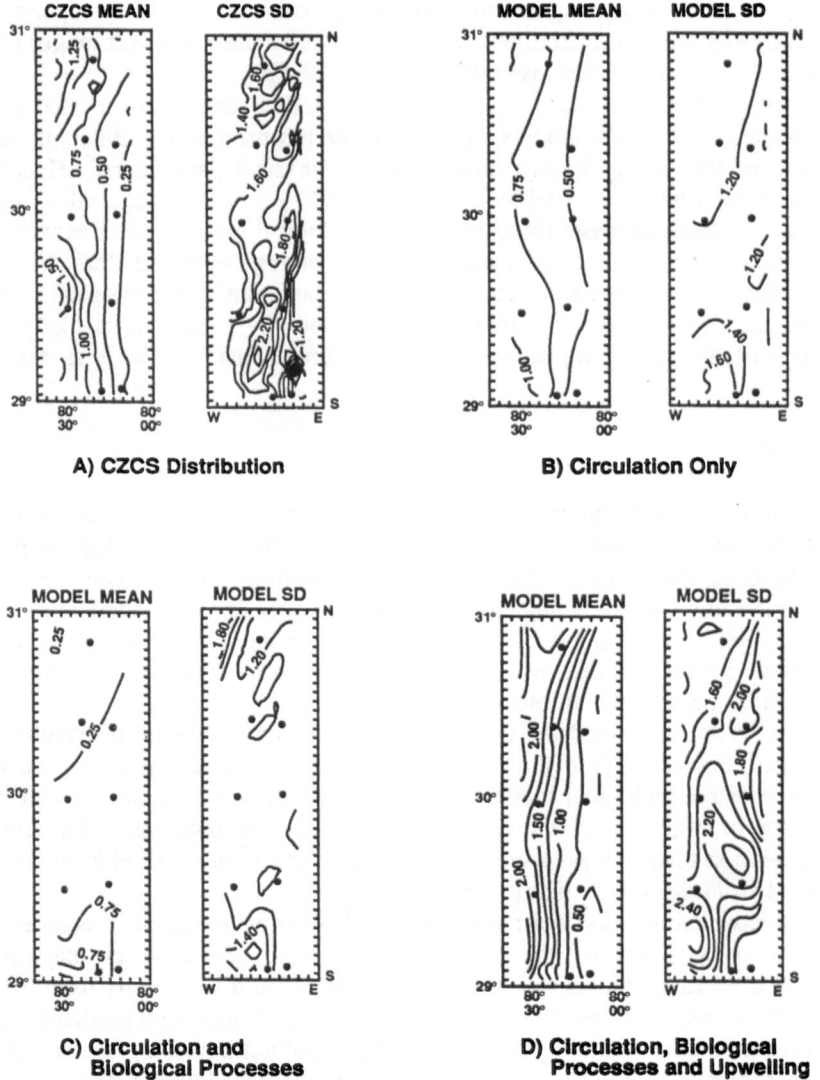

Figure 8. (A) Mean and standard deviation fields calculated from a series of CZSC images from April 1980. (B) Mean and standard deviation distributions calculated from simulated chlorophyll fields that include circulation effects only. (C) Mean and standard deviation distributions calculated from simulated chlorophyll fields that include circulation and biological processes. (D) Mean and standard deviation distributions calculated from simulated chlorophyll fields that include circulation, biological, and upwelling effects. Contour intervals for the mean and standard deviation fields are 0.25 and 0.2 μg chl a^{-1}, respectively. Figure is adapted from Ishizaka (1990b).

continental shelf must contain accurate representations of horizontal and vertical advective processes as well as a good representation of the food web of the lower trophic levels.

Model Upgrades

Another potential use for CZCS measurements is to upgrade physical-biological models. Upgrade means an improvement in model capability through assimilation of an available data set, i.e., CZCS measurements. However, assimilating data into an ecosystem model presents difficulties; if one portion of the ecosystem is adjusted, all other portions of the ecosystem must also be adjusted. For example, if the phytoplankton field is adjusted with a CZCS field, then the other ecosystem components such as zooplankton must also be adjusted so that they are in balance with the new phytoplankton field. If data are available for all ecosystem components, this is not a problem. However, such data are almost never available. Hence, relationships between ecosystem components must be used to adjust the fields.

Ishizaka (1990c) developed three approaches for adjusting other ecosystem components following adjustment of the phytoplankton component with CZCS data. The first technique assumed constant values for the inorganic nitrate (N) and zooplankton (Z) components. Detrital nitrogen (D) was estimated by the difference in the sum of the P-Z-N nitrogen and the total nitrogen estimated from a nitrate-temperature relationship. The second technique was similar except that the model distributions were used to estimate the values for N and Z; they were not assumed constant. The third technique assumed a ratio between N:Z:D in the model distributions. The new N-Z-D distributions were calculated from the difference between the total nitrogen estimated from temperature and the phytoplankton concentration obtained from the CZCS. The effect of the various methods on improvement of the model solutions was quantified by calculating the COR and RMSE between the simulated and CZCS-derived phytoplankton distributions.

The three methods described above all resulted in improvements of the model performance relative to the original (no input of CZCS data) model (Figure 9). Comparisons among the three methods indicate that they give about the same improvement in model capability. The more interesting result is that the upgraded model solution returns to the original solution within one to two days. The implication is that the phytoplankton fields need to be upgraded frequently. This same result is true even when the model is upgraded at all times when CZCS data are available (Ishizaka 1990c).

The outer southeastern U. S. continental shelf system is influenced by a strong boundary current. Therefore, horizontal advection is the primary process controlling biological distributions in this region (McClain et al. 1990). As a result, information input through model upgrades with CZCS measurements is removed from the model region in just a few days. In other regions, where the biological effects are stronger or the physical forcing is weaker, the choice of model upgrade technique could potentially be more important in determining the distributions obtained from a physical-biological model.

SUMMARY

1. Physical-biological models are useful for addressing problems in biological oceanography.

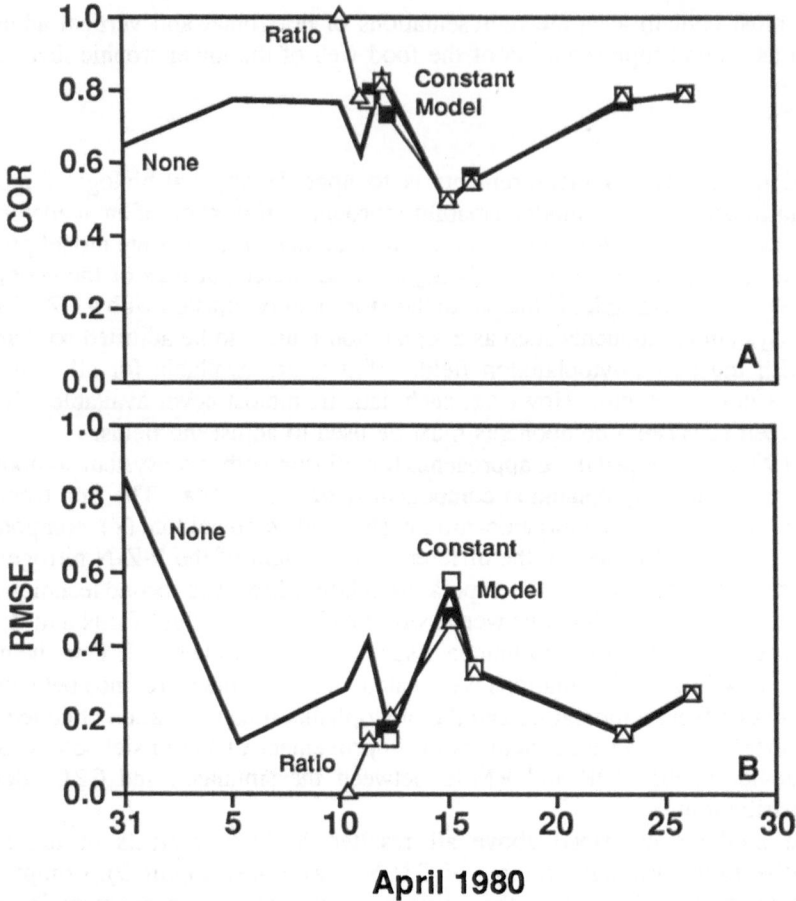

Figure 9. Time variations in the COR and RMSE for the data assimilation experiments. Heavy line indicates the non-upgraded model (none). Other lines indicate the results obtained from the model that was upgraded with the April 10, 1980, CZCD data. Those results obtained with a constant 0.1 μM nutrient and zooplankton value are indicated by the white square (constant). Results obtained with the model nutrient and zooplankton distributions are indicated by the solid square (model). Results obtained with a ratio of nutrient and zooplankton are indicated by the cross (ratio). Figure is from Ishizaka (1990c).

2. Once a circulation field is available, Lagrangian experiments can provide considerable insight into the processes that control biological distributions. This is a type of modeling that should be exploited by biological oceanographers, especially since several high-resolution, sophisticated circulation models are now available.

3. Physical-biological models provide a mechanism for synthesizing large amounts of data. The simulated distributions from Eulerian models can be analyzed using the same approaches and techniques that are used to analyze field distributions.

4. Assimilation of biological data (e.g., CZCS distributions) into ecosystem models requires that all portions of the modeled ecosystem be updated. Techniques to do this need development.

5. Quantitative comparisons with distributions derived from ocean color provide an approach for adjusting poorly known parameters/processes in marine ecosystem models. It also provides an approach for separating the contribution of physical and biological effects on phytoplankton distributions.

ACKNOWLEDGMENTS

Support for this research was provided by the Office of Naval Research and the National Aeronautics and Space Administration. I thank Lisa Hommel, Margaret Dekshenieks, Cathy Lascara, Ana Martins, John Moisan, and John Steele for their comments on the manuscript.

REFERENCES

Atkinson, L. P., D. W. Menzel, and K. A. Bush (eds.). 1985. *Oceanography of the Southeastern U. S. Continental Shelf*, Estuarine and Coastal Sciences 2, American Geophysical Union, 156 pp.

Blanton, J. O., 1972. Exchange of Gulf Stream water with North Carolina shelf water in Onslow Bay during stratified conditions. *Deep-Sea Res.* 18:167-178.

Bryan, K., and J. L. Sarmiento. 1985. Modeling ocean circulation. In: S. Manabe (ed.). Issues in Atmospheric and Oceanic Modeling, Part A, Climate Dynamics. *Advances in Geophysics.* Vol. 28, Academic Press, 433-459.

Bretherton, F. D., R. E. Davis, and C. D. Fandry. 1976. A technique for objective analysis and design oceanographic experiments applied to MODE-73. *Deep-Sea Res.* 23:559-582.

Bucklin, A., M. M. Rienecker, and C. N. K. Mooers. 1989. Genetic tracers of zooplankton transport in coastal filaments off northern California. *J. Geophys. Res.* 94:8277-8288.

Church, J. A., and A. M. G. Forbes. 1981. Non-linear model of tides in the Gulf of Carpentaria. *Aust. J. Mar. Freshwater Res.*, 32:685-698.

Ekman, V. W. 1906. On the influence of the earth's rotation on ocean currents. *Arkiv. f Matematik Astronomi. Fysik*, Bd 2, No, 11.

Haidvogel, D. B., and A. R. Robinson (eds.). 1989. *Data Assimilation, Dynamics of Atmospheres and Oceans.* Vol. 13, nos. 3-4, 518 pp.

Haidvogel, D. B., A. Beckmann, and K. S. Hedstrom. 1991a. Dynamical simulations of filament formation and evolution in the Coastal Transition Zone. *J. Geophys. Res.* 96:15,017-15,040.

Haidvogel, D. B., J. L. Wilkin, and R. Young. 1991b. A semi-spectral primitive equation circulation model using vertical sigma and orthogonal curvilinear horizontal coordinates. *J. Comp. Phys.* 94:151-185.

Hedstrom, K. S. 1990. User's manual for a semi-spectral primitive equation regional ocean-circulation model. Version 3.0, Institute for Naval Oceanography Technical Note FY90-2, 90 pp.

Hofmann, E. E. 1988. Plankton dynamics on the outer southeastern U. S. continental shelf. Part III: A coupled physical-biological model. *J. Mar. Res.* 46:919-946.

Hofmann, E. E., and J. W. Ambler. 1988. Plankton dynamics on the outer southeastern U. S. continental shelf. Part II: A time-dependent biological model. *J. Mar. Res.* 46:883-917.

Hofmann, E. E., K. S. Hedstr, J. R. Moisan, D. B. Haidvogel, and D. L. Mackas. 1991. The use of simulated drifter tracks to investigate general transport patterns and residence times in the Coastal Transition Zone. *J. Geophys. Res.* 96:15,041-15,052.

Holland, W. R. 1985. Simulation of mesoscale ocean variability in mid-latitude gyres. In: S. Manabe (ed.). *Issues in Atmospheric and Oceanic Modeling, Part A, Climate Dynamics.* Advances in Geophysics, Vol. 28, Academic Press, 479-523.

Ishizaka, J. 1990a. Coupling of Coastal Zone Color Scanner data to a physical-biological model of the southeastern U. S. continental shelf ecosystem 1. CZCS data description and Lagrangian particle tracing experiments. *J. Geophys. Res.* 95:20,167-20,181.

_____. 1990b. Coupling of Coastal Zone Color Scanner data to a physical-biological model of the southeastern U. S. continental shelf ecosystem 2. An Eulerian Model. *J. Geophys. Res.* 95: 20,183-20,199.

_____. 1990c. Coupling of Coastal Zone Color Scanner data to a physical-biological model of the southeastern U. S. continental shelf ecosystem 3. Nutrient and phytoplankton fluxes and CZCS data assimilation. *J. Geophys. Res.* 95:20,201-20,212.

Ishizaka, J., and E. E. Hofmann. 1988. Plankton dynamics on the outer southeastern U. S. continental shelf. Part I: Lagrangian particle tracing experiments. *J. Mar. Res.* 46:853-882.

Lee, T. N., L. P. Atkinson, and R. Legeckis. 1981. Observations of a Gulf Stream frontal eddy on the Georgia continental shelf, April 1977. *Deep-Sea Res.* 28:347-378.

Lewis, M. R., J. J. Cullen, and T. Platt. 1983. Phytoplankton and thermal structure in the upper ocean: Consequences of nonuniformity in chlorophyll profile. *J. Geophys. Res.* 88:2365-2570.

Mackas, D. L., L. Washburn, and S. L. Smith. 1991. Zooplankton community pattern associated with a California Current cold filament. *J. Geophys. Res.* 96:14,781-14,797.

Malone, T. C. 1975. Environmental control of phytoplankton cell size. *Limnol. Oceanogr.* 20:490.

McClain, C. R., J. Ishizaka, and E. E. Hofmann. 1990. Estimation of the processes controlling variability in phytoplankton pigment distributions on the southeastern U. S. continental shelf. *J. Geophys. Res.* 95:20,213-20,235.

Mills, E. L. 1989. *Biological Oceanography: An Early History.* 1870-1960. Cornell University Press, Ithaca and London, 378 pp.

Peterson, W. T., C. B. Miller, and A. Hutchinson. 1979. Zonation and maintenance of copepod populations in the Oregon upwelling zone. *Deep-Sea Res.* 26:467-494.

Riley, G. A. 1946. Factors controlling phytoplankton populations on Georges Bank. *J. Mar. Res.* 6:54-73.

Rothlisberg, P. C. 1982. Vertical migration and its effect on dispersal of penaeid shrimp larvae in the Gulf of Carpentaria, Australia. *U. S. Fish. Bull.* 80:541-554.

Rothlisberg, P. C., J. A. Church, and A. M. G. Forbes. 1983. Modelling the advection of vertically migrating shrimp larvae. *J. Mar. Res.* 41:511-538.

Semtner, A. J., Jr. 1974. An oceanic general circulation model with bottom topography. *Numerical Simulation of Weather and Climate,* Technical Report No. 9, Department of Meteorology, University of California, Los Angeles.

_____. 1986. Finite-difference formulation of a World Ocean model. In: J. J. O'Brien (ed.). *Advanced Physical Oceanography Numerical Modelling.* D. Riedel Publishing Company, Philadelphia, pp. 187-202.

Smith, W. O., G. W. Heburn, R. T. Barber, and J. J. O'Brien. 1983. Regulation of phytoplankton communities by physical processes in upwelling ecosystems. *J. Mar. Res.* 4,41:539-556.

Staples, D. J. 1979. Seasonal migration patterns of postlarval and juvenile banana prawns, *Penaeus merguiensis de Man,* in the major rivers of the Gulf of Carpentaria, Australia. *Aust. J. Mar. Freshwater Res.* 30: 143-157.

Thompson, J. D. 1974. The coastal upwelling cycle on a beta plane: Hydrodynamics and thermodynamics. Ph.D. Dissertation, Florida State University, 141 pp.

Walsh, J. J. 1972. Implications of a systems approach to oceanography. *Science* 176:969-975.

161

_____. 1975. A spatial simulation model of the Peru upwelling ecosystem. *Deep-Sea Res*. 22:201-236.

Walsh, J. J., D. A. Dieterle, and M. B. Meyers. 1988. A simulation analysis of the fate of phytoplankton within the Mid-Atlantic Bight. *Cont. Shelf Res*. 8:757-787.

Wroblewski, J. S. 1977. A model of phytoplankton plume formation during variable Oregon upwelling. *J. Mar. Res.*, 35:357-393.

_____. 1982. Interaction of currents and vertical migration in maintaining *Calanus marshallae* in the Oregon upwelling zone—a simulation. *Deep-Sea Res*. 29:665-686.

Wroblewski, J. S., and E. E. Hofmann. 1989. U. S. interdisciplinary modeling studies of coastal-offshore exchange processes: Past and future. *Prog. Oceanog*. 23:65-99.

11

An Invitation to Structured (Meta)Population Models

Odo Diekmann

Introduction

The contribution of an individual to population growth and interaction depends, as a rule, on various characteristics related to its physiology and its spatial position. Structured population models take this observation seriously and start the modeling process at the individual level, (i-level for short). First an i-state space Ω is specified and the movement of individuals through Ω, in dependence on the state of the environment (E-state), is described. Also described are the dependence on i-state and E-state of reproduction, death, and influence on the environment. These ingredients at the i-level completely determine the deterministic formulation at the p-level (p for population): simple bookkeeping principles tell us how the p-equations should look. (Admittedly, however, the appearance of the p-equations depends somewhat on our choice of p-state space: either $L_1(\Omega)$, if we expect that the population distribution over Ω has a nice density or $M(\Omega)$, if we expect that the distribution may contain measures concentrated on subsets of Ω).

Here I shall restrict myself to models that, though having stochastic components at the i-level (in particular related to death and reproduction), are deterministic at the p-level. Concerning the biology-mathematics interface (the "true" modeling), I shall be rather abstract, general, and vague, but references are provided to published material where one can find concrete and detailed elaborations. My aim is to introduce briefly some techniques that are helpful in deriving biologically meaningful conclusions from an analysis of the infinite dimensional p-dynamical systems. These techniques range from positive operator theory and numerical analysis to formal time scale arguments.

The first part of this chapter is concerned with the simplifications that arise, both in model formulation and in mathematical analysis, from the assumption that the environment is constant. A definition of the basic reproduction ratio R_0 is provided and computational aspects are discussed. Next the notion of "asymptotic speed of propagation" for homogeneous and isotropic infinite spatial domains is explained. References for the application of these ideas to the spread of infectious diseases and the invasion of new species are given.

The second part introduces energy budget models for Daphnia (and other organisms) as an example of structured models incorporating variable environments, where feedback through the environment leads to nonlinearity. A numerical method is briefly described.

Finally, the chapter considers metapopulation models, where local populations are considered as "individuals." It is shown that the formalism of structured population models is rich enough to incorporate many biological mechanisms, but that the resulting mathematical problem at the p-level is rather formidable. Using simplifications, based on quasi-steady-state assumptions, the neglect of delays or special choices of model ingredients, more tractable approximations are derived from which one can obtain qualitative conclusions.

INVASION IN A CONSTANT ENVIRONMENT

The Definition of R_0

Consider a population living under constant environmental conditions. The key question is: Will the population grow or decline? In order to answer the question, we can count either on a generational basis or at real time intervals as the basis for comparison (but of course the answer should not depend on our choice). We start by looking from one generation to the next.

Let $B(\tau, \xi, \eta)$ denote the expected number of offspring with state ξ at birth, produced per unit of time by an individual of age τ that was born with state η. We claim that B gives all the information we need to settle the growth-or-decline question. To substantiate this claim we introduce the so-called *next-generation operator K*:

$$(K\phi)(\xi) = \int_{\Omega} (\int_0^{\infty} B(\tau, \xi, \eta) d\tau) \phi(\eta) d\eta \tag{1}$$

The idea here is that ϕ describes both how many members a certain generation has (viz., $\int_{\Omega} \phi(\eta) d\eta$) and how these are distributed with respect to their state at birth. Then $K\phi$ does exactly the same for the next generation, and that explains how K arrived at its name. Our key question now translates into the following: What happens when we iterate K? It seems that, in principle, the answer could depend on the initial condition, the zero'th generation. In general (that is, under minor conditions on B), however, it does not. The reason is positivity. To serve as a meaningful model, both ϕ and B should be nonnegative. Hence K is a positive operator and, under the condition that the spectral radius R_0 of K, defined by

$$R_0 = \lim_{n \to \infty} \|K^n\|^{1/n}, \tag{2}$$

is a *strictly* dominant eigenvalue of K (i.e., the other spectral values of K lie strictly inside the disk of radius R_0 in the complex plane), one finds that

$$K^n \phi \sim c(\phi) R_0^n \phi_d, \quad n \to \infty, \tag{3}$$

where ϕ_d is the eigenvector corresponding to R_0 and all dependence on the zero'th generation ϕ is through a scalar quantity c. We conclude that the answer to the key question is completely determined by R_0: When $R_0 > 1$ the population will grow, when $R_0 < 1$ it will decline. In particular, invasion by a new species will be successful if and only if $R_0 > 1$.

Remarks. 1) Here we restrict ourselves to the deterministic setting at the population level. The stochastic formulation leads to multitype branching processes (Jagers and Nerman 1984, Mode 1971). Although $R_0 > 1$ is then still a necessary condition for exponential growth, it is no longer a sufficient condition, since even in the supercritical case there exists a non-zero probability that extinction occurs.

2) Let $u(t, \xi)$ be the rate at which, at time t, individuals are born with state ξ. Then

$$u(t, \xi) = \int_\Omega \int_0^\infty B(\tau, \xi, \eta) u(t - \tau, \eta) d\tau d\eta \tag{4}$$

If we substitute $u(t, \xi) = e^{\lambda t} \psi(\xi)$, we find that ψ has to be an eigenvector corresponding to the eigenvalue one for the operator K_λ defined by

$$(K_\lambda \phi)(\xi) = \int_\Omega (\int_0^\infty B(\tau, \xi, \eta) e^{-\lambda \tau} d\tau) \phi(\eta) d\eta . \tag{5}$$

Monotonicity arguments imply that

$$R_0 > 1 \Leftrightarrow r > 0 \tag{6}$$

where r is the unique *real* value of λ for which K_λ has dominant eigenvalue one. The population will grow, asymptotically for large time, at the rate r (in other words, r is the intrinsic rate of natural increase). Note that the time dependence in B has no influence on R_0, but that it does influence the value of r (in particular, the value of $R_0 - 1$ does not by itself give any information about the magnitude of r).

Computational Aspects

We discuss three special cases in which the determination of R_0 can be reduced to a manageable problem (Diekmann et al. 1990, Diekmann, 1991).

Suppose that the probability distribution for state at birth is independent of the state at birth of the mother or, more precisely, assume that

$$\int_0^\infty B(\tau, \xi, \eta) d\tau = a(\xi) b(\eta) . \tag{7}$$

Then K has one-dimensional range spanned by a (i.e., $K\phi$ is a multiple of a, no matter what ϕ we take) and consequently

$$R_0 = \int_\Omega b(\eta) a(\eta) d\eta. \tag{8}$$

So the independence allows us to obtain R_0 by averaging the life-time fertility b with respect to the probability distribution a of state at birth (assume, for the sake of the interpretation, that $\int_\Omega a(\eta) d\eta = 1$ is used to normalize a).

Next, consider the case

$$\int_0^\infty B(\tau, \xi, \eta) d\tau = a(\xi) b(\eta) + c(\eta) \delta(\eta - \xi) \tag{9}$$

where δ is Dirac's "function." As an example of a biological situation where such an assumption might be an appropriate idealization, think of i-state as geographical position and two modes of seed dispersal: either seeds fall to the ground more or less directly or they are blown into the air, after which they can land essentially everywhere, irrespective of where they came from. Some straightforward algebraic manipulations lead to the conclusion that R_0 is the largest real root of the characteristic equation $f(\lambda)=1$, where, by definition,

$$f(\lambda) = \int_\Omega \frac{b(\eta)a(\eta)}{\lambda-c(\eta)} d\eta. \tag{10}$$

Since f is a monotone function of λ, finding its one point is rather easy.

Finally, consider the case where the state has a discrete component, i.e., $\xi=(i,\xi_i)$, and

$$\int_0^\infty B(\tau,(i,\xi_i), (j,\xi_j))d\tau = a_i(\xi_i)b_{ij}(\xi_j) \tag{11}$$

The relation (11) is a conditional independence statement: conditional on the first component being i, the probability distribution for state-at-birth-component ξ_i is fixed and independent of the state-at-birth of the parent. In the application to sexually transmitted diseases (where begetting offspring corresponds to infecting a susceptible), one may think of i as distinguishing between men and women, while ξ_i is an indicator of sexual activity. If

$$m_{ij} = \int_{\Omega_i} b_{ij}(\xi_j)a_j(\xi_j)d\xi_j \, , \tag{12}$$

then R_0 is the dominant eigenvalue of the positive matrix with entries m_{ij}. Thus we have reduced an infinite dimensional problem to a finite dimensional one, for which standard software is available.

Submodels for B

In many situations B has the representation

$$B(\tau,\xi,\eta) = h(\tau)\int_\Omega c(\xi,\zeta)P(\tau,\zeta,\eta)d\zeta \, , \tag{13}$$

where h is the expected intrinsic fecundity at age τ, c describes the distribution of the birth state ξ, given the *current* state ζ of the mother, while P is the probability that the current state of an individual is ζ, given that it started its life τ units of time ago with state η. Deterministic examples are

$$P(\tau,\zeta,\eta) = \delta(\zeta-\eta) \tag{14}$$

for a static i-state and

$$P(\tau,\zeta,\eta) = \delta(\zeta-\eta-\tau) \tag{15}$$

for the case where i-state equals age and $\Omega=[0,\infty)$. When Ω is discrete, say $\{1,2,...,n\}$, and changes of i-state follow a Markov process with transition matrix G, we find

$$P(\tau,i,j) = (e^{G\tau})_{ij} \, ; \tag{16}$$

and when h is constant (and absorbed into c), we can easily perform the τ-integration to obtain

$$\int_0^\infty B(\tau,i,j)d\tau = \sum_{k=1}^n c(i,k)(-G^{-1})_{kj} \,. \tag{17}$$

The specification of the coupling c will depend heavily on the meaning of the i-state variable, and nothing can be said in generality. The function h may be deduced from data, from an energy budget consideration (see the section below on variable environments) or from some stochastic submodel. For instance, in the context of infectious diseases, one considers compartmental models and probabilities per unit of time to go from one compartment to another, while infectivity depends on the compartment one is in. Such a situation is described by a d-state (d from disease), which takes values $1,2,...,n$; a vector b with $\Sigma_{i=1}^n b_i = 1$, which gives the probability distribution for d state at the moment immediately following infection; a $n \times n$ matrix Σ of transition rates; and a vector q of infectivities. In terms of these ingredients, one has

$$h(\tau) = <q,e^{\Sigma\tau}b> \,. \tag{18}$$

In epidemic models, it is useful to factorize

$$B(\tau,\xi,\eta) = \Lambda(S)(\xi)A(\tau,\xi,\eta) \,, \tag{19}$$

where $\Lambda(S)(\xi)$ describes how B depends on the demographic steady state density of susceptibles S, as distributed with respect to i-state ξ. The strong form of the Law of Mass Action would yield

$$\Lambda(S)(\xi) = S(\xi) \tag{20}$$

but, for diseases transmitted during sexual intercourse or by biting mosquitoes, saturation effects lead to different functional forms for Λ (Heesterbeek and Metz, submitted).

Epidemic models clearly illustrate the untenability of our assumption that environmental conditions stay constant: the density of susceptibles will actually decrease as the disease makes victims (at least when the disease leads to immunity or death and the epidemic time scale is much faster than the demographic time scale). The assumption is an idealization appropriate for the initial phase of population expansion or disease spread, when density dependence can still be ignored (that's why we speak about invasion). Mathematically it amounts to linearization. Note, however, that the specification of a full nonlinear model may be a much more formidable task and that our idea to focus on B directly has a certain economy of thought and parameterization. The main point of this section, then, has been to explicitly emphasize that B may be composed from various building blocks, each of which may be related to a submodel for certain (biological) mechanisms.

When we have to iterate many times in (3) before the generations adopt the stable distribution as described by ϕ_d, we may be at variance with the restriction to the initial phase of population expansion, as is implicit in the linearization giving constancy of environmental conditions. And indeed, the next section is devoted to a situation in which, though the sign of $R_0 - 1$ still decides whether or not invasion will be successful, r does not give any information about the rate at which the invading organism will increase in numbers during the initial phase.

The Asymptotic Speed of Propagation

When we consider a fungal plant disease that spreads in a very large field by spore dispersal, we may take

$$B(\tau,\xi,\eta) = h(\tau)V(|\xi-\eta|) \tag{21}$$

for $\tau \geq 0$ and $\xi,\eta \in \mathbb{R}^2$. What happens when we introduce the disease at a localized spot? First there is a transient phase in which the details of the initial condition matter. Next there is an intermediate asymptotic phase in which the disease propagates in all directions in a wave-like manner with a constant velocity c_0. When these waves reach the boundary, a final phase begins in which the details of the geographic situation matter. In a truly infinite field, the third phase does not occur; and after the initial phase, similarity solutions, which are plane waves traveling with velocity c_0, take over. (Diekmann 1986, van den Bosch 1990, van den Bosch et al. 1990).

The ingredients h and V determine the velocity c_0 as follows. For the linear birth-rate equation (4), with B given by (21), the travelling plane wave "ansatz"

$$u(t,\xi) = e^{\lambda(t+c\xi_1)} \tag{22}$$

(where ξ_1 denotes the first component of ξ) leads to the equation

$$L(c,\lambda) = 1 , \tag{23}$$

where, by definition,

$$L(c,\lambda) = \int_0^\infty h(\tau)e^{-\lambda\tau}d\tau \int_{\mathbb{R}^2} V(|\eta|)e^{-\lambda c\eta_1} d\eta . \tag{24}$$

Provided $R_0 = L(0,0) > 1$ there exist solutions $\lambda > 0$ whenever c exceeds a critical number, which we call c_0. Hence c_0 is characterized by the simultaneous equations

$$L(c_0,\lambda_0) = 1, \quad \frac{\partial L}{\partial \lambda}(c_0,\lambda_0) = 0 \tag{25}$$

In his thesis, van den Bosch (1990) has worked through the following program:

- determine reasonable h and V from (quasi-) mechanistic submodels
- calculate c_0, either from (25) or by using approximation formulae
- compare with observed speeds for plant diseases and animal invasions

An account of the outcome of this program is given by van den Bosch et al. (1990).

Operationalization of R_0: Some Examples

In the last section it was indicated that one can use the general and somewhat abstract definition of the asymptotic speed c_0 as a tool in the quantitative analysis of concrete biological invasions. With R_0 one can do exactly the same thing. For a certain type of structure, one can develop algorithms which compute R_0 from a specification of the various ingredients. De Jong and Diekmann (1992) give an example in the context of veterinary epidemiology; and work in progress by de Jong, Heesterbeek, and Diekmann elaborates on this by allowing for a combination of the barn structure of a swine production farm and the stage structure of the animal population, while exploiting the conditional independence conditions discussed above to reduce the computer computations to finite dimensions. Ultimately this piece of bio-industrial mathematics should yield a valuable tool for the assessment of the efficacy of vaccination and other control measures.

When offspring is produced only in pairs that form a stable configuration for non-negligible periods of time, one cannot, in general, work with an age-representation to describe fecundity. This is particularly relevant for models for sexually transmitted diseases that take pair formation into account. But, as demonstrated in Diekmann et al. (1991), our general methodology easily leads to the matrix that represents the next generation operator and that has R_0 as its dominant eigenvalue. In work in progress, Dietz, Heesterbeek, and Tudor analyse how, for a disease of given virulence, R_0 depends on the parameters of the pair formation and dissociation process.

VARIABLE ENVIRONMENTS (AND NONLINEARITY BY FEEDBACK THROUGH THE ENVIRONMENT)

Energy Budget Models

Let the individuals of a population be distinguished from one another by their size x. Assume that all are born with birth size x_b and that they grow at a rate g, which depends on their size and on the prevailing food concentration S. Assume that individuals have a probability of dying per unit of time $\mu = \mu(x,S)$ and that they produce offspring at a rate $\lambda = \lambda(x,S)$ and consume food at a rate $\gamma = \gamma(x,S)$.

Let the density function n describe the size and composition of the population, i.e.,

$$\int_{x_1}^{x_2} n(t,\xi)d\xi$$

is the number of individuals, at time t, with size between x_1 and x_2. Then one can follow either Lagrange or Euler in the derivation of the bookkeeping equations (see Metz and Diekmann (eds.) 1986):

$$\begin{cases} \dfrac{\partial n}{\partial t} + \dfrac{\partial}{\partial x}(gn) = -\mu n \\[2em] g(x_b,S)n(t,x_b) = \displaystyle\int_{x_b}^{\infty} \lambda(\xi,S)n(t,\xi)d\xi \end{cases} \tag{26}$$

These equations tell us how processes at the individual level express themselves at the population level. We can think of S either as a given, time-dependent variable (for instance, when we deal with controlled experiments) or as a dynamic variable by itself (think of a field situation) that is governed by a differential equation such as

$$\frac{dS}{dt} = \alpha S(1 - \frac{S}{K}) - \int_{x_b}^{\infty} \gamma(\xi,S)n(\cdot,\xi)d\xi. \tag{27}$$

In the latter case, the combined problem (26) - (27) is nonlinear by feedback through the environment. Note, moreover, the *non-local* character (in particular, in the boundary condition describing the appearance of newborns).

When the vital rates μ and λ depend not only on size but also on age, equations (26) should be replaced by

$$
\left\{
\begin{aligned}
&\frac{\partial n}{\partial t} + \frac{\partial n}{\partial a} + \frac{\partial}{\partial x}(gn) = -\mu n \\[2mm]
&n(t,0,x) = \int_0^\infty \int_0^\infty \pi(y)(x)\lambda(\alpha,y,S)n(t,\alpha,y)\,d\alpha\,dy ,
\end{aligned}
\right.
\tag{28}
$$

where $\pi(y)(x)$ is the probability that offspring produced by a mother of size y will be born with size x.

An *energy budget model* consists of the specification of g,λ,π,μ and γ as functions of the i-state (either size x alone or size x and age a) and the environmental state (here the food concentration S) (see Kooijman 1986). Note that modeling is concerned with processes at the individual level and that straightforward mathematical bookkeeping suffices to "lift" the model to the population level.

The word "straightforward" in the last sentence above refers to the *formulation* of the model. To *analyze* the model and to arrive at biologically relevant conclusions is far less straightforward, unfortunately. Nevertheless, some conclusions have been obtained for the age-size structured prey-predator model sketched above.

Apart from the familiar "paradox-of-enrichment" prey-predator cycles, the model can exhibit "demographic" cycles, in which time periods in which there are only a few large predators alternate with periods in which there are many small predators, in such a way that the total prey consumption is nearly constant. This dynamic regime manifests itself for small values of prey carrying capacity K and for small values of the predator death rate μ. It does explain experimental observations on Daphnia: Murdoch and McCauley (1987) found that sometimes numbers of Daphnia oscillate wildly while the algae concentration remains nearly constant.

The dynamic behaviour described above was found by de Roos and co-workers (Metz et al. 1988, de Roos et al. 1990, 1992). A key tool in the analysis is a numerical technique, called the escalator box car train, to integrate first order partial differential equations like (26) and (28), involving non-local boundary conditions. The technique is based on an approximation of the density function n by cohorts (or, in more mathematical jargon, by concentrated measures), which reduces the partial differential equation to a finite system of ordinary differential equations for which numerical methods are readily available. Existing cohorts change in number due to death

$$
\frac{dn_i}{dt} = -\mu(\xi_i,S)n_i
\tag{29}
$$

and in mean size ξ due to growth

$$
\frac{d\xi_i}{dt} = g(\xi_i,S)
\tag{30}
$$

(and in mean age, when age is another i-state variable). Here i refers to the cohort number, a purely administrative index which we may adjust at regular time intervals, just as in the discretized age Lesley matrix model, to keep the range of i-values within bounds. The only non-trivial element of the technique is the derivation of ordinary differential equations for "cohorts in creation," i.e., for cohorts that are gradually formed from newborns. (There is just one such cohort in the case of (26), but there may be several of them needed along the boundary $a = 0$ in the case of (28)). A detailed exposition of this aspect is presented in de Roos et al. (1992). I conclude this part by emphasizing that, in my opinion, an important attractive feature of this technique is that the resulting system of ordinary differential equations has a clear interpretation as a biological model itself! In fact, in de Roos et al. (1992). the equations are derived directly, and not as a numerical approximation to a partial differential equation.

Finally, I refer to work by Kooijman (1986) on models that take energy reserves into account.

METAPOPULATION MODELS

Local Populations Considered as Individuals

When we want to describe an ensemble of local patches, we may conceive of a local population as an individual, characterized by its local population size. When these patches are arenas for interaction, there are several sizes to take into account, and the i-state space becomes higher-than-one-dimensional. Here we concentrate on the interaction of phytophages (spider mites) with their food source (plants) and their natural enemies (predator mites). So a local colony is characterized by:

1. the amount of food (leaf area) z
2. the number of phytophages/prey x
3. the number of predators y.

Our basic assumptions are that:

-   new prey colonies are founded by prey emigrating from existing colonies
-   prey colonies come to an end when the local food source is over-exploited
-   predator invasion in a prey colony will, after a while, lead to extermination of the prey followed by dispersion of the remaining predators.

So what we need are submodels for the local tritrophic interaction:

-   food production
-   prey increase/food decline
-   predator increase/prey decline

and for the non-local process of dispersion:

-   tendency to emigrate
-   dynamics of "aerial plankton" (starvation, search)
-   foundation of prey colonies
-   predator invasion in prey colonies.

Here one should note that geographical structure is not explicitly taken into account (no "distance"), but only implicitly through the fact that we acknowledge. the existence of *local* colonies. But what exactly qualifies to be called a local colony? In the real world we see the same phenomena at different spatial (and temporal) scales (leaves, twigs, branches, shrubs, etc.). Once we decide about the scale on which we want to concentrate, we can be more precise about the definition of a local colony. (The structure of the equations introduced below does not depend on our choice, but of course the numerical values of the parameters do). The key feature is that movement within a patch should be easy and frequent relative to movement from one patch to another.

In the next section we shall formulate mathematical models that are in the spirit of the above considerations, but still relatively simple. For instance, we shall assume that all patches are equal as far as food is concerned (but see Hanski and Gyllenberg, preprint), and we shall not model the food dynamics. Neither shall we take into account that parameters do depend on temperature and other meteorological variables.

Deriving the Population Balance Laws

Let x denote prey colony size and let the density function n be such that

$$\int_{x_1}^{x_2} n(t,\xi)d\xi$$

is the number of prey patches at time t of a size between x_1 and x_2. Let $v(x)$ be the rate of growth of x. Concerning the rate at which predators invade a prey colony of size x, we assume that it factorizes as a product of a vulnerability index $\eta(x)$ and the prevailing density of predator aerial plankton $Q(t)$. Then the time evolution of the population state n is described by the partial differential equation (pde)

$$\frac{\partial}{\partial t}n(t,x)+\frac{\partial}{\partial x}(v(x)n(t,x)) = -\eta(x)Q(t)n(t,x) \tag{31}$$

together with the boundary condition

$$v(1)n(t,1)=\zeta n_0(t)P(t) , \tag{32}$$

where n_0 is the number of suitable "free" patches, P is the density of prey aerial plankton, and ζ is a "reaction" constant. There is a strange but deliberate inconsistency here. Whereas prey colony growth is described by a differential equation for a continuous variable x, the founding of the colony is described by a discrete change $x=0 \to x=1$. The argument here is that, after a very short initial phase, population growth is deterministic to a good approximation, with further immigration negligible relative to reproduction.

Let y denote the size of the predator population in a certain patch and let the density function m be such that

$$\int_{x_1}^{x_2} \int_{y_1}^{y_2} m(t,x,y)dydx$$

is the number of predator patches at time t with prey population size between x_1 and x_2 and predator population size between y_1 and y_2. The local predator-prey interaction is assumed to be described by the system of ordinary differential equations (ode's)

$$\begin{aligned}\frac{dx}{dt}&=g(x,y)\\[1ex]\frac{dy}{dt}&=h(x,y)\end{aligned} \tag{33}$$

where consistency requires that $v(x) = g(x, 0)$. As a consequence, we find at the population level the balance law

$$\frac{\partial}{\partial t}m(t,x,y) + \frac{\partial}{\partial x}(g(x,y)m(t,x,y)) + \frac{\partial}{\partial y}(h(x,y)m(t,x,y)) = 0 \tag{34}$$

with the boundary condition

$$h(x, 1)m(t,x, 1) = \eta(x)Q(t)n(t,x) \tag{35}$$

describing the transmutation of prey patches into predator patches upon invasion.

We assume that the ode system (33) is such that orbits starting at $(x, 1)$ reach the boundary $x = 0$ after finite time or, in biological words, that predators do exterminate the local prey population. So the ode system determines somewhat implicitly both the interaction period and the predator yield at the end of that period. Note, finally, that preys may emigrate from predator-invaded patches before extermination is a fact.

In the present context, the "environment" consists of the aerial plankton P and Q. We assume that dispersing prey and predators die with probability per unit of time, respectively, μ and ν. The production of prey plankton is described by the emigration rate $\pi(x,y)$ and we obtain as the differential equation for P:

$$\frac{dP}{dt}(t) = \int\limits_{1}^{\infty} \pi(x, 0)n(t,x)dx + \int\limits_{1}^{\infty} \int\limits_{0}^{\infty} \pi(x,y)m(t,x,y)dydx - \mu P(t) \tag{36}$$

For predator plankton, on the other hand, production is related to the massive dispersal from patches in which the prey population is exterminated. So we have to multiply the flux through the boundary $x = 0$ by the yield y and sum over y. This leads to

$$\frac{dQ}{dt}(t) = - \int\limits_{0}^{\infty} yg(0,y)m(t, 0,y)dy - \nu Q(t) . \tag{37}$$

Along the same lines as followed above, one may introduce the food (leaf area) in a given patch as another steady-state variable z and work with densities $n_0(t,z)$, $n(t,x,z)$, and $m(t,x,y,z)$. In doing so, one increases not only the generality but also the complexity of the model considerably.

Our ideal now is to understand the global dynamical behaviour of the nonlinear infinite dimensional system described by (31), (32), and (34) through (37) and how this behaviour depends on the various ingredients (submodels and parameters). Unfortunately, this is an impossible task. The next section describes some techniques that allow us to achieve less ambitious goals. The main message of this section is simply that the *formalism* of structured population models is rich enough to incorporate a great number of biological mechanisms.

Some Limiting Cases

The processes of dispersal, prey colony growth, and local prey-predator interaction all have their characteristic time scale, and these need not be the same. If some of these scales are widely different, we may either use quasi-steady-state approximation or neglect some of the delays between cause and effect to obtain less complicated models. Moreover, even if these time scales are actually not very different, one may still adopt the sound mathematical strategy of studying limiting special cases first before tackling the full problem. Most of the time, insight obtained from special simplified cases is of much help in the analysis of the general case. Last but not least, any qualitative under-standing of close relatives of complicated models can be a key factor in the design and sensitivity analysis of computer experiments.

Instantaneous Prey Extermination After Predator Invasion. When the time between predator invasion and prey extermination is negligible compared to the average time of dispersal and prey colony growth, we may forget about (34) and (35), drop the term involving m in (37), and describe the dynamics of Q by

$$\frac{dQ}{dt}(t) = \delta Q(t) \int_1^\infty x\eta(x)n(t,x)dx - \nu Q \tag{38}$$

where δ is the prey-to-predator conversion factor. As a further simplification we may consider the special case where n_0 is constant. The point of this is that one obtains a

closed system of ode's for $O, P,$ and Q where, by definition,

$$O(t) = \int_1^\infty xn(t,x)dx . \tag{39}$$

A straightforward analysis (see Diekmann et al. 1988) shows that the ode system has an asymptotically stable steady state. Comparison with the neutral stability of the Volterra-Lotka system then leads to the conclusion that

a prey dispersal phase of non-negligible duration has a stabilizing effect on the global prey-predator interaction.

The prey dispersal phase acts as a temporary refuge. In contrast (Diekmann et al. 1988),

a predator dispersal phase acts as a destabilizing delay.

So if we treat the free patches as a dynamic variable, things become more complicated, since the herbivores consume (i.e., act as predator towards) plants but are prey for their predators. Sabelis et al. (1991) found that

a dispersal phase for the middle level in a tritrophic system has a destabilizing effect.

Constant Interaction Time, Predator Yield, Vulnerability, and Prey Dispersal Rate. The prey-predator interaction time (i.e., the time between predator invasion and prey exter-mination) as well as the predator yield at the end of the interaction depend on the prey colony size at the time of predator invasion. The precise form of this dependence is determined by the solutions of the ode system (33). The pde (31) adds to this no more and no less than the bookkeeping of the number of patches. So, if we make alternative assumptions concerning the interaction time and the predator yield we may forget about (33) and (31). In this section we shall assume that both are constant, i.e., independent of the prey colony size at the time of predator invasion.

Under this assumption, prey colony size is still relevant, since it determines the vulnerability η, the prey dispersal rate π, and, finally, the probability that the patch will crash. But if we assume that all these parameters are, in fact, independent of prey colony size, we may dispose of n and work with the total number of prey patches

$$N(t) = \int\limits_{1}^{\infty} n(t,x)dx \tag{40}$$

instead. One then arrives at a system of differential delay equations (which can still be further simplified by adopting quasi-steady-state approximations for P and Q). Standard steady-state stability analysis yields the following conclusions:

the founding of new prey colonies by prey dispersing from predator patches is a stabilizing mechanism (Sabelis and Diekmann 1988, Sabelis et al. 1991)

and

postponement of predator dispersion to the end of the interaction period is a destabilizing mechanism.

Instantaneous Host Plant Destruction, Possibly Defeated by Predator Invasion. Whenever the prey exhaust their host plant very quickly, we may employ a somewhat more sophisticated time scale argument. In the absence of predators, the founding of a prey colony leads instantaneously to the production of new searching prey. When predators are around they may invade, and then the instantaneous yield consists of predators rather than prey. The probability of predator invasion as well as the predator yield after invasion depend on predator aerial plankton density Q in a manner to be derived from the limiting procedure (see Diekmann et al. 1988, Appendix). The end result is a system of three ode's for n_0, P, and Q. Remarkably, the system exhibits bistability in certain regions of parameter space (see Diekmann et al. 1989), whence the conclusion:

for successful biological control one possibly needs to introduce many predators.

ACKNOWLEDGMENTS

The writing of this paper was funded in part by the U.S. Army Research Office and was initiated during a three-weeks visit to the Mathematical Sciences Institute, Cornell University. I would like to thank Si Levin and Rick Durrett for their hospitality and stimulating discussions.

REFERENCES

van den Bosch, F. 1990. The Velocity of Spatial Population Expansion. Ph.D. thesis, Leiden University, Leiden, The Netherlands.
van den Bosch, F., J.A.J. Metz, and O. Diekmann. 1990. The velocity of spatial population expansion, *J. Math. Biol.* 28:529-556.
Diekmann, O. 1991. Modelling Infectious Diseases in Structured Populations. In: B.D. Sleeman and R.J. Jarvis, (eds.). *Ordinary and Partial Differential Equations*, Vol. III. Pitman Research Notes in Mathematics 254: 67-79, Longman, Harlow.
Diekmann, O. 1986. Dynamics in bio-mathematical perspective. In: M. Hazewinkel, J.K. Lenstra, and L.G.L.T. Meertens, (eds.). *Mathematics and Computer Science II.* CWI Monograph 4:23-50, North-Holland, Amsterdam.

Diekmann, O., K. Dietz, and J.A.P. Heesterbeek. 1991. The basic reproduction ratio R_0 for sexually transmitted diseases, part I: Theoretical considerations. *Math. Biosc.* 107:325-339.

Diekmann, O., J.A.P. Heesterbeek, and J.A.J. Metz. 1990. On the definition and the computation of the basic reproduction ratio R_0 in models for infectious diseases in heterogeneous populations, *J. Math. Biol.* 28:365-382.

Diekmann, O., J.A.J. Metz, and M.W. Sabelis. 1988. Mathematical models of predator-prey-plant interaction in a patchy environment. *Experimental and Applied Acarology.* 5:319-342.

Diekmann, O., J.A.J. Metz, and M.W. Sabelis. 1989. Reflections and calculations on a prey-predator-patch problem. *Acta Applicandae Mathematicae.* 14:23-35.

Gyllenberg, M., and I. Hanski. In press. Single-species metapopulation dynamics: A structured model. *Theor. Pop. Biol.*

Hanski, I., and M. Gyllenberg. Preprint 1991. Two general metapopulation models and the core-satellite species hypothesis. Lulea University, Lulea, Sweden.

Heesterbeek, J.A.P., and J.A.J Metz. Submitted. The saturating contact rate in marriage- and epidemic models.

Jagers, P., and O. Nerman. 1984. The growth and composition of branching populations. *Adv. Appl. Prob.* 16:221-259.

de Jong, M.C.M., and O. Diekmann. 1992. A method to calculate - for computer-simulated infections - the threshold value, R_0, that predicts whether or not the infection will spread. *Prev. Vet. Med.* 12:269-285

Kooijman, S.A.L.M. 1986. Population dynamics on basis of budgets. In: J.A.J. Metz and O. Diekmann (eds.). *The Dynamics of Physiologically Structured Populations.* Lecture Notes in Biomathematics 68. Springer-Verlag, Berlin, pp. 453-473.

McCauley, E., and W.W. Murdoch. 1987. Cyclic and stable populations: Plankton as a paradigm. *Amer. Nat.* 129:97-121.

Metz, J.A.J., and O. Diekmann. (eds.). *The Dynamics of Physiologically Structured Populations.* Lecture Notes in Biomathematics 68. Springer-Verlag.

Metz, J.A.J., and O. Diekmann. 1991. Exact finite dimensional representations of models for physiologically structured populations. I. The abstract foundations of linear chain trickery. In: J.A. Goldstein, F. Kappel, and W. Schappacher (eds.). *Differential Equations with Applications in Biology, Physics and Engineering.* Lecture Notes in Pure and Applied Mathematics 133. Marcel Dekker, New York pp. 269-289.

Metz, J.A.J., A.M. de Roos, and F. van den Bosch. 1988. Population models incorporating physiological structure: A quick survey of the basic concepts and an application to size-structured population dynamics in waterfleas. In: B. Ebenman, and L. Persson (eds.). *Size-Structured Populations: Ecology and Evolution.* Springer, Berlin, pp. 106-124.

Mode, C.J. 1971. *Multitype Branching Processes: Theory and Applications.* Elsevier, New York, NY.

de Roos, A.M., J.A.J. Metz, E. Evers, and A. Leipoldt. 1990. A size-dependent predator-prey interaction: Who pursues whom? *J. Math.Biol.* 28:609-643.

de Roos, A.M., O. Diekmann, and J.A.J. Metz. 1992. Studying the dynamics of structured population models: A versatile technique and its application to Daphnia *Amer.Nat.* 139:123-147

Sabelis, M.W., and O. Diekmann. 1988. Overall population stability despite local extinction: The stabilizing influence of prey dispersal from predator invaded patches, *Theor.Pop.Biol.* 34:169-176.

Sabelis, M.W., O. Diekmann, and V.A.A. Jansen. 1991. Metapopulation persistence despite local extinction: Predator-prey patch models of the Lotka-Volterra type. *Biol. J. Linnean Soc.* 42:267-283.

12
STOCHASTIC MODELS OF GROWTH AND COMPETITION

Richard Durrett

INTRODUCTION

The purpose of this chapter is to give an introduction to interacting particle systems by describing the behavior of several examples. In each system there is a collection of spatial locations called sites, which in all our examples will be the d-dimensional integer lattice, $\mathbf{Z}^d$, that is, the points in d-dimensional space with all integer coordinates. At each time $t \in [0, \infty)$, each site can be in one of a finite set of states, F, so the state of the process at time t is a function $\xi_t : \mathbf{Z}^d \to F$. The time evolution is described by declaring that each site changes its state at a rate that depends upon the states of a finite number of neighboring sites. Here, we say that something happens at rate r if the probability of an occurrence between times t and $t + h$ is $\sim rh$ as $h \to 0$ is small; that is, when divided by h, the probability converges to r as $h \to 0$. Historically the first example of such a system studied was

THE STOCHASTIC ISING MODEL

In this model $\xi_t : \mathbf{Z}^d \to \{-1, 1\}$. We think of having an iron atom at each point $x \in \mathbf{Z}^d$, which can be in one of two states: spin up (1) or spin down (-1). To describe the time evolution, we let $\mathcal{N} = \{(1, 0), (0, 1), (-1, 0), (0, -1)\}$ be the four nearest neighbors of 0 and define the energy at x to be

$$H_x(\xi) = -\beta \sum_{y : y - x \in \mathcal{N}} \xi(x)\xi(y).$$

To explain the definition, note that if $\xi(x) = \xi(y)$ for all y in the sum, then $H_x(\xi) = -4\beta$, and this minimizes the energy at x. Let $c(x, \xi) = \exp(H_x(\xi))$. In words, $c(x, \xi)$ is the flip rate at x when the configuration is ξ, or, to be more precise,

$$P(\xi_{t+s}(x) = -\xi_t(x)|\xi_t) \sim c(x, \xi_t)s \quad \text{as } s \to 0.$$

There are good reasons for choosing this definition of the $c(x, \xi)$, which are explained in Liggett 1985, Chapter 3. We will content ourselves here to observe that flips occcur slowly when the energy is small.

The basic question concerning the Ising model and all other interacting particle systems is What happens as $t \to \infty$? An important first step in answering this question for the Ising model and many other systems is the observation that the flip rates defined above are *attractive*, a technical term that means that if $\xi(x) \leq \xi'(x)$ for all x, then

$$c(x, \xi) \geq c(x, \xi') \quad \text{when } \xi(x) = \xi'(x) = 1$$
$$c(x, \xi) \leq c(x, \xi') \quad \text{when } \xi(x) = \xi'(x) = -1$$

In words, increasing the number of 1's, makes 1's flip to -1's at a smaller rate and -1's flip to 1's at a larger rate. Two important consequences of this property are that: (a) given initial configurations with $\xi(x) \leq \xi'(x)$ for all x, we can construct the time evolutions in such a way that $\xi_t(x) \leq \xi'_t(x)$ for all x and all $t \geq 0$; and (b) if we start from $\xi_0^+(x) = 1$ for all x then $\xi_t^+(x) \Rightarrow \xi_\infty^+(x)$ as $t \to \infty$. Here $\Rightarrow$ is short for *converges in distribution*, which means that for any sequence $x_1, \ldots, x_n$ of points in $\mathbf{Z}^d$ and sequence of values $i_1, \ldots, i_n \in \{1, -1\}$,

$$P(\xi_t^+(x_1) = i_1, \ldots \xi_t^+(x_n) = i_n) \to P(\xi_\infty^+(x_1) = i_1, \ldots \xi_\infty^+(x_n) = i_n)$$

Symmetry implies that if we start from $\xi_0^-(x) = -1$ for all x then $\xi_t^- \Rightarrow \xi_\infty^-$ and ξ_∞^- has the same distribution as $-\xi_\infty^+$. General results (see Liggett 1985, Chapter 1) imply that ξ_∞^+ and ξ_∞^- are stationary distributions for the stochastic Ising model. That is, if we start with a random initial state that has this distribution, then the process has this distribution at any time $t > 0$. The basic fact about the stochastic Ising model is that there is a critical value $\beta_c = \{\log(1+\sqrt{2})\}/2$, so that if $\beta \leq \beta_c$, $P(\xi_\infty^+(x) = 1) = 1/2$ and there is a unique stationary distribution; while if $\beta > \beta_c$, $P(\xi_\infty^+(x) = 1) > 1/2$, so ξ_∞^+ and ξ_∞^- are two different stationary distributions.

Physically, the parameter β is proportional to $1/T$, where T is the temperature and the nonuniqueness of the stationary distribution for low temperature corresponds to the experimental fact that it is possibile to magnetize an iron bar when the temperature is low, but not possible if the temperature is too high. There is a moral in this example that is important for the other systems we will introduce below. The dynamics are a drastic oversimplification of what goes on in a ferromagnet: spins can point in any direction and interact with more than just the nearest neighbors. However, the fact that we are able to reproduce the observed phenomena indicates that we have identified the important qualitative features of the interaction (in this case, the competition between the thermal fluctuations and the tendency of the spins to align).

1. The Basic Contact Process

In this model $\xi_t : \mathbf{Z}^d \to \{0, 1\}$, we think of 0 as vacant and 1 as occupied by a "particle," and the system evolves as follows:

(i) Particles die at rate one, give birth at rate β.

(ii) A particle born at x is sent to a y chosen at random from the $2d$ nearest neighbors $\{y : \|x - y\|_1 = 1\}$.

(iii) If y is occupied then the birth is suppressed.

Rule (iii) says that there can be at most one particle per site. This is a reasonable constraint if you are thinking of the spread of a plant species, but this realism makes the model very difficult to analyze. Let ξ_t^A be the state at time t when initially $\xi_0^A(x) = 1$ if and only if $x \in A$, and let $\tau^A = \inf\{t : \xi_t^A \equiv 0\}$. If there are no particles then none can be born, so $\xi_t^A \equiv 0$ for all $t \geq \tau^A$. In words, the "all 0" state is an *absorbing state* and we say the system *dies out* at time τ^A.

The first question to be addressed is When does the system have positive probability of not dying out starting from a single occupied site? or When is $P(\tau^{\{0\}} = \infty) > 0$? It suffices to use a single occupied site as an initial configuration since $P(\tau^{\{0\}} = \infty) = 0$ implies $P(\tau^A = \infty) = 0$ for all finite A. Now, increasing β improves the chances for survival, so it should be clear that there is a critical value

$$\beta_c = \inf\{\beta : P(\xi_t^0 \not\equiv 0 \text{ for all } t) > 0\}.$$

If we delete rule (iii) from the definition, the resulting system is called a branching random walk and has $\beta_c = 1$. That is, in order for a branching random walk to survive, it is sufficient to have a birth rate larger than the death rate. Since in the contact process some of the birth rate will be wasted on occupied sites, this proves the easy half of the following result.

Theorem 1A. $1 < \beta_c(\mathbf{Z}^d) \leq 4$

The lower bound is due to Harris 1974, the upper bound to Holley and Liggett 1978. Both bounds are reasonably accurate. Numerical results (see Brower et al 1978) suggest that $\beta_c(\mathbf{Z}) \approx 3.299$ and $\beta_c(\mathbf{Z}^2) \approx 1.645$, and it has been shown (see Holley and Liggett 1981 or Griffeath 1983) that $\beta_c(\mathbf{Z}^d) \to 1$ as $d \to \infty$.

Once it was established that $\beta_c \in (0, \infty)$, attention turned to What does the process look like when it does not die out? To answer this question, we modify slightly our approach to the Ising model: if $\xi(x) \leq \xi'(x)$ the birth rates are higher for ξ' and the death rates are smaller for ξ' than in ξ, i.e., the system is attractive. Hence (a) given initial states with $\xi_0(x) \leq \xi_0'(x)$ for all x, we can construct the time evolutions in such a way that $\xi_t(x) \leq \xi_t'(x)$ for all x and all $t \geq 0$, and (b) if we start from $\xi_0^1(x) = 1$ for all x, then as $t \to \infty$ $\xi_t^1(x) \Rightarrow \xi_\infty^1(x)$, and ξ_∞^1 is a stationary distribution for the process.

At the other extreme, the point mass on the "all 0" state, δ_0, is a trivial stationary distribution. There is no guarantee that $\xi_\infty^1 \neq \delta_0$ and indeed this will happen if β is too small. However the "duality relation" for the contact process implies

$$P(\xi_\infty^1(y) = 0) = P(\tau_t^{\{y\}} < \infty),$$

so $\xi_\infty^1 = \delta_0$ if the contact process dies out, but is a nontrivial stationary distribution if the contact process survives. The next result, called the *complete convergence theorem*, implies that ξ_∞^1 is the only nontrivial stationary distribution.

Theorem 1B. $\xi_t^A \Rightarrow P(\tau^A < \infty)\,\delta_0 + P(\tau^A = \infty)\,\xi_\infty^1$

In words, when the process dies out it looks dead, but when it survives and t is large it looks like the system starting from all sites occupied.

The last result took fifteen years to evolve to its current form. Harris 1974, Griffeath 1978, Durrett 1980, Durrett and Griffeath 1982, and Durrett and Schonmann 1987 proved increasingly more general results before Bezuidenhout and Grimmett 1990 finished the problem and in addition proved

Theorem 1C. *When $\beta = \beta_c$, $P(\tau^{\{0\}} = \infty) = 0$.*

In words, the contact process dies out at the critical value. For applications (including some we will make in this paper), it is worthwhile to note that all the results in this section hold if (ii) is replaced by

(ii) A particle born at x is sent to a y chosen at random from $x + \mathcal{N}$.

and if we assume $\mathcal{N}$ is (a) *symmetric* with respect to reflection in any coordinate plane, and (b) *irreducible*, i.e., the group generated by $\mathcal{N}$ is $\mathbf{Z}^d$.

2. MULTITYPE CONTACT PROCESSES

It is well known, even to mathematicians, that there is more than one type of plant, so it is natural to generalize the contact process to have two (or more) types of particles. In this model, the state at time t $\xi_t : \mathbf{Z}^d \rightarrow \{0, 1, 2\}$ and we think of 0 as vacant and 1 and 2 as occupied by pine and maple trees, respectively. With this in mind, we formulate the evolution as follows:

(i) Particles of type i die at rate one, give birth at rate β_i.

(ii) A particle born at x is sent to a y chosen at random from $x + \mathcal{N}$ where $\mathcal{N}$ is symmetric and irreducible.

(iii) If y is occupied, then the birth is suppressed.

When only one type of particle is present the system reduces to the basic contact process, so if $\beta_1, \beta_2 > \beta_c(\mathbf{Z}^d)$, then there are three trivial equilibria: δ_0, μ_1 and μ_2, where μ_i is the limit starting from $\xi_t(x) \equiv i$. The main question to be answered about the new system is: Is there a nontrivial stationary distribution?, i.e., one that concentrates on configurations that contain both 1's and 2's. The first result is a negative one.

Theorem 2A. *If $\beta_1 > \beta_2$, then there are no nontrivial translation invariant stationary distributions.*

Here translation invariant means that that the distribution is invariant under spatial shifts. This result and the others in this section are from Claudia Neuhauser's 1990 thesis. We conjecture that Theorem 2A holds without the assumption of translation invariance, but that assumption is often difficult to remove. Note that Harris proved Theorem 1B for translation invariant initial distributions in 1974, but the general case was settled 15 years later.

Restricting our attention now to the special case $\beta_1 = \beta_2 > \beta_c(\mathbf{Z}^d)$, we have

Theorem 2B. *In dimensions $d \leq 2$, for any initial configuration, we have $P(\xi_t(x) = 1, \xi_t(y) = 2) \rightarrow 0$ for all $x, y \in \mathbf{Z}^d$, so all stationary distributions are trivial.*

Theorem 2C. *In dimensions $d \geq 3$, there is a one–parameter family of stationary distributions ν_θ, $\theta \in [0, 1]$, and all translation–invariant stationary distributions are convex combinations of the ν_θ.*

As in the voter model, (see Liggett 1985, Chapter 5, or Durrett 1988, Chapter 2), the dichotomy between the behavior in $d \leq 2$ and $d \geq 3$ comes from the fact that random walks are recurrent in the first case and transient in the second. The stationary distributions are constructed by starting the system from an initial product measure in which 1's have density θ and 2's have density $1 - \theta$, i.e., $\xi_0(x)$ are independent and take values 1 and 2 with probabilities θ and $1 - \theta$. The reader should note that while the basic contact process has a single nontrivial stationary distribution, the two–color version has a one–parameter family in $d \geq 3$.

3. SUCCESSIONAL DYNAMICS

In this model we again have $\xi_t : \mathbf{Z}^d \to \{0, 1, 2\}$ but this time we think of 0 as vacant and 1 and 2 as occupied by a bush or tree, respectively. With this interpretation in mind, the dynamics are formulated as follows:

(i) Particles of type i die at rate one, give birth at rate β_i.

(ii) A particle born at x is sent to a y chosen at random from $\{y : \|x - y\|_1 \leq M\}$, where M is an integer.

(iii) If $\xi_t(y) \geq \xi_t(x)$, then the birth is suppressed.

In words, trees can give birth onto sites occupied by bushes, but not conversely. In biological terms, the two species are part of a successional sequence. When only one type of particle is present, the system reduces, as in the last example, to a contact process, so if $\beta_1, \beta_2 > \beta_c$, then there are three trivial equilibria: δ_0, μ_1, and μ_2, where μ_i is the limit starting from $\xi_t(x) \equiv i$.

Again, the main question to be answered is: Are there nontrivial stationary distributions? or, more briefly: Is coexistence possible? Our first answer is

Theorem 3A. *If $d = 1$ and $M = 1$, then for any initial configuration we have $P(\xi_t(x) = 1, \xi_t(y) = 2) \to 0$ as $t \to \infty$ for all $x, y \in \mathbf{Z}$, so there is no coexistence.*

This result can be proved by drawing a picture of a "typical" realization of the process starting with a single 2

$$0\,0\,1\,0\,1\,0\,2\,0\,2\,2\,0\,2\,0\,0\,0\,2\,1\,0\,1\,0\,0\,1$$

and checking that, since $M = 1$, there can never be a 1 between the leftmost and rightmost 2's. If the 2's do not die out, then the ends of the interval of 2's go to $-\infty$ and ∞, respectively (see Durrett 1980) and the 1's get crowded out. In general, either (a) all the 2's die out, or (b) some 2 starts an interval that grows forever. In either case, $P(\xi_t(x) = 1, \xi_t(y) = 2) \to 0$ as $t \to \infty$.

We believe that coexistence is possible in all other cases.

Conjecture 3A. *If $d > 1$ or $M > 1$ then coexistence is possible when $\beta_2 = \beta_c + \epsilon$ and β_1 is large.*

The main trouble with proving this conjecture is that coexistence can occur only near the critical value. It is not hard to show that if $\beta_2 > \beta(d, M)$, then there is no coexistence for any $\beta_1 \leq \infty$. Somewhat surprisingly, this problem, which is difficult to solve when $d = 1$ and $M = 2$, or when $d = 2$ and $M = 1$, turns out to be more tractable when M is large. In addition to proving Theorem 3A, Durrett and Swindle 1990 have shown

Theorem 3B. *If $\beta_1 > \beta_2^2 > 1$, then coexistence occurs for large M.*

To explain the last conclusion, we need to introduce the long–range contact process, a modification of the basic contact process in which (ii) is changed to:

(ii) A particle born at x is sent to a y chosen at random from $\{y : \|x - y\|_1 \leq M\}$.

If we write $\beta_c(M)$ to indicate the dependence of the critical value on M and use ξ_∞^1 to denote the limit starting from all 1's, then we have

Theorem 3C. *As* $M \to \infty$, $\beta_c(M) \to 1$. *Furthermore, if* $\beta > 1$, *then* ξ_∞^1 *converges weakly to a product measure with density* $(\beta - 1)/\beta$.

This result (for the neighborhood $\{y : \|x - y\|_\infty \leq M\}$) was proved by Bramson et al 1989 who identified the rate at which $\beta_c(M)$ approached 1. A simpler and more general proof, which does not give the right rate, can be found in Durrett 1989.

To explain the condition in Theorem 3B, observe that $\eta_t = \{x : \xi_t(x) = 2\}$ is a long–range contact process, so if M is large and we are in equilibrium, η_t is approximately a product measure with density $(\beta_2 - 1)/\beta_2$. If the 2's were exactly that product measure, a 1 would die at rate $1 + \frac{\beta_2 - 1}{\beta_2}\beta_2$ (the second term representing births onto the site by 2's) and give birth at rate β_1/β_2 (the site must not be occupied by a 2 for a successful birth to occur). So for coexistence to occur, we need $1 + \frac{\beta_2 - 1}{\beta_2}\beta_2 < \beta_1/\beta_2$ or $\beta_1 > \beta_2^2$. The careful reader will have noted that we have just argued the condition is necessary, while Theorem 3B proves it is sufficient. Having faith in the heurisitc argument, we make

Conjecture 3B. *If* $\beta_1 < \beta_2^2$ *then there is no coexistence for large M.*

Remark: The heuristic argument generalizes easily to show that if the two particles die at different rates, then we need

$$\delta_1 + \frac{\beta_2 - \delta_2}{\beta_2}\beta_2 > \beta_1 \frac{\delta_2}{\beta_2}$$

and the proof of Theorem 3B generalizes to show that this condition is sufficient. It is natural to generalize the multitype contact process in this way, but we do not know how to prove any results in that generality. The naive guess is that $\beta_1/\delta_1 > \beta_2/\delta_2$ is the right hypothesis for Theorem 2A. We believe this is correct but have no idea how to prove it.

Having discussed the existence of nontrivial stationary distributions, we turn to the question of uniqueness. Durrett and Møller 1991 have proved a "complete convergence theorem." To state their result, let δ_0, μ_1, and μ_2 be the trivial stationary distributions mentioned at the beginning of this section. Let μ_{12} be the nontrivial stationary distribution constructed in Theorem 3B. Let $\eta_t = \{x : \xi_t(x) = 1\}$, $\zeta_t = \{x : \xi_t(x) = 2\}$, $\tau_1 = \inf\{t : \eta_t = \emptyset\}$, and $\tau_2 = \inf\{t : \zeta_t = \emptyset\}$.

Theorem 3C. *If* $\beta_1 > \beta_2^2 > 1$ *and M is large, then*

$$\xi_t \Rightarrow P(\tau_1 < \infty, \tau_2 < \infty)\,\delta_0 + P(\tau_1 = \infty, \tau_2 < \infty)\,\mu_1$$
$$+ P(\tau_1 < \infty, \tau_2 = \infty)\,\mu_2 + P(\tau_1 = \infty, \tau_2 = \infty)\,\mu_{12}$$

In words, if the 1's and/or 2's die out, we end up with a trivial stationary distribution in which one or zero types of particles are present. If both the 1's and 2's survive and t

is large, the system looks like μ_{12}, so that is the only nontrivial stationary distribution. The value of M required for Theorem 3C is larger than that for Theorem 3B, which is enormous. With more work this difference might be eliminated, but the interesting problem is to show

Conjecture 3C. The complete convergence theorem holds whenever coexistence occurs.

4. AN EPIDEMIC MODEL

Our fourth system is a process $\xi_t : \mathbf{Z}^2 \to \{0, 1, 2\}$ that has been used to model the spread of epidemics and forest fires. In the epidemic interpretation, $0 = $ healthy, $1 = $ infected, $2 = $ removed $ = $ immune or dead. In the forest fire interpretation, $0 = $ alive, $1 = $ on fire, and $2 = $ burnt. With these interpretations in mind, we formulate the dynamics as follows:

(i) A burning tree sends out sparks at rate β.

(ii) A spark emitted from x flies to one of the four nearest neighbors $\{y : \|y - x\|_1 = 1\}$ chosen at random. If the spark hits a live tree, the tree catches fire and begins immediately to emit sparks.

(iii) A tree remains on fire for an exponential amount of time with mean 1 then becomes burnt.

(iv) Burnt trees come back to life at rate α.

At first glance, the spontaneous re-appearance of trees may not seem reasonable. In the epidemic interpretation this is quite natural, however. Consider a disease like measles that confers lifetime immunity upon recovery. New susceptibles are born and immune individuals die. We combine the two transitions into the one in (iv) to keep a constant population size.

When $\alpha = \infty$, sites change instantaneously from 2 to 0 and the result is the contact process. At the other extreme, $\alpha = 0$, is the so-called "spatial epidemic with removal" in which regrowth is impossible. We begin by considering the behavior of our processes starting with a single burning tree at the origin in the midst of an otherwise virgin forest, i.e., $\xi_0^0(0) = 1$, $\xi_0^0(x) = 0$ for $x \neq 0$. Let $\eta_t^0 = \{x : \xi_t^0(x) = 1\}$, let $\zeta_t^0 = \{x : \xi_t^0(x) = 2\}$, and define a critical value by

$$\beta_c(\alpha) = \inf\{\beta : P(\zeta_t^0 \neq \emptyset \text{ for all } t) > 0\}.$$

Cox and Durrett 1988 considered the case $\alpha = 0$ and showed

Theorem 4A. *If* $\beta > \beta_c(0)$, *then there is a nonrandom convex set* D, *so that on* $\{\eta_t \neq \emptyset \text{ for all } t\}$ *we have* $\zeta_t^0 \approx \zeta_\infty^0 \cap tG$, *and* $\eta_t^0 \approx t\partial G$. *To be precise, for any* $\epsilon > 0$ *the following inequalities hold for large* t

$$\zeta_\infty^0 \cap (1 - \epsilon)tG \subset \zeta_t^0 \subset (1 + \epsilon)tG$$
$$\eta_t^0 \subset (1 + \epsilon)tG - (1 - \epsilon)tG$$

In words, this result says that the fire expands linearly and has an asymptotic shape. The statement is made contorted by the fact that the set of trees that will ever burn,

ζ^0_∞, is not all of $\mathbf{Z}^d$. Thus what we prove is that when t is large, ζ^0_t is contained in $(1 + \epsilon)tG$ and (if nonempty) contains all the points of ζ^0_∞ in $(1 - \epsilon)tG$.

When $\alpha = 0$, the system cannot have a nontrivial stationary distribution but Durrett and Neuhauser 1991 have shown

Theorem 4B. *If $\beta > \beta_c(0)$ and $\alpha > 0$, then there is a nontrivial stationary distribution, i.e., one that assigns no mass to "all healthy" state.*

The last result illustrates some of the frustrations in "applied probability." The proof is intricate and required several months to put down on paper, but we have been repeatedly told by physicists and biologists that the conclusion is obvious. In view of our difficulties in proving existence, the reader should not be surprised to learn that we have little to say about uniqueness.

Conjecture 4C. *If $\beta > \beta_c(\alpha)$, then there is a unique nontrivial stationary distribution.*

In the first three examples we have had varying degrees of success in identifying the set of stationary distributions. In each of those cases however there is a useful "duality equation" and we have not been able to find one here.

REFERENCES

Bezuidenhout, C. and G. Grimmett 1990 : The critical contact process dies out. *Ann. Probab.* 18, 1462–1482.

Bramson, M., R. Durrett, and G. Swindle 1989 : Statistical mechanics of crabgrass. *Ann. Probab.* 17, 444–481.

Brower, R.C., M.A. Furman, and M. Moshe 1978 : Critical exponents for the Reggeon quantum spin model. *Physics Letters* 76B: 213–219.

Cox, J.T. and R. Durrett 1988 : Limit theorems for the spread of epidemics and forest fires. *Stoch. Processes Appl.* 30: 171–191.

Durrett, R. 1980: On the growth of one-dimensional contact processes. *Ann. Probab.* 8 : 890–907.

Durrett, R. 1988: *Lecture Notes on Particle Systems and Percolation.* Wadsworth Pub. Co., Pacific Grove, CA.

Durrett, R. 1991: A new method for proving the existence of phase transitions. *Spatial Stochastic Processes* edited by K. Alexander and J. Watkins, Birkhauser, Boston

Durrett, R. and D. Griffeath 1982: Contact processes in several dimensions. *Z. Warsch. Verw. Gebiete* 59 : 535–552.

Durrett, R. and A.M. Møller 1991: Complete convergence theorem for a competition model. *Prob. Th. Rel. Fields* 88 : 121–136.

Durrett, R. and C. Neuhauser 1991: Epidemics with recovery in d=2. *Adv. in Applied Probab.* 1: 189–206

Durrett, R. and R.H. Schonmann 1987: Stochastic growth models. *Percolation Theory and the Ergodic Theory of Interacting Particle Systems.* Springer, New York.

Durrett, R. and G. Swindle 1991: Are there bushes in a forest? *Stoch. Processes Appl.* 37, 19–31.

Griffeath, D. 1978: Limit theorems for non-ergodic set-valued Markov processes. *Ann. Probab.* 6: 379–387.

Griffeath, D. 1983: The binary contact path process. *Ann. Probab.* 11 : 692–705.

Harris, T.E. 1974: Contact interactions on a lattice. *Ann. Probab.* 2: 969–988.

Holley, R. and T.M. Liggett 1978: The survival of contact processes. *Ann. Probab.* 6 : 198–206.

Holley, R. and T.M. Liggett 1981: Generalized potlatch and smoothing processes. *Z. Warsch. verw. Gebiete* 55: 165–195.

Liggett, T.M. 1985: *Interacting Particle Systems.* Springer, New York.

Neuhauser, C. 1990: Ergodic theorems for the multitype contact process. Ph.D. Thesis, Cornell University, Ithaca, New York.

13
MECHANISMS OF PATCH FORMATION

Douglas H. Deutschman, Gay A. Bradshaw, W. Michael Childress, Kendra L. Daly, Daniel Grünbaum, Mercedes Pascual, Nathan H. Schumaker, and Jianguo Wu

INTRODUCTION

Many mechanisms both physical (e.g., light, temperature, ocean currents, density gradients, topography) and biological (e.g., allelopathy, competition, predation, selective foraging) are considered responsible for patch formation. Wiens (1976) presented an excellent review of population responses to environmental patchiness. He identified localized random disturbances (e.g., fire, erosion, tree windfalls), predation, selective herbivory, and vegetational patterns as potential causes of patch formation. Roughgarden (1977) discussed five general mechanisms that are responsible for patchiness: resource distribution, dispersal, aggregation behavior, competition, and reaction-diffusion.

Patch-forming mechanisms operate at different spatial and temporal scales. Different mechanisms may predominate on one or more characteristic scales. For instance, fires are considered a dominant mechanism producing conspicuous patchiness at relatively large scales in many terrestrial systems (e.g., deciduous forests, grasslands). In contrast, vegetative propagation creates patchiness at relatively small scales. In marine plankton, reproductive population growth may dominate at large scales while behavioral adaptations of individuals tend to dominate at small scales. Further, the scale at which this transition occurs will vary with the size, longevity, and mobility of the organism (Mackas et al. 1985). Variability of environmental conditions (e.g., temperature) and the diversity of an organism's behavioral and physiological responses contribute to patchiness at many different scales in both terrestrial and marine systems (see Pickett and White 1985, Kolasa and Pickett 1991).

To understand the consequences of patchiness, we need to understand:

1)   On what spatial and temporal scales do organisms respond to patches in their environment?

2)   How do processes scale up and scale down?

3)   How do simple patch-forming mechanisms interact in space and time to give rise to more complex patterns?

In this chapter, we present brief discussions of our attempts to integrate the scale of patch-forming processes. The first section focuses on the importance of identifying the characteristic scales of the organism and its environment in order to understand how organisms perceive and potentially react to patchiness. We address the question: What can be determined about small-scale (= high resolution) spatial patterns from large-scale (= low resolution) data such as satellite imagery and aerial photographs? Finally, we present an integrated empirical and theoretical investigation of the multiple scales over which patchiness is observed in the distribution of the Antarctic krill, *Euphausia superba*. This first section demonstrates the need to consider the implicit scales of the organism, the environment, and the data.

The second section is devoted to demonstrating how interactions of relatively simple processes can generate very complex patterns of patchiness. This is approached through the use of two different spatial simulation models. The first model examines the results of random movement interacting with predator-prey cycles on a one-dimensional gradient. In the second model, the environment consists of a two-dimensional array of cells of uniform quality. In both cases, the interaction of movement with the predator-prey dynamics leads to complex spatial patterns. While these studies reflect only a narrow range of possible examples, they demonstrate how both the scaling and the interactions of patch-forming mechanisms are critical to an understanding of patchiness.

SCALE

Relative Scales

Despite its prevalence in recent ecological literature, the definition of a patch has remained controversial. Early definitions distinguished a patch from spatial variability by its relatively discrete boundary and internal homogeneity (Levin and Paine 1974, Wiens 1976). Kotliar and Wiens (1990) stressed that a patch is simply a surface area differing from its surroundings and that patches form a hierarchical mosaic over a broad range of scales. In the marine context, this range of scales is portrayed as a continuum with methods like power spectra (Platt and Denman 1975). This dichotomy corresponds to the difference between terrestrial and aquatic patterns. Thus, it would not be fruitful to attempt a general definition of patchiness because patches cannot be defined in the abstract. They must be defined relative to the investigation in order to render the term meaningful (for more discussion, see the chapter by García-Moliner et al.).

Investigations into the role of patchiness must consider the spatial and temporal scales that are relevant to the organism under study (Steele 1978, Wiens and Milne 1989, Downes 1990). It is clear that a sessile organism will respond to phenomena at spatial scales quite distinct from those to which a motile organism responds. Similarly, the adult stage of an organism may experience vastly different environmental patches than a larva or juvenile of the same species. Insects that undergo metamorphosis and benthic organisms with pelagic larvae are extreme examples of this phenomenon. Other features of an organism's life history such as size, dispersal, dormancy, and foraging strategies will affect its perception of scales.

Addicott et al. (1987) formalized the notion of relevant scales with the definition of "ecological neighborhoods." The ecological neighborhood is defined in three sequential steps: 1) choosing the organism; 2) identifying the ecological process; and 3) determining the appropriate temporal and spatial scales. This approach has a graphical analog similar to the

Stommel diagram (Haury et al. 1978, after Stommel 1963; for more discussion see the chapter by Marquet et al.), which involves the generation of a log-log plot of the time and space components of the organism's ecological neighborhood (Figure 1a).

The graphical approach is useful in exploring interactions between a specific organism and its environment. Environmental events (e.g., seasonal and diel changes, disturbance) can be plotted onto the same axes (Figure 1b). Spatial scale indicates the range of the event, while the temporal scale is based on the frequency of the event. Predictable events (i.e., diel, lunar, seasonal, and annual) are indicated by discrete lines. Less predictable events are represented by confidence ellipses indicating the amount and direction of the spatial and temporal variability.

The position of these events relative to the organism will influence how the organism responds to each event (see Hutchinson 1961). Events occurring over spatial and temporal scales similar to those of the organism itself are likely to have the most pronounced effect. Processes that are larger in spatial or temporal extent than the organism's ecological neighborhood would not be experienced as patches or events by the organism; the effect would more likely be at the population level. Events occurring over smaller scales or shorter time period (below the "grain," *sensu* Kotliar and Wiens 1990) are often averaged out by the behavior and physiology of the organism.

Resolving Small-scale Spatial Patterns and Processes from Large-scale Patterns: Grid-based Computer Simulations

Although comprehensive in extent, aerial and satellite imagery have fixed minimum resolutions well above the scales at which individual organisms can be resolved. Because many important ecological interactions occur below the pixel level, it is necessary to develop methods that can derive information about small-scale spatial processes and patterns from large-scale spatial patterns. Assuming that ecological processes at a particular scale can cause patterns at other (especially larger) scales, we consider the following question: What can be determined about high-resolution spatial patterns and processes from spatial patterns observed at lower resolutions? As a heuristic example, a two-dimensional grid cell spatial model is presented and used as a means to investigate the general relationships among spatial patterns in grids of different scales.

Methods. Grids and spatial patterns were generated using computer simulations similar to those in Milne (1991). Each high-resolution grid was a 100 x 100 cell matrix wherein each cell could be in one of two states, ON or OFF. These binary states are considered equivalent to the presence or absence of an individual organism at a particular location in a given landscape. Different grids were generated in which the proportion f of ON cells varied from 0.05 to 0.95, and the location of ON cells across the grid was random, clustered, or dispersed.

The initial high-resolution grid was generated in a two-step procedure, beginning with an initial scan of the grid to "seed" the ON cells. Here, 10% of the number of ON cells were placed at random in an empty grid. Subsequent scans and placement of ON cells were performed until the desired proportion of ON cells was achieved. In each scan and for each OFF cell, a random number was compared with $0.02fg^n$, where g is an aggregation factor and n is the number of neighboring cells (from 0 to 8) that are ON. This calculation allowed weighting the likelihood that a particular cell would be turned ON by the presence of other

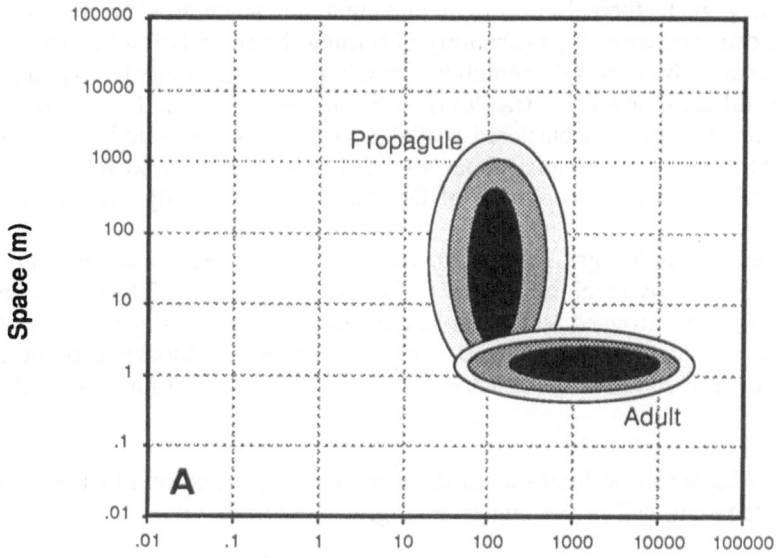

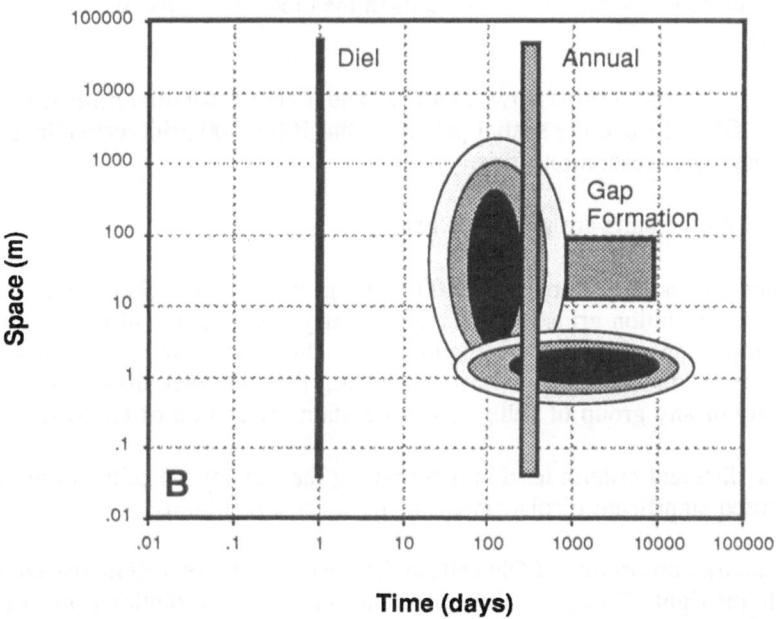

Figure 1. Diagram of the spatial and temporal scales for the propagule and adult stages of a hypothetical tree. (a) Shaded confidence ellipses show the variability in the "species." (b) Environmental processes that can affect tree growth and survival. Diel and annual environmental change occur across all spatial scales. Gap formation is more variable but has a characteristic frequency and size distribution.

ON cells: when $g = 1$, there is no effect of neighbors; when $g < 1$, the presence of neighboring ON cells decreases the probability of turning the cell ON; and when $g > 1$, the probability is increased by each ON neighbor. In this study, the amount of aggregation (g) was set at three different levels: 1 (random), 0.5 (spread), and 1.5 (clustered). By this procedure, random, spread, and clustered spatial patterns were generated. The coefficient 0.02 ensured that a number of scans (about 40 - 50) was conducted so that patterns were gradually built and the desired proportion of ON cells was not greatly overshot by the last scan.

Three lower-resolution grids (20 x 20, 10 x 10, and 5 x 5 total pixels) were derived from the 100 x 100 grid by combining blocks of cells. Separate 5 x 5 blocks of cells from the 100 x 100 grid were grouped to form a single cell in the 20 x 20 grid; 10 x 10 blocks were combined for each cell in the 10 x 10 grid; and 20 x 20 cell blocks made up each cell in the 5 x 5 grid. The state for the lower-resolution cells was determined by one of three methods:

a) Presence/Absence: if at least one of the cells in the aggregation block was ON, then the corresponding cell in the higher-level grid was turned ON.

b) 30% threshold: if at least 30% of the cells in the block were ON, then the corresponding cell was turned ON.

c) 50% threshold: if at least 50% of the cells in the block were ON, then the corresponding cell was turned ON.

Evaluations of relationships between lower- and higher-resolution grids were based on the proportion of ON cells and the spatial pattern in the 100 x 100 grid versus the proportion of ON cells in the lower-resolution grids.

Results. Three results are presented here.

1) For a random spatial pattern in the 100 x 100 grid, the proportion of ON cells in the three lower-resolution grids increased nonlinearly and at different rates with increase in the proportion of ON cells (Figure 2a). This particular result can be derived analytically. Since each cell is independent, the binomial distribution gives the probability of any group of cells having a certain proportion of ON cells.

2) The three different criteria used in determining the state of the cells in low-resolution grids caused significant displacements of these curves (Figure 2b).

3) As the required proportion of ON cells in the lower-resolution cell increased, the curve shifted to the right. There was little difference in the low-resolution grids for different spatial patterns (random, spread, or clustered) in the 100 x 100 grid (for an example, see Figure 2c).

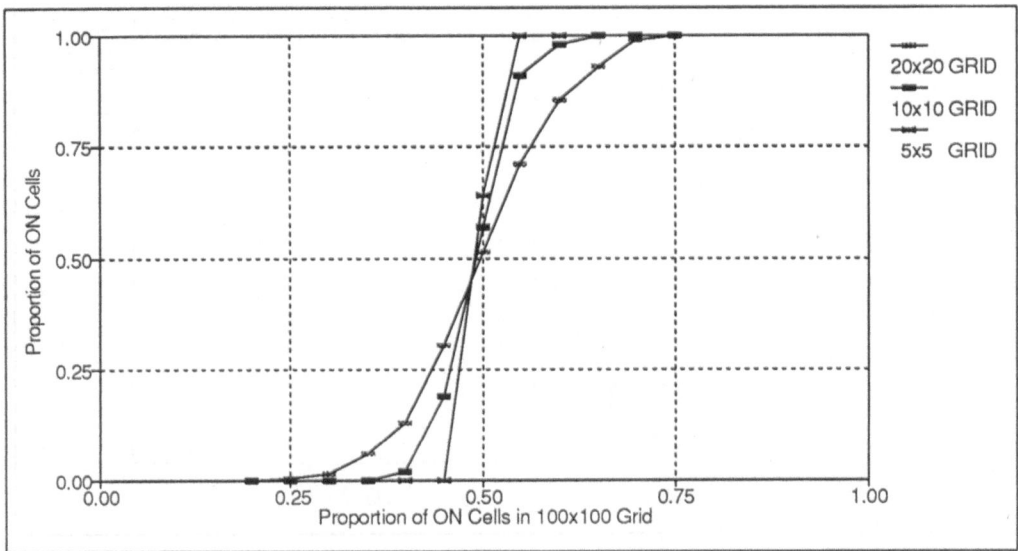

Figure 2a. Proportion of ON cells in the three higher-level aggregated grids at various proportions of ON cells in random pattern 100 x 100 low-level grid. The threshold criterion for aggregation of low-level cells to high-level cells was 50% ON in the cell block.

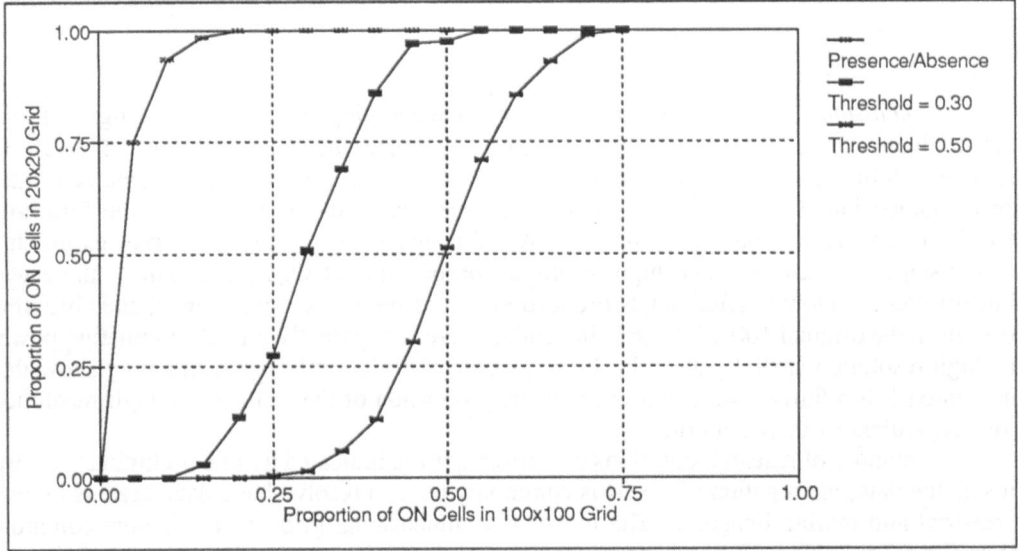

Figure 2b. Proportion of ON cells in the 20 x 20 cell grid using different criteria for aggregating low-level cell states into higher-level cell states. Cells in the low-level 100 x 100 grid were distributed in a random pattern.

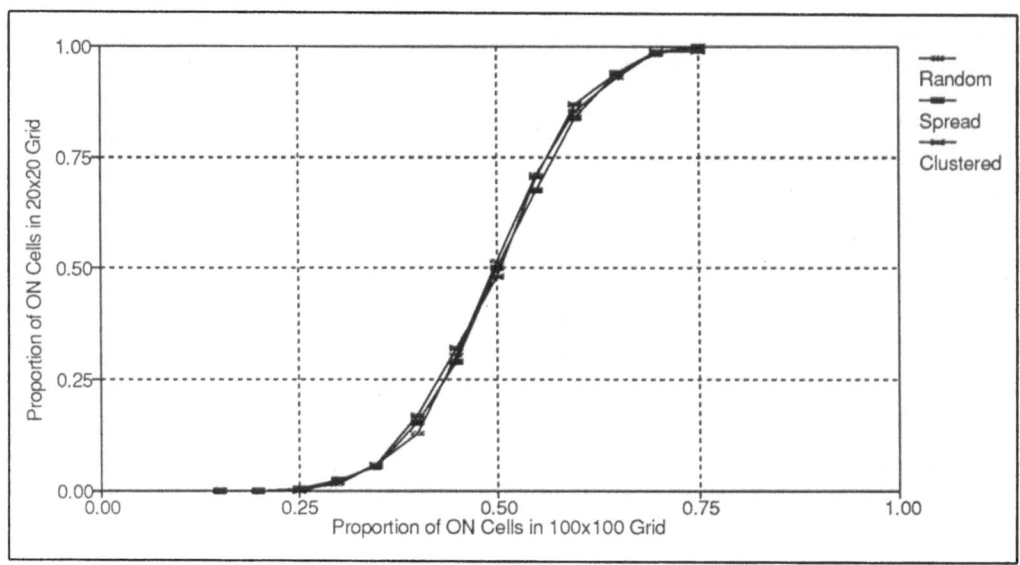

Figure 2c. Proportion of ON cells in the 20 x 20 grid for different spatial patterns of ON cells in the low-level 100 x 100 grid. The aggregation criterion was an ON threshold of 50% for cells in the aggregation block.

Conclusions. These results are both encouraging and disappointing. First, low-resolution grids can apparently be used to estimate the proportion of ON cells in higher-resolution grids. However, this requires some knowledge of how individual cell states are combined into larger cells. Second, it appears that both the scales and the threshold criteria for low-resolution grids strongly affect the shape and location of curves expressing relationships between low- and high-resolution proportions of ON cells. Third, there was little difference in low-resolution ON proportion curves among random, spread, and clustered patterns in the original 100 x 100 grid, indicating there is apparently little discriminating power for high-resolution spatial pattern in this approach. On the other hand, this may show that the approach is a fairly robust indication of the proportion of ON cells in the high-resolution grid, regardless of their pattern.

A number of research questions and directions are indicated by this preliminary effort. In satellite data, aggregation of pixels is commonly used to resolve huge data streams in both terrestrial and marine imagery. The use of two-dimensional grid models is now common. Thus the relationship of these image-processing techniques to patch dynamic simulations is worth further study.

Multiscale Patterns: Modeling and Empirical Data on Krill Aggregations

Spatial patchiness in physical and biological phenomena has been observed at virtually every scale investigated (Haury et al. 1978, Steele 1978), but little is known about the roles of physical vs. biological or induced vs. autonomous mechanisms that cause heterogeneity. Analytical techniques that elucidate the predominance or interactions of these mechanisms, therefore, are urgently needed.

We compare the results of analyses of field data from the Southern Ocean using two techniques: spectral analysis, a method often applied in marine systems (e.g., Platt and Denman 1975), and wavelet analysis, a relatively new statistical procedure recently applied to ecological and physical data (Argoul et al. 1989, Bradshaw 1991, Bradshaw and Spies, in press). The wavelet analysis is an alternative method for quantifying spatial pattern. These methods are employed in the present analysis to examine the characteristic scales of variability of a physical property (temperature), primary production (chlorophyll fluorescence), and an important herbivore (the Antarctic krill, *Euphausia superba*) in order to develop hypotheses about mechanisms that generate multi-scale patterns (see the chapter by García-Moliner et al. for a description of wavelet analysis).

Previous Antarctic investigations have concluded that observed variability in phytoplankton distribution is governed primarily by physical processes, because the power spectra of temperature and chlorophyll were consistent with those of passively advected scalars (Weber et al. 1986, Levin et al. 1989). In contrast, the spectra of krill biomass distribution showed greater variability at small spatial scales than those of temperature and phytoplankton. We hypothesize that this difference arises, in part, from the social behavior of krill, and we examine this hypothesis with the use of a model of an idealized aggregating species (Grünbaum 1991). Results from both the field data and the model suggest that behavior is an important mechanism contributing to the heterogeneous distribution of krill at small scales.

E. superba is an important link in the Antarctic marine food web and also the subject of a commercial fishery. On the scale of this large marine ecosystem, the spatial patterns of krill are probably controlled primarily by physical processes (Sahrhage 1988). On smaller scales, however, biological factors including behavior may influence krill spatial patterns (Price et al. 1988, Daly and Macaulay 1991).

Methods. Sea-surface temperature and chlorophyll-*a* fluorescence (provided by O. Holm-Hansen and W. Helbling, Scripps Institution of Oceanography; and A. Amos, University of Texas at Austin) and vertically integrated acoustic biomass of krill (provided by M. C. Macaulay, University of Washington) were collected along a 120 km transect north of Elephant Island near the Antarctic Peninsula during two surveys, approximately one month apart, in mid-summer (Rosenberg and Hewitt 1991).

In the present study, we analyzed temperature, chlorophyll, and acoustic biomass data to identify possible spatial relationships between krill and physical-biological factors by determining (1) the characteristic distances over which significant changes in each variable occurred, and (2) the spatial correlation among the three variables. The sampling resolution was 100 m for acoustic data and 800 m for temperature and chlorophyll data. Thus, the smallest scale compared for correlation analyses was 800 m (These scales complement the analysis of krill data in the chapter by García-Moliner et al.).

Field Results. The field data showed spatial patterns typical of other results (e.g., Mackas et al. 1985, Weber et al. 1986). Sea-surface temperature showed little structure and changed gradually with distance; chlorophyll was similar but showed more spatial pattern; and krill density showed complex spatial structure (Figure 3). The spatial pattern of krill also was not uniform along the transect; patch size and abundance varied with location. The power spectra were similar to those of Weber et al. (1986) and indicated that the distribution of krill on both transects was highly variable at spatial scales smaller then 1000 m. The wavelet transform revealed that all three variables had spatial structure at different scales. Sea-surface temperature (Figure 4a) and chlorophyll concentration (Figure 4b) each had one characteristic scale, ca. 15 km and 5 - 10 km, respectively, while krill distribution (Figures 4c,d) was characterized by patches on several scales, ca. 25 km, 10 km, and < 200m. These patches had a hierarchical structure, with fine-scale features nested in larger features, as shown by the increasing complexity of structure and detail from large (Figure 4c) to small scales (Figure 4d). The special feature of the wavelet method is to display locational information. Wavelet analysis indicated that krill were aggregated into three large patches along the transect, 20 - 25 km in width; that there were few krill between 30 - 45 km; and that fine-scale structure was less pronounced between 0 - 10 and 100 - 120 km (Figures 3 and 4c).

Wavelet cross-covariance analysis did not indicate a strong association between any of the variables at zero lag. The spatial patterns of temperature and chlorophyll, however, were associated with an offset of 2 km (Figure 5a). In other words, 10 km patches of relatively high phytoplankton density were offset by 2 km from 10 km patches of cold temperature water. Chlorophyll and krill distributions also were negatively correlated at a scale of about 20 km, with an offset of 4 km (Figure 5b).

Model Results. To provide additional insight for understanding spatial patterns of krill, a model of an idealized aggregating species is used to examine the qualitative effects of schooling or swarming behavior on spatial distribution and spectral density. This continuum model of density-dependent aggregation is derived directly from a stochastic model of aggregating individuals (Grünbaum 1991).

In the stochastic model, individuals form aggregations by swimming toward or away from neighbors in search of a desired "target" density. The premise of the model is that aggregating individuals make decisions on the direction in which to move based on a limited amount of information about the distribution of other individuals around them. This limitation is incorporated in the model in the form of a "sensing range," the distance at which individuals can detect one another. Movement based on density-dependent decisions can only be in response to a sample of neighbors within this range. Instantaneous measurements of this relatively small number of individuals, therefore, are random quantities, which only statistically describe the average local population density and density gradient.

The response of a species to a continuously forcing velocity field can be compared in the presence and absence of aggregation behavior. To make this comparison as explicit as possible, it is convenient to have a velocity field that has a single characteristic length scale. One possible choice of such a velocity field is simply a traveling sine wave. At any position this velocity field results in a succession of converging and diverging velocity fields.

A comparison between the density distributions and the spectra of aggregating and non-aggregating species for a typical choice of parameters is shown in Figure 6. Here, the characteristic length scale of the velocity perturbation is much larger than the sensing distance. The density distribution (Figure 6a) of the non-aggregating species shows variation primarily

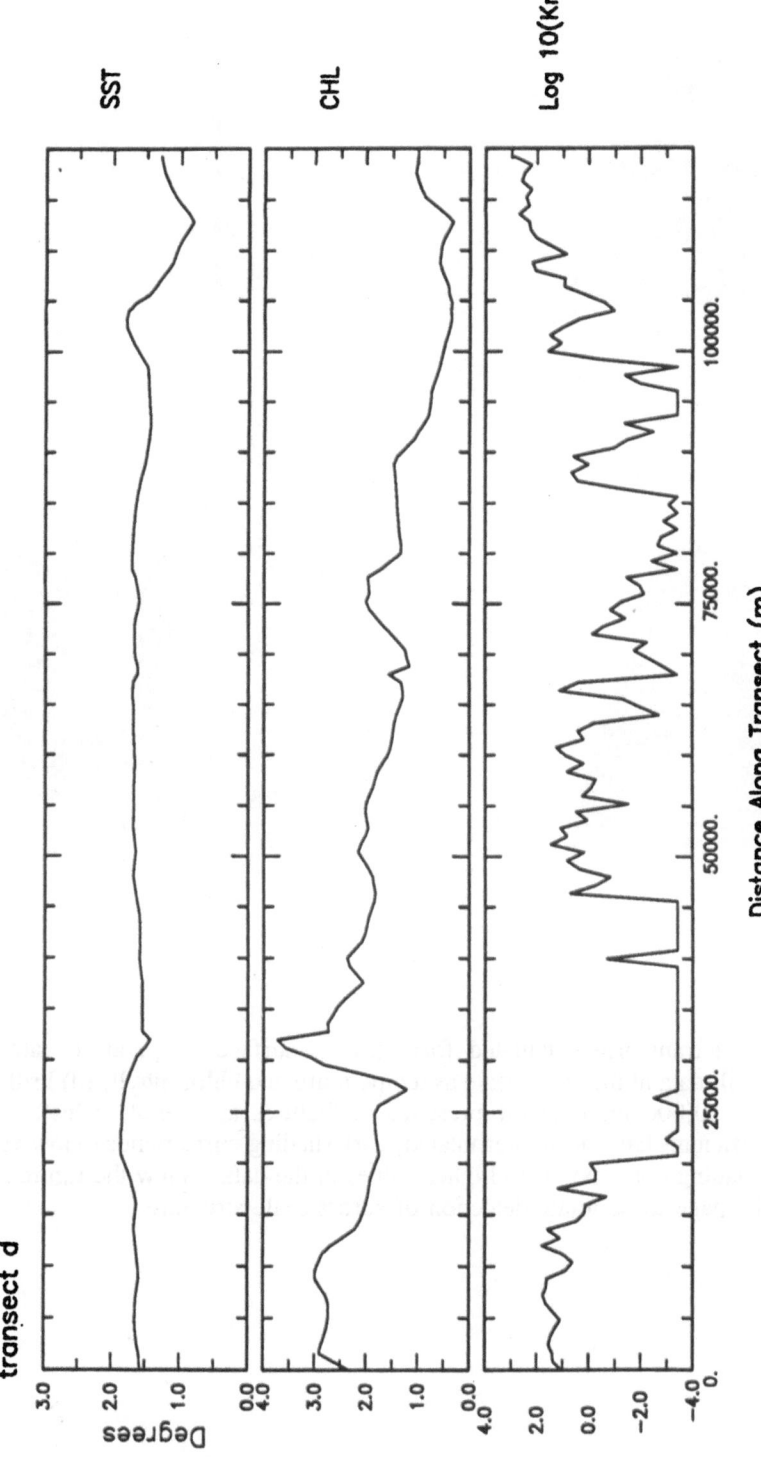

Figure 3. Field data for transect d illustrating different spatial patterns for sea-surface temperature (SST, °C) and chlorophyll-*a* concentration (CHL, mg l⁻¹), and acoustic biomass of krill integrated 6 - 250 m (log₁₀ krill, g m⁻²). X-axis denotes distance along the transect in meters; transect runs approximately west to east.

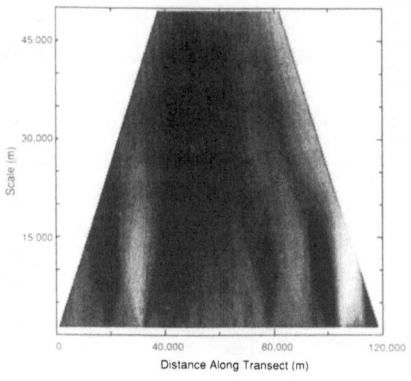

a

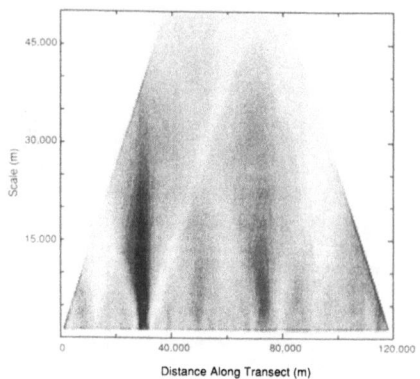

b

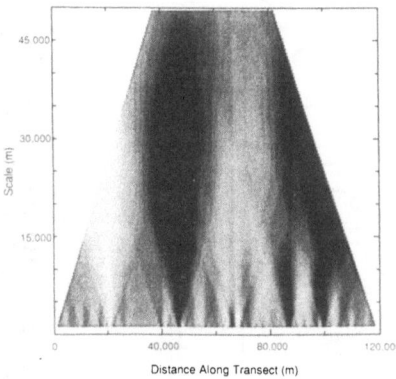

c

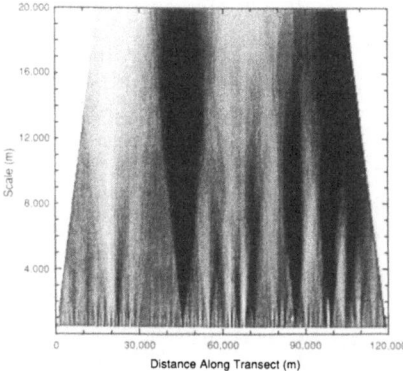

d

Figure 4. The wavelet transform calculated for: (a) sea-surface temperature data; (b) chlorophyll data; (c) krill data at the same scale as temperature and chlorophyll; (d) krill data at a finer scale (100 - 20,000 m; note the presence of features at several scales). The wavelet transform coefficients have been interpolated; dark shading corresponds to low values in the data and light shading corresponds to higher values in the data. View the figure near-parallel to plane of the page to facilitate detection of across-scale structure.

SST - Chlorophyll

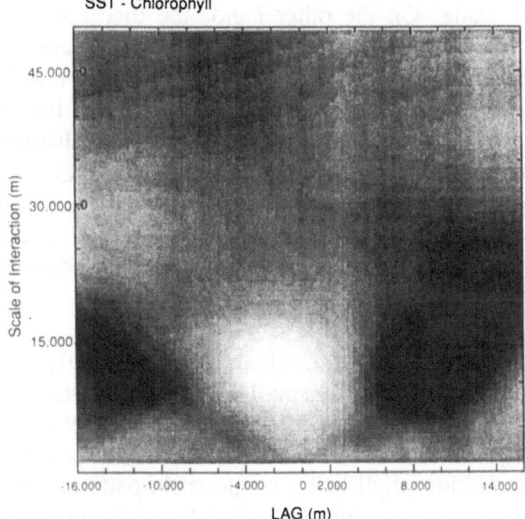

a

Krill - Chlorophyll

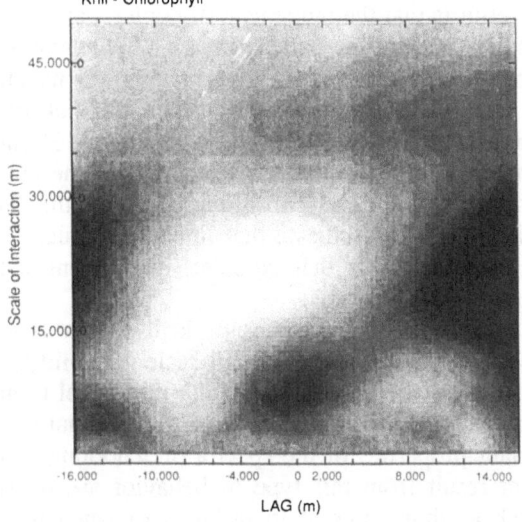

b

Figure 5. The wavelet cross-covariance function for: (a) sea-surface temperature; (b) chlorophyll. The grey-scale bar at the base of the figure identifies the magnitude of the wavelet cross-covariance; white corresponds to negative values and dark grey to positive values.

corresponding to this length scale. On the other hand, the distribution of the aggregating species shows a shorter characteristic length scale that is autonomously generated by density-dependent social behavior. More difficult to discern in the density distribution, but evident in the associated power spectra (Figure 6b), is the result that the large-scale variation in the density distribution of the aggregating species is nearly unchanged from that of the purely diffusing species. Substantial changes in the spectra, however, occur at higher wave numbers (i.e., small scales), where the addition of aggregating behavior to a randomly dispersing species increased the spectral energy at small length scales. This trend is qualitatively consistent with the trends observed in the empirical temperature, chlorophyll, and krill biomass distributions.

Conclusions. The results of the spectral and wavelet analyses confirm that krill biomass distribution has more variability at small length scales than the distributions of temperature and chlorophyll, consistent with earlier investigations (Weber et al. 1986, Levin et al. 1989). Wavelet analysis also demonstrated that krill had a multiscale aggregation pattern while temperature and chlorophyll had a single-scale pattern within the limits of the data resolution. The large-scale aggregations of krill (10 and 25 km range) appeared to be composed of many small patches or swarms and were not the result of a diffuse, homogeneous aggregation of krill over a wide area. This implies that the mechanisms that controlled temperature and phytoplankton spatial patterns were different from those that controlled krill.

The wavelet cross-covariance analyses indicated that temperature and chlorophyll were negatively correlated but spatially offset by some distance. One possible hypothesis that might be explored based on this result is that the movement of cooler water masses at a scale of 10 km produces a corresponding 10 km sub-pattern in the phytoplankton community. The cross-covariance of chlorophyll and krill spatial patterns also was negatively correlated and spatially offset. Grazing by krill may have contributed to the lack of coherence between temperature and chlorophyll distributions. The field data (Figure 3) suggest a downstream decrease in phytoplankton concentration, with the exception of the chlorophyll maximum, which occurred in the same location as the krill biomass minimum. The lack of coupling between chlorophyll and krill may also indicate that important scales of interaction may be smaller than the sampling resolution or that complex interactions between physical and biological processes obscure the mechanisms.

Nevertheless, the numerous small patches of krill within larger aggregations are evidence of the importance of social behavior to small-scale variability. The implications of this behavior are demonstrated by the model. Although the model tremendously simplifies the true behavior of Antarctic krill in a complex, three-dimensional environment, it can be used to show some important properties of the spectra of schooling and swarming species. First, the aggregations that result from this type of behavior are resistant to break-up by velocity perturbations (such as those present in turbulence) when the length scale of the aggregation is up to an order of magnitude larger than the sensing distance. These aggregations are characterized by high variability at small length scales. Thus, aggregation behavior can account for, in principle, the observed increase of krill spectral density at small length scales relative to those of a passively convected scalar.

Second, at length scales much larger (several orders of magnitude) than the sensing distance, aggregations resulting from density-dependent behavior have very little resilience. At large length scales, the distribution patterns are determined by the external advection field, even when the characteristic external velocity is small compared to the characteristic

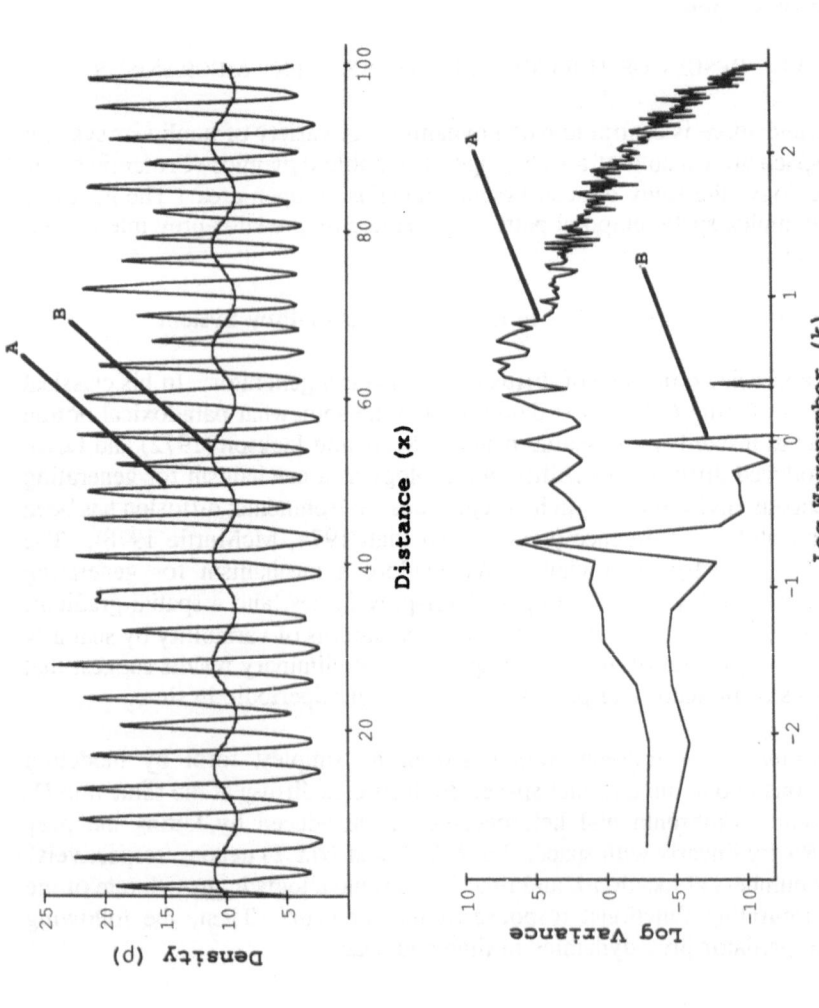

Figure 6. (a) Plots of population density vs. distance for aggregating (A) and non-aggregating (B) species in an environment with a continuous forcing velocity field of the form $v(x,t) = v_o \sin(k_o (x - ct))$. The advection parameters are $v_o = 0.5$, $k_o = 0.16$ n, and $c = 0.05$. The diffusion coefficient, aggregation velocity, and target density are $D = 0.25$, $\tau = 0.3535$, and $\mu = 20$, respectively. (b) Power spectra (log-log plots) for density distributions of the aggregating (A) and non-aggregating (B) species in Figure 6a. The peak in both plots at wave number $k = -0.688$ corresponds directly to the sinusoidal forcing frequency. At higher wave numbers, the pure diffusion case (B) shows some additional energy peaks at subharmonics of the...

aggregation velocities of individuals. Hence, the effect of aggregation behavior on spectral density is strongly length-scale specific: the model predicts an approximate length-scale threshold, below which distribution is determined by social behavior and above which distribution derives from physical oceanographic processes.

These preliminary results illustrate the benefits of a combined analytical and empirical approach to studying pattern-generating mechanisms. Both statistical analyses of field data and theoretical modeling suggest that behavior may be more important than physical processes in determining meso-scale spatial patterns of krill. A number of underlying mechanisms may produce similar spatial patterns; hence, an investigation of an ecosystem must combine knowledge about the spatial distribution patterns of physical and biological variables with experimental and theoretical studies.

PATCH FORMATION AS A RESULT OF THE COUPLING OF DIFFERENT MECHANISMS

The interplay of time and space is central to patch dynamics. A variety of nonlinear systems when extended into space are capable of a rich range of possible dynamics (Crutchfield and Kaneko 1987). In ecology, the study of such systems remains an open area. The next two models illustrate the complex spatiotemporal patterns generated by predator-prey interactions when coupled by dispersal.

Diffusion-Induced Chaos or Quasiperiodicity in a Spatial Predator-Prey System

Ecologists have long appreciated the role of diffusion as a pattern generator. In his classical work on morphogenesis, Turing (1952) first demonstrated the somewhat paradoxical notion that diffusion may lead to instability and spatial pattern. Segel and Jackson (1972) and Levin and Segel (1976) introduced diffusion instability into ecology as a mechanism for generating patchiness in homogeneous environments. In heterogeneous environments, diffusion has been generally viewed as a stabilizing influence (Comins and Blatt 1974, McMurtie 1978). The following work argues against such a view. We explore a mechanism for generating patchiness involving the interplay of diffusion, predator-prey cycles, and a spatial gradient. The model demonstrates that diffusion can lead to complex patterns of variability by spatially coupling local predator-prey cycles of different frequencies. Preliminary results suggest that the spatial distributions of predator and prey are chaotic or quasiperiodic in time.

Model description. The problem was posed in its simplest form by modeling predator-prey interactions in one-dimensional space. Both species diffuse at the same rate D_0 along a spatial gradient. Environmental heterogeneity is introduced by letting the prey intrinsic growth rate R vary linearly with space. Let $P(X,T)$ and $H(X,T)$ denote, respectively, the prey and predator numbers at location X and time T. Assume a logistic growth rate of the prey and a Type II saturating functional response of the predator. Then, the following equations describe the predator-prey dynamics in time and space:

$$\frac{\partial P}{\partial T} = RP(1-\frac{P}{K}) - \frac{bC_1P}{C_2+P}H + D_0\frac{\partial^2 P}{\partial X^2}$$

$$\frac{\partial H}{\partial T} = \frac{C_1P}{C_2+P}H - d_0H + D_0\frac{\partial^2 H}{\partial X^2}$$

(1)

The parameters K, d_0, and $1/_b$ denote the carrying capacity of the prey, the death rate of the predator, and the conversion rate of prey ingested to predator growth, respectively. The constants C_1 and C_2 set the parameters for the saturating functional response.

Also, no flux is assumed at the boundaries, and hence, at $X = 0$ and $X = L$,

$$\frac{\partial P}{\partial X} = \frac{\partial H}{\partial X} = 0$$

With the dimensionless variables $p = P/K$, $h = bH/K$, $x = X/L$, and $t = RT$ where $R = R(X_0)$ for some X_0 in $(0,L)$, system (1) becomes

$$\frac{\partial p}{\partial t} = rp(1-p) - \frac{Ap}{1 + Bp}h + D\frac{\partial^2 p}{\partial x^2}$$

$$\frac{\partial h}{\partial t} = \frac{Ap}{1+Bp}h - dh + D\frac{\partial^2 h}{\partial x^2}$$

(2)

where

$$r(x) = \frac{R}{\overline{R}} = m + nx, \quad A = \frac{C_1K}{C_2\overline{R}}, \quad B = \frac{K}{C_2}, \quad d = \frac{d_0}{\overline{R}}, \quad D = \frac{D_0}{L^2\overline{R}}.$$

At the boundaries, $x = 0$ and $x = 1$

$$\frac{\partial p}{\partial x} = \frac{\partial h}{\partial x} = 0$$

The dynamics of system (2) were investigated numerically with a finite difference integration scheme using 100 spatial grid sites.

Results. The following results illustrate the dynamics of equations (2) for a single parameter set (A = 5, B = 5, d = 0.6, m = 1.8 and n = -1.4), chosen to obtain limit cycles at each fixed location along the gradient in the absence of diffusion. When diffusion is added, $D = 10^{-4}$, the system exhibits sharply different dynamics. Changes in behavior occur in both space and time (Figures 7a,b,c), and trajectories show sensitivity to initial conditions. The resulting spatial distributions of both species along the gradient are then aperiodic, either quasiperiodic or chaotic in time.

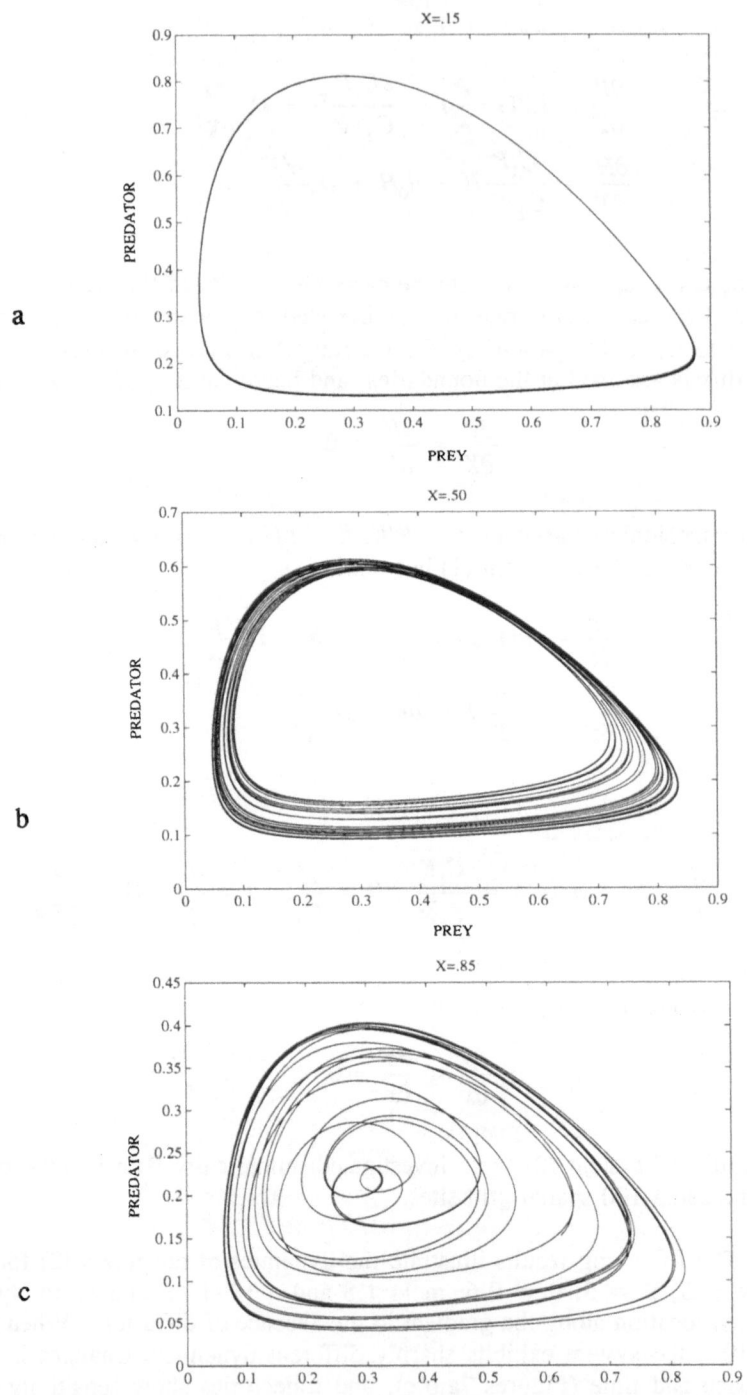

Figure 7. Phase plots of predator vs. prey numbers at three different locations along the spatial gradient. (a) x = 0.1, (b) x = 0.5, (c) x = 0.85. (Transients have been removed).

Hence, diffusion may yield an asymptotic state that changes continuously in space and time. A synoptic observer would then see patches changing continuously in form and location (Figure 8). Preliminary exploration of the parameter space reveals that complex dynamics occur only at intermediate values of the diffusion rate. Also, as the diffusion rate is decreased, the transition to temporal chaos may occur through a quasiperiodic route (Schaffer 1988). The steepness of the environmental gradient also plays a critical role. At the limit of a uniform environment, cycles are again the rule. We are presently working on a more definite characterization of the dynamics of the system.

Conclusions. Predator-prey systems, by virtue of their nonlinearity, are known to exhibit complex dynamics. Previous work has focused on the temporal interplay of oscillations (Gilpin 1979, Hastings and Powell 1991, Kot et al. 1991). These preliminary results add the spatial dimension and show that diffusion may drive an otherwise periodic predator-prey system into aperiodic behavior. Vastano et al. (1990) describe a similar phenomenon for a chemical system. Kot (1989) demonstrates that diffusion can lead to period-doubling bifurcation and chaos in discrete predator-prey models. However, these models are already capable of complex dynamics with no added spatial dimension. We have presented a mechanism that in spite of its simplicity creates a dynamic patchiness in the distributions of predator and prey.

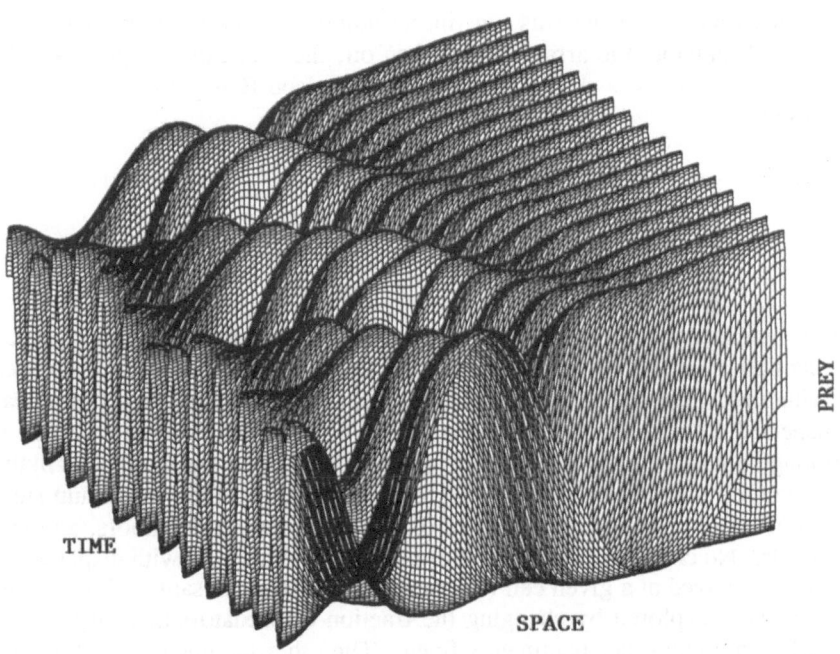

Figure 8. Spatiotemporal variation in prey numbers.

Dispersal of a Predator and Its Prey in a Two-Dimensional Spatial Grid

Environmental heterogeneity can function as an external or induced mechanism generating patchy distributions of organisms (Kareiva 1990, Doak et al. in press, Hilborn 1975, Robinson and Quinn 1988). Behavioral strategies that promote aggregation can be expected to create clumpy distributions of individuals in homogeneous landscapes (Grünbaum, 1991); such behavior would typify an internal or autonomous mechanism. However, the presence of non-uniform distributions of species need not imply the existence of such explicit patch-forming mechanisms are at work. In this section, a model is presented in which random movement, in the form of dispersal, interacting with a predator-prey model, leads to the generation of patchy distributions of individuals in a landscape consisting of discrete but equal patches.

Model description. The model consists of a two-dimensional square array of cells. Every cell in the array can support a fixed population of a prey species; i.e., a carrying capacity for the cells is defined. In addition, a predator species can exist in each cell. At any one time, a cell is classified as being in one of four possible states: empty, only prey present, only predators present, and both prey and predator present. Because the predator cannot exist in the absence of prey, the state with only predators is necessarily considered to be transient. Cells in the array are linked through the dispersal of both the prey and the predator species. The array that we used for this investigation was 50 x 50 cells and had wrapping boundary conditions.

Each time the model was run, initial numbers of prey and predators were randomly distributed throughout the array. From then on, the predator and prey populations in each cell were governed by a density-dependent Nicholson-Bailey model (Hassell 1976). The model used was:

$$N_{t+1} = N_t \exp\left[r(1 - \frac{N_t}{K} - aP_t\right]$$

$$P_{t+1} = \alpha N_t\{1 - \exp[-aP_t]\}$$

where N indicates the prey and P the predator. If the population of either species fell to less than one-half (in one cell), it was truncated to zero. Spatial interactions were defined by specifying maximum step sizes for both species, and by fixing the fraction of each population that dispersed from a cell at every time step. The fraction dispersing was the same for each cell (though not necessarily the same for prey and predator), and dispersing individuals moved in a random direction a distance ranging between zero and their maximum step size. A step of size one would take an individual to one of its 8 nearest neighbors, and so on for larger movements. No cost (e.g., increased mortality) was associated with dispersal. All dispersers successfully arrived at a given cell (dispersal to and from the same cell was allowed). Model dynamics were explored by changing the fraction of predators that dispersed (from 0.05 to 0.15) while holding other parameters fixed. The other parameters used for each model run were: fraction of prey that disperse = 0.01; prey maximum step size = 1; predator maximum step size = 5; $r = 2$, $K = 100$, $a = 0.5$, and $\alpha = 1.00$. In a single cell, these parameters led to the predators' eliminating the prey in a relatively short time.

To investigate further the dynamics of the spatial predator-prey system, a patch-counting algorithm was constructed. A patch was defined as any collection of cells that were connected by sides or corners. The inclusion of patches connected only by a corner is consistent with our dispersal mechanism as described above. This design eliminates the bias toward orthogonal directions that is otherwise present in grid models. The patch-counting algorithm was used to generate a frequency distribution of patch sizes. Each frequency distribution was constructed by allowing the model to run for 100 time steps to eliminate transient effects. Then, prey patches, consisting of cells with prey or cells with predators and prey, were counted for 750 subsequent time steps. The shape of the patch frequency distribution became fully expressed after at most 500 steps.

Results. The limiting distribution of prey and predators was quite sensitive to the rate at which predators dispersed. The distribution of prey appears increasingly clumpy as the fraction of dispersing predators grows (Figure 9). The mean numbers of prey and predators decrease as the predator dispersal rate increases. However, the predators were unable to persist if significantly less than 5% dispersed. Predators would eliminate the prey in their vicinity and then die off before finding new patches. At 15% predator dispersal, the prey were frequently reduced nearly to zero, which would precipitate a crash of the predator population. Large outbreaks of prey were common in this scenario, as a few cells always escaped the onslaught and managed to grow into large new patches before the predator population recovered. Predator dispersal rates greater than 15% led to a complete extermination of the prey.

Patch frequencies resulting from a completely random process were also generated. When the Nicholson-Bailey model was run, the number of grid cells containing prey was tabulated for every step. A patch size distribution for a random process was then obtained by placing an equal number of cells randomly throughout the array for the same number of steps and counting up the patches that resulted. This was done to allow the frequency distributions obtained from the Nicholson-Bailey model to be contrasted with those obtained from a random placement of cells.

The "envelopes" that contained the patch size distributions for the Nicholson-Bailey and the random process model runs are outlined to facilitate inspection (Figure 10). The distributions for the two processes represent the same total number of cells of prey in each case. A sudden loss of large patches of prey takes place as the fraction of predators dispersing increases from 5% to 9%. The presence of large patches of prey when 5% of the predators disperse can be explained in part by a tendency for percolating networks to form as the number of cells containing prey increases in a grid of fixed size. Applications from results of percolation theory suggest that, as the frequency of cells of one type grows, the likelihood that extremely large clusters of these cells will form increases dramatically (Gardner and O'Neill 1991). With this model, the effect of predator dispersal reaches a threshold at around 8%, and beyond this percolating networks rapidly disappear.

As the fraction of dispersing predators increases, the Nicholson-Bailey patch size distribution deviates from that of a random process (Figure 10). The average number of cells containing prey declines rapidly as predator dispersal grows, and this is reflected in the patch size distribution for the random process containing increasing numbers of patches in the range of 1 to 10, with the difference made up in diminishing numbers of large patches. The shape of the Nicholson-Bailey patch size distribution changes abruptly at the ends of the range of dispersal parameters investigated, and more smoothly in the middle.

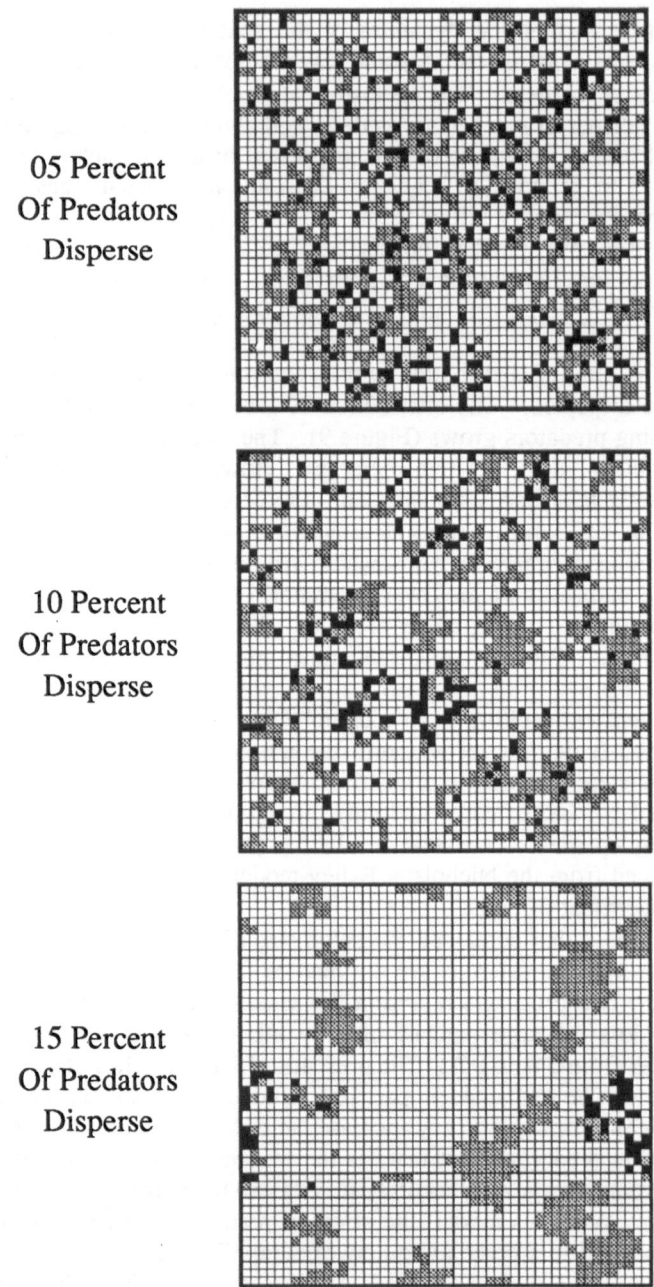

05 Percent
Of Predators
Disperse

10 Percent
Of Predators
Disperse

15 Percent
Of Predators
Disperse

Figure 9. Snapshots in time of the array of cells for three different fractions of predators dispersing. Grey cells contain only the prey species and black cells contain both prey and predators. Empty cells have neither prey nor predators in them. The value of 10% predator dispersal was not used in the model runs, but is displayed here for clarity of presentation. The grid is 50 x 50 cells.

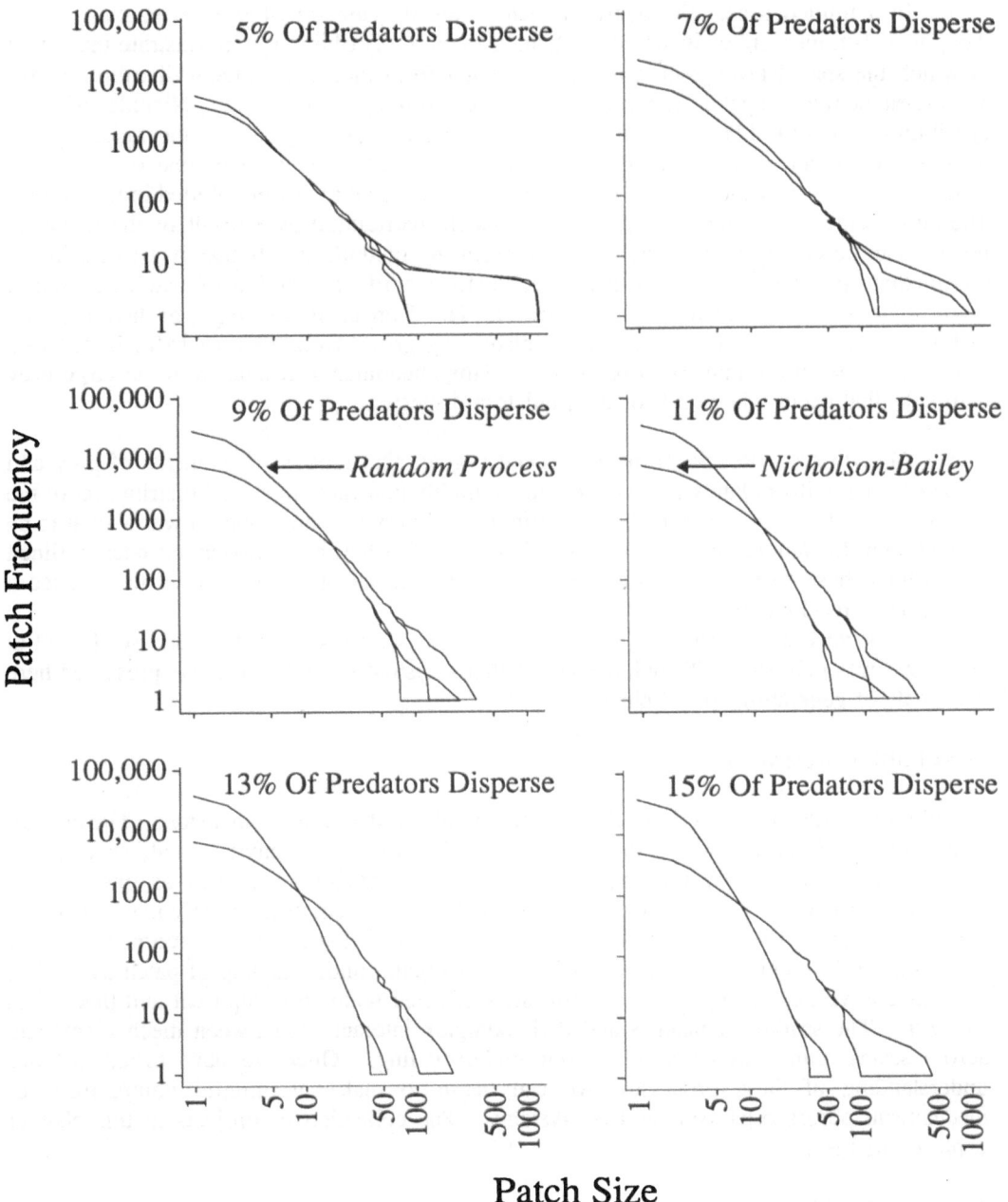

Figure 10. Envelopes containing the patch size distributions for the Nicholson-Bailey and random process models. All of the data points fell within the regions plotted in the figure, and the envelopes were drawn so that overlapping areas could be seen more clearly. Each graph corresponds to a different fraction of predators dispersing.

In a limited sense, the degree to which the distribution of prey is patchy can be thought of as a measure of its deviation from a random distribution. To illustrate the extent to which the spatial arrangement of prey deviated from that of a random distribution, we subtracted the random patch frequency distributions from the corresponding Nicholson-Bailey distributions (Figure 11). Our intent in this analysis was to focus on the tendency for increased clumping of prey in the spatial Nicholson-Bailey model over that of a random process, and thus only the portions of the graphs greater than zero were plotted in this figure. The enhancement of a narrow region of the patch distribution as a result of the dispersal process can be clearly seen. The deviation from the random distribution is greatest in the intermediate patch sizes, and these data serve to quantify the notion of patch generation invoked from the series of plots (see Figure 9). The decrease in the height of the maximum in Figure 11, as the fraction of predators dispersing grows from 13% to 15%, is due to a sudden increase in the maximum patch size. This phenomenon results from the large prey outbreaks that occur when 15% of the predators disperse.

Conclusions. Results from this model suggest that random movements of prey and predators in a uniform landscape can generate a highly non-random spatial distribution of the prey species. The degree to which the distribution of prey was non-random was shown to be a function of the number of predators that dispersed. Further, an approach has been outlined with which these spatial distributions can be quantified and the results used to enhance a definition of patchiness.

This work was carried out and the section was written before the work by Hassell et al. (1991) was published. They have shown that a system similar to the one presented here is capable of generating spatial chaos.

CONCLUDING REMARKS

By definition, patch dynamics combines the temporal and the spatial dimensions. The explicit consideration of space leads to the explicit consideration of the environment, such as the transport of organisms by the physics in the ocean. Dichotomies such as physical vs. biological, or induced vs. autonomous, have developed in attempts to classify mechanisms of patch formation. However, a focus on the interaction between the environment and the organisms at different spatial scales would lead to a better understanding of patch dynamics. It is imperative that ecologists avoid arbitrarily selected scales for empirical and theoretical research. The scaling of patches and their complex interactions between mechanisms and across scales remain as obstacles to our understanding. Once we have broadened our understanding of these processes, we can begin to make meaningful comparisons of mechanisms across organisms and ecosystems. The collection of projects in this chapter explore this idea.

ACKNOWLEDGMENTS

This chapter has benefitted from discussions with many of the participants in the Patch Dynamics Workshop. In particular, we wish to thank Simon Levin, Tom Powell, and John Steele, Graciela García-Moliner, and Pablo Marquet. This chapter also was improved by comments from Jonathon Dushoff and Mary Schumacher.

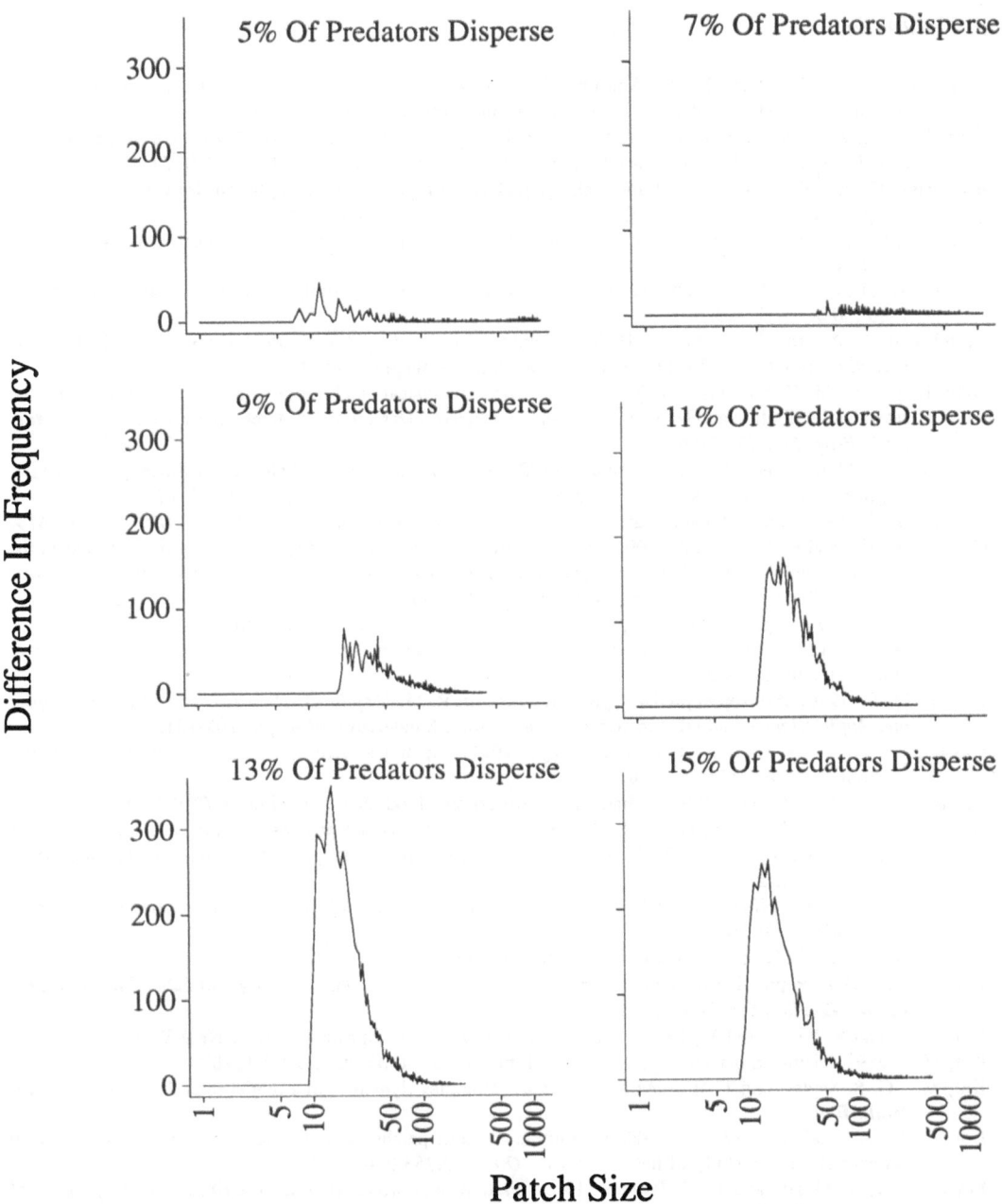

Figure 11. The amount by which the Nicolson-Bailey patch size distribution exceeded the random distribution. This graph was generated by subtracting the frequency distributions for the random process from those of the Nicolson-Bailey process in Figure 10 (when the random distribution exceeded the Nicholson-Bailey distribution, the results were set to zero). Each graph corresponds to a different fraction of predators dispersing.

208

REFERENCES

Addicott, J. F., J. M. Aho, M. F. Antolin, D. K. Padilla, J. S. Richardson and D. A. Soluk. 1987. Ecological neighborhoods: Scaling environmental patterns. *Oikos* 49:340-346.

Argoul, F., A. Arneodo, G. Grasseau, Y. Gagne, E. J. Hopfinger, and U. Frisch. 1989. Wavelet analysis of turbulence reveals the multi-fractal nature of the Richardson cascade. *Nature* 338:51-53.

Bradshaw, G. A. 1991. Analysis of hierarchical pattern and process in Douglas-fir forests using wavelet analysis. Ph. D. dissertation, Oregon State University, Corvallis, OR.

Bradshaw, G. A., and T. A. Spies. In press. Characterizing canopy gap structure in forests using the wavelet transform. *J. Ecol.*

Comins, H. N., and D. W. E. Blatt. 1974. Prey-predator models in spatially heterogenous environments. *J. Theor. Biol.* 48:75-83.

Crutchfield, J. P., and K. Kaneko. 1987. Phenomenology of spatio-temporal chaos. In: H. Bai-Lin, *Directions in Chaos*. World Scientific Publishing, Singapore. pp. 272-353.

Daly, K. L., and M. C. Macaulay. 1991. The influence of physical and biological mesoscale dynamics on the seasonal distribution and behavior of *Euphausia superba* Dana in the Antarctic marginal ice zone. *Mar. Ecol. Prog. Ser.* 79:37-66.

Doak D., P. Marino, and P. M. Kareiva. In Press. Spatial scale mediates the influence of habitat fragmentation on dispersal success: Implications for conservation. *Theor. Popul. Biol.*

Downes, B. J. 1990. Patch dynamics and mobility of fauna in streams and other habitats. *Oikos* 59:411-413.

Gardner, R. H., and R. V. O'Neill. 1991. Pattern, process, and predictability: The use of neutral models for landscape analysis. In: M. G. Turner and R. H. Gardner (eds.). *Quantitative Methods in Landscape Ecology*. Springer-Verlag, New York, NY, pp. 289-307.

Gilpin, M. E. 1979. Spiral chaos in a predator-prey model. *Am. Nat.* 107:306-308.

Grünbaum, D. 1991. Three unrelated projects in mathematical biology. Ph. D. dissertation, Cornell University, Ithaca, NY.

Hassell, M. P. 1976. Arthropod predator-prey systems. In: R. M. May (ed.). *Theoretical Ecology: Principles and Applications* - 2nd ed. Sinauer Associates, Inc., Sunderland, MA, pp. 105-131.

Hassell, M. P., H. N. Comins, and R. M. May. 1991. Spatial structure and chaos in insect population dynamics. *Nature* 353:255-258.

Hastings, A., and T. Powell. 1991. Chaos in a three-species food chain. *Ecology* 72(3):896-903.

Haury, L. R., J. A. McGowan, and P. H. Wiebe. 1978. Patterns and processes in the time-space scales of plankton distributions. In: J. H. Steele (ed.). *Spatial Pattern in Plankton Communities*. Plenum Press, New York, pp. 277-327.

Hilborn R. 1975. The effect of spatial heterogeneity on the persistence of predator-prey interactions. *Theor. Popul. Biol.* 8(3):346-355

Hutchinson, G. E. 1961. The paradox of the plankton. *Am. Nat.* 95:137-145.

Kareiva, P. 1990. Population dynamics in spatially complex environments: Theory and data. *Philos. Trans. R. Soc. London*, Ser. B 330:175-190.

Kolasa, J., and S. Pickett (eds). 1991. *Ecological Heterogeneity*. Springer-Verlag, New York.

Kot, M. 1989. Diffusion-driven period-doubling bifurcations. *BioSystems* 22:279-287.

Kot, M., G. S. Sayler, and T. W. Schultz. In press. Complex dynamics in a model microbial system. *Bull. Math. Biol.*

Kotliar, N. B., and J. A. Wiens. 1990. Multiple scales of patchiness and patch structure: A hierarchical framework for the study of heterogeneity. *Oikos* 59:253-260.

Levin, S. A., A. Morin, and T. M. Powell. 1989. Patterns and processes in the distribution and dynamics of Antarctic krill. CCAMLR Scient. Rep. VII/BG/20:281-296.

Levin, S. A., and R. T. Paine. 1974. Disturbance, patch formation and community structure. *Proc. Nat. Acad. Sci. USA* 71(7):2744-2747.

Levin, S., and L. A. Segel. 1976. Hypothesis for origin of planktonic patchiness. *Nature* 259:659.

Mackas, D. C., K. L. Denman, and M. R. Abbott. 1985. Plankton patchiness: Biology in the physical vernacular. *Bull. Mar. Sci.* 37:652-674.

McMurtie, R. 1978. Persistence and stability of single-species and prey-predator systems in spatially heterogeneous environments. *Math. Biosci.* 39:11-51.

Milne, B. T. 1991. Renormalization relations for spatial models. Paper presented at the Annual Meeting, Ecological Society of America, San Antonio, TX.

Platt, T., and K. L. Denman. 1975. Spectral analysis in ecology. *Ann. Rev. Ecol. Syst.* 6:189-210.

Pickett, S. T. A., and P. S. White (eds). 1985. *The Ecology of Natural Disturbances and Patch Dynamics.* Academic Press, San Diego, CA.

Price, H. J., K. R. Boyd, and C. M. Boyd. 1988. Omnivorous feeding behavior of the Antarctic krill *Euphausia superba*. *Mar. Biol.* 97:67-77.

Robinson G. R., and J. F. Quinn. 1988. Extinction, turnover and species diversity in an experimentally fragmented California (USA) grassland. *Oecologia* (Berlin) 76:71-86.

Rosenberg, J. and R. Hewitt. 1991. AMLR 1990/1991 field season report, Objectives, accomplishments and tentative conclusions. Administrative report LJ-91-18, NOAA/NMFS/Southwest Fisheries Science Center, La Jolla, California.

Roughgarden, J. 1977. Patchiness in the spatial distribution of a population caused by stochastic fluctuations in resources. *Oikos* 29:52-59.

Sahrhage, D. 1988. Summary and conclusions. In: D. Sahrhage, (ed.). *Antarctic Ocean and Resources Variability*. Springer-Verlag, Berlin, pp. 297-300.

Schaffer, W. M. 1988. Perceiving order in the chaos of nature. In: M. S. Boyce (ed.). *Evolution of Life Histories of Mammals: Theory and Pattern*. Yale University Press, New Haven and London, pp. 313-350.

Segel, L. A., and J. L. Jackson. 1972. Dissipative structure: An explanation and an ecological example. *J. Theor. Biol.* 37:545-559.

Steele, J. H. 1978. Some comments on plankton patches. In: J. H. Steele (ed.). *Spatial Pattern in Plankton Communities*. Plenum Press, New York, pp. 1-20.

Stommel, H. 1963. Varieties of oceanographic experience. *Science* 139:572-576.

Turing, A. M. 1952. The chemical basis of morphogenesis. *Philos. Trans. R. Soc. London*, Ser. B 237:37-72.

Vastano, J. A., T. Russo, and H. L. Swinney. 1990. Bifurcation to spatially induced chaos in a reaction-diffusion system. *Physica D* 46:23-42.

Weber, L. H., S. Z. El-Sayed, and I. Hampton. 1986. The variance spectra of phytoplankton, krill and water temperature in the Antarctic Ocean south of Africa. *Deep-Sea Res.* 33:1327-1343.

Wiens, J. A. 1976. Population responses to patchy environments. *Ann. Rev. of Ecol. Syst.* 7:81-120.

Wiens, J. A., and B. T. Milne. 1989. Scaling of 'landscapes' in landscape ecology, or landscape ecology from a beetle's perspective. *Landscape Ecology* 3:87-96.

PART IV

ECOLOGICAL AND EVOLUTIONARY CONSEQUENCES: AN OVERVIEW

Simon A. Levin

The description of spatial and temporal pattern is only the first step in understanding its biological importance. Observing pattern, one is impelled to wonder about its determinants and consequences, ecological and evolutionary. The implications are great, and have become central issues in conservation biology, biodiversity, and the study of global change. The chapters in this section begin to explore these issues, with an emphasis on problems of scale and pattern in ecological and evolutionary time.

Much of our understanding of patchiness and scale has come from studies of marine systems, especially regarding the spatial distribution of plankton (Steele 1978). In particular, as John Steele has emphasized, too often the scales of description are chosen as matters of convenience rather than biological necessity, biasing our understanding and perceptions. Denman (Chapter 14) emphasizes the importance of rising above these limitations if we are to address problems of global scales. In particular, both in marine and in terrestrial systems, the need exists to relate phenomena on broad global scales to processes that are operating on scales many orders of magnitude smaller. Indeed, the cycling of carbon through the ocean is determined both by physical processes, generally operating at large scales, and by biological processes, including fine-scale processes such as the production of phytoplankton detritus and zooplankton fecal pellets. Denman reviews the major marine mechanisms mediating the global carbon cycle, focusing especially on the pelagic planktonic ecosystem. His analysis is a valuable introduction to cross-scale problems, which must be faced in all ecosystems.

In any system of classification, many aspects of description become simpler if we can discretize, organizing entities into clusters according to what often are somewhat arbitrary criteria. In evolutionary biology, for example, we must recognize that the biological species is a fuzzy set, whose members have varying degrees of membership; nonetheless, the concept of biological species, reinforced by operational criteria, becomes an essential organizing notion, providing order to the complex systematic relationships among individuals. The same principle applies in describing spatial and temporal pattern. In the earlier chapters, a variety of continuum methods have been presented for describing the degree of spatial and temporal structure in an ecosystem. Things become simpler, however, if we can recognize discrete clusters or patches, either in the distribution of organisms, or in the environmental parameters that influence their distribution. That patches come in all sizes and shapes is no deterrent; we can simply include as part of our formulation explicit consideration of the dynamics of the patch population (Levin and Paine 1974, Paine and Levin 1981). This has been what has made the patch dynamic paradigm (Levin and Pain 1974, Pickett and White 1985) so seductive, and has led to the increasing popularity of metapopulation approaches such as those discussed in several of the chapters that follow.

In a wide-ranging paper (Chapter 15), Clark builds a bridge between the theory of metapopulations and perennial plant dynamics. The metapopulation approach (Levins 1970, Levin and Paine 1974, Hanski and Gilpin 1991) treats the landscape as a mosaic of patches, explicitly considering both the demography of the patch population and the internal dynamics of the patch. Clark, building on his earlier work, creates a foundation for his investigations through a relationship between the density (x) of trees in a stand and the area (A) per individual. Clark in effect assumes

$$x = \kappa/(A + c),$$

where κ and c are constants; more precisely, he makes the equivalent assumption that

$$d(\log x)/da = -F^* (d(\log A)/da),$$

where $F = xA/\kappa$ and a is stand age. As a increases, so does A; in the limit, the measure of crowding (F) tends to unity, and hence gap area tends to zero. Clark thus develops a theory relating gap dynamics to the internal dynamics of the canopy, and uses the approach to determine optimal life history strategies in these disturbance-mediated environments.

Gap information of the type described in Clark's paper may be the dominant process in certain systems, at least on short time scales. But broader scale processes, such as fire, also play a critical role in the dynamics of forests. The modeling of the spread of a fire, built largely on physical models (Rothermel 1972), has occupied a central role in the literature for two decades. In REFIRES, a regional fire regime simulation model, Davis and Burrows (Chapter 16) complete the loop by interfacing models of fire spread with ignition models and geographical information systems. The result is a tool for relating fire spread to stand succession models, and to aspects of landscape fragmentation and heterogeneity, which influence the length of the period between fires.

Holt and Gaines (Chapter 17) show the power of the patch dynamic approach in dealing with a fragmented old-field system, an experimental archipelago of habitat islands separated by mowed areas. By careful controlled experiments, involving a variety of distributions of patch sizes and configurations, Holt and Gaines investigate the successful dynamics of patches and their interrelationships via dispersal, as well as patch utilization by small mammals. Studies of this sort provide a realization of the original hope for the theory of island biogeography, allowing investigation of the importance of the size and spatial distribution of patches.

Finally, in Chapter 18, Marquet et al. provide an insightful review and synthesis of some of the consequences of patchiness, integrating concepts and data that were presented in earlier chapters and providing an introduction to the effects of patchiness on gene flow and life history theory. The Stommel diagram, which has been an influential but usually abstract guide to the consideration of multi-scale phenomena, is given substance in this paper through a serious attempt to quantify key marine and terrestrial scales. This provides a solid basis for cross-systems comparisons across the marine-terrestrial boundary. The resultant review is an extremely useful companion to the literature.

REFERENCES

Gilpin, M., and I. Hanski (eds.). 1991. Metapopulation dynamics: Empirical and theoretical investigations. Reprinted from the *Biological Journal of the Linnean Society* 42(1 & 2). Academic Press, Orlando, FL.

Levin, S. A., and R. T. Paine. 1974. Disturbance, patch formation, and community structure. *Proceedings of the National Academy of Sciences* 71(7):2744-2747.

Levins, R. 1970. Extinction. In: M. Gerstenhaber (ed.). *Some Mathematical Problems in Biology*. American Mathematical Society, Providence, RI, pp. 77-107.

Paine, R. T., and S. A. Levin. 1981. Intertidal landscapes: disturbance and the dynamics of pattern. *Ecological Monographs* 51(2):145-178.

Pickett, S. T. A., and P. S. White (eds.). 1985. *The Ecology of Natural Disturbance and Patch Dynamics*. Academic Press, Orlando, FL.

Rothermel, R. C. 1972. A mathematical model for predicting fire spread in wildland fuels. USDA Forest Services, Research paper INT-115, Inter-mountain Forest Range Experiment Station, Ogden, UT.

Steele, J. H. (ed.). 1978. *Spatial Pattern in Plankton Communities*. Plenum, New York, NY.

14

THE OCEAN CARBON CYCLE AND CLIMATE CHANGE:
AN ANALYSIS OF INTERCONNECTED SCALES

Kenneth L. Denman

INTRODUCTION

Many studies of patch dynamics develop from provocative observations: hence, the scales of interest are those at which observations were practical. If further work suggests that patchiness at scales outside the range observed may be important, then the observation capabilities may be expanded into these ranges of scales. Recently, oceanographers have taken on a daunting challenge where the choice of scale selection has been removed. The ocean is important to climate change and global warming—as a storer and transporter of heat and carbon—but our understanding of the operative processes is inadequate to make predictions with the required skill. We cannot choose the observational "window" where we are most capable: we must address all scales that contribute to the global climate. In particular, to assess the role of the marine ecosystem in the ocean carbon cycle, we have had initially to extrapolate to ocean basin scales (10^5 km) from, for example, a few tens of sediment traps (1 m diameter) or water samples (10 cm, based on a 1 L sample). How do we bridge over 9 orders of magnitude to address problems of global scale from water samples typically of 1 L volume?

In this chapter I want to analyze the possible role of the pelagic planktonic ecosystem—the so-called "biological pump"—in the global carbon cycle. The analysis will necessarily involve an evaluation of the relative importance of processes operating at various space and time scales. I take it as axiomatic, a result of two decades of scientific research, that the role of the marine ecosystem in the ocean carbon cycle cannot be addressed without inclusion of physics and geochemistry. In the following sections, I will: briefly review the ocean's role in the global CO_2 cycle and climate change; describe the importance of mesoscale processes in the ocean; examine the main pathways of carbon exchange between the ocean surface layer (in contact with the atmosphere) and the ocean interior; analyze the scales of the physical processes involved in nutrient limitation of the "biological pump"; and review the implications for climate change.

THE OCEAN'S ROLE IN CLIMATE CHANGE

Humankind is conducting a massive geophysical experiment, and we do not have much confidence in our ability to predict the outcome (see IPCC 1990, and references cited therein).

therein). We are pumping gases into the atmosphere that allow short wave solar radiation to reach the earth's surface but reduce the amount of heat radiated back to outer space. A steady state greenhouse effect maintains the surface of the earth at a temperature warm enough for life to exist, but any imbalance that lasts long enough can result in climate change. An increased heat retention should result in warming of the earth-ocean-atmosphere system. One gas, carbon dioxide (CO_2), resulting from fossil fuel and biomass burning, is responsible for more than 50% of the increase in the greenhouse effect caused by anthropogenic inputs (Ramanathan et al. 1985), with a potentially greater contribution in the future because of its long lifetime in the atmosphere. In the last century, we have released so much CO_2 into the atmosphere—from burning fossil fuels and from burning down forests—that the level of CO_2 in the atmosphere (Figure 1) is apparently higher than at any time during the last 160,000 years (Figure 2). The lower panel of Figure 2 shows the atmospheric CO_2 concentration for the last 160,000 years, reconstructed from an ice core from Vostok, Antarctica (Barnola et al. 1987). The upper panel shows regional air temperature, estimated from deuterium isotope results from the same ice core. Clearly, for that period, air temperatures are closely correlated with atmospheric CO_2 concentration. Peak-to-peak glacial-interglacial warming appears to be about 10 C degrees. Predictions suggest that, because of the current buildup of atmospheric CO_2, we can expect a global warming of 2 - 4 C degrees by the year 2050.

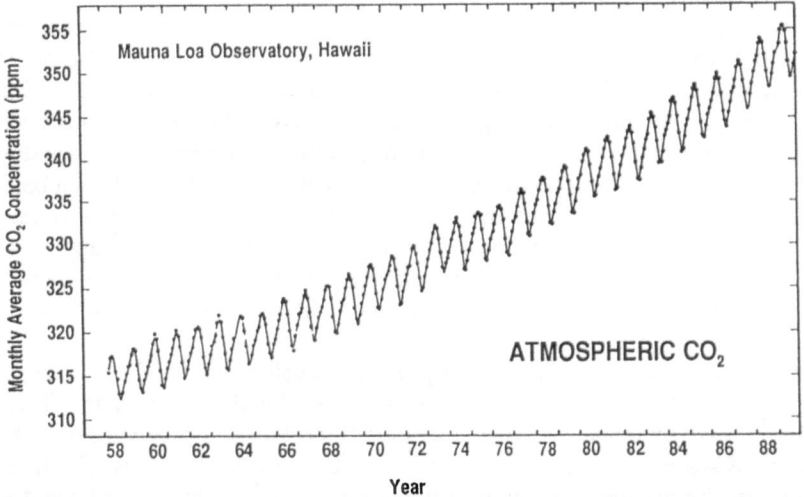

Figure 1.   Change in CO_2 concentration between 1958 and 1988 at the Mauna Loa Observatory, Hawaii (from Keeling 1989).

The ocean stores and transports vast quantities of heat and CO_2. Figure 3 (redrawn and corrected from the IPCC report, based on personal communication with U. Siegenthaler) shows a current estimate of the global reservoirs and fluxes of carbon. There is more than 50 times the CO_2 or equivalent in the ocean (not including that stored in the sediments) compared with that in the atmosphere, i.e., about 39000 Gt in the ocean versus 750 Gt in the atmosphere (1 Gt = 10^{12} kg). Therefore, if the same amount of CO_2 were injected into the atmosphere and the ocean, the average change in the ocean would be about 50 times smaller than in the atmosphere—not easy to detect. In fact, only 60% or less of the CO_2 released into the atmosphere from fossil fuel burning and deforestation remains in the atmosphere; most of the rest is hypothesized to end up in the ocean. Global estimates of net CO_2 input to the ocean from the atmosphere, based on partial pressure differences at the air-sea interface, are consistent with this hypothesis, but their spatial distribution is inconsistent with that inferred from atmospheric models calibrated against the distributions of radioactive tracers (Tans, Fung and Takahashi, 1990). The relative storage capacity of the ocean for heat is even greater: the heat capacity of the ocean is at least 1000 times that of the whole atmosphere. Thus, heat that penetrates into the deep ocean from the atmosphere can readily be stored and reduces the rate of increase of atmospheric temperature that might arise due to the greenhouse effect. The heat required to cause the projected 4 C degrees increase in the atmosphere would, if added to the ocean, cause only an increase of 4 one-thousandths of a C degree.

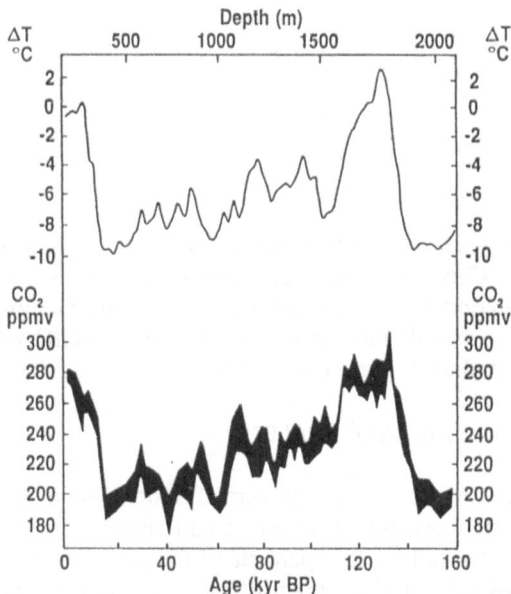

Figure 2. CO_2 concentrations (bottom) and estimated temperature changes (top) during the past 160,000 years, as determined from the ice core from Vostok, Antarctica (Barnola et al. 1987). Temperature changes estimated from measured deuterium concentrations (from IPCC 1990).

The ocean, then, is buying us time in this projected warming by moderating the buildup of CO_2 and heat in the atmosphere. In the context of CO_2 as presented in Figures 1 - 3, we must ask the critical question: How long can the ocean continue to accept more CO_2 from the atmosphere than it gives back to the atmosphere—10 years, 100 years, or 1000 years? At present, our understanding of how the ocean stores, transports, and transforms carbon, and how it stores and transports heat, is inadequate to allow us to predict future climate change with any acceptable level of confidence.

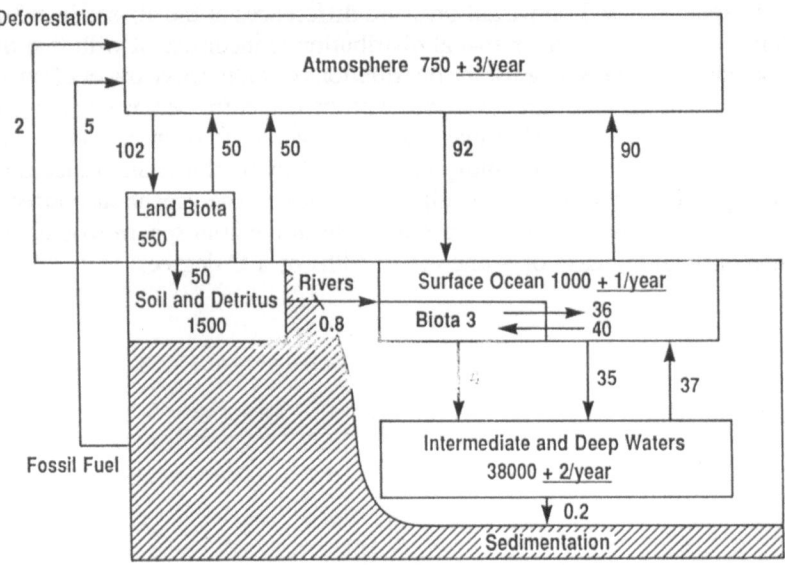

Figure 3. Global carbon reservoirs and fluxes, representative of typical literature values for the present-day situation. Fluxes are gross annual exchanges, and numbers underlined indicate net annual CO_2 accumulation due to human activity. Units are gigatons of carbon (GtC; 1 Gt = 10^9 metric tonnes = 10^{12} kg) for reservoir sizes and GtC yr^{-1} for fluxes. Redrawn with correction (U. Siegenthaler, pers. comm.) from IPCC (1990).

THE OCEAN IS IN MOTIONS, AT ALL SCALES

The global carbon cycle, in its present perturbed state shown in Figure 3, is almost in balance: the net fluxes between the atmosphere and surface ocean or between the surface ocean and ocean interior are only a few percent of the gross fluxes in either direction. To compound the problem of determining small differences, the gross and net fluxes are patchy in both time and space (e.g. Tans et al. 1990). Moreover, carbon undergoes many transformations while in the ocean (e.g., Broecker and Peng 1982, Berger et al. 1989). The roughly 4 GtC yr^{-1} that exits the surface ocean biota box (Figure 3) leaves as particles, but, as they sink into deeper water, they are constantly being "remineralized" by bacteria into dissolved inorganic forms. The 35 GtC yr^{-1} leaving and the 37 GtC yr^{-1} returning to the surface ocean box is dissolved matter, mostly inorganic. These large dissolved fluxes are transported by physical advective and mixing processes.

To improve prediction of climate change, it is necessary to understand the physical transport processes well enough to develop quantitative models. Achieving this goal would seem to be straightforward, given an averaged knowledge of currents on ocean basin scales, as pictured for the Pacific Ocean in Figure 3 of Tabata (1975). There are gyres and boundary currents, but it should be possible to estimate the flux of any substance by calculating the

217

vector product of the concentration and the current at an array of sites. Vertical fluxes could be estimated from convergences and divergences of horizontal fluxes by continuity.

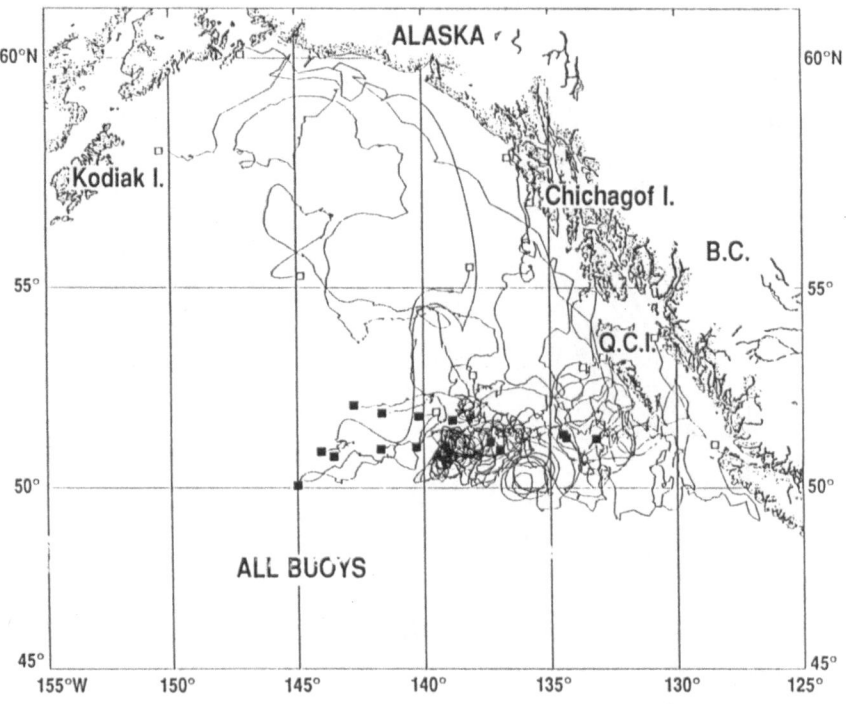

Figure 4. Daily positions of 24 satellite-tracked drifters (drogued at 100 to 120 m) deployed in the northeast Pacific Ocean in June and October 1987. Solid symbols denote start locations, open symbols end locations. Tracks cover period up to 1 April 1989 (from Thomson et al. 1990).

The gyres and boundary currents exist in an average sense, but their direct measurement is non-trivial. Figure 4 (from Thomson et al. 1990) shows the tracks of 24 satellite-tracked drifters, drogued at 100 to 120 m depth, in the northeast Pacific Ocean. Many of the drifters did eventually follow the anticlockwise Alaskan Current, but the dominant characteristics of the tracks are the eddies and meanderings on scales of about a degree latitude (~ 100 km). Mean currents around the eddies were 13 cm s^{-1}, mean propagation speed of the eddies was 1.5 cm s^{-1}, and mean currents on scales larger than the eddies, except in the coastal boundary currents, were less than 5 cm s^{-1}, consistent with mean currents derived from average geopotential fields (e.g., Dodimead et al. 1963). Speeds within eddies are clearly larger than the mean flows: mean currents can be observed only with appropriate averaging and can be modeled correctly only by resolution or appropriate parameterization (as yet undeveloped) of these eddies. These eddies do not result from wind forcing: Figure 5 shows a similar eddylike nature in drifter tracks at 700 m in the North Atlantic. The tracks are smoother, probably due to the absence of wind effects, but the eddy motions with length scales around 100 km still dominate. The eddies or gyres penetrate vertically for hundreds of meters, but they are highly baroclinic with the result that surface motions are not necessarily well-correlated with deep motions.

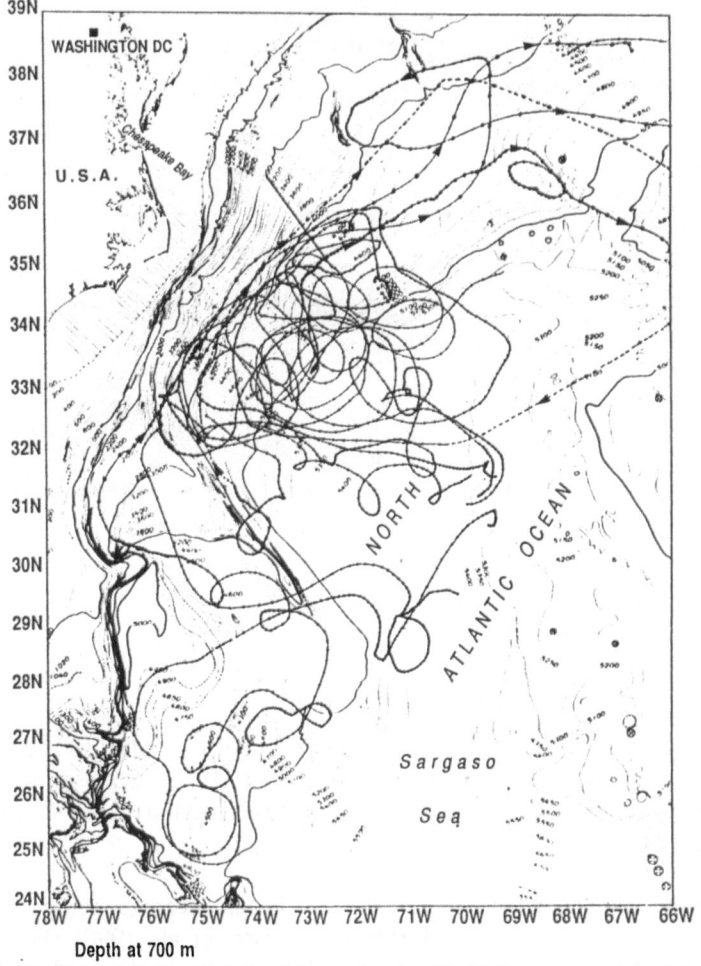

Depth at 700 m

Figure 5. Trajectories of floats that had been in the Gulf Stream and that lie at a depth of about 700 m (from Rossby et al. 1983).

In analogy with weather patterns in the atmosphere, we refer to these motions with spatial scales of 100 km and time scales of 10 - 20 days as the ocean mesoscale. Atmospheric mesoscale motions have a length scale of about 1000 km and a time scale of several days. To observe and model ocean mesoscale motions requires horizontal spatial resolution on a much finer scale than for the atmosphere. Ocean general circulation models that resolve mesoscale eddy structure require a grid spacing of 1/4 to 1/2 degree latitude (e.g., Cox 1985, Semtner and Chervin 1988; Webb et al. 1991) versus 3 - 5 degrees in the atmosphere. In two dimensions, ocean models then require about 100 times as many grid points to resolve the mesoscale structure. If the time step can be 10 times as large, then ocean models appear to need about 10 times the computing power. However, the spinup time for ocean models to approach equilibrium behavior is usually longer than for atmospheric models. Embedding equations into advective models for carbon conservation and transformations will require additional computing power, and preliminary modeling is underway (Fasham et al. 1990, Sarmiento et al., in press).

PATHWAYS FOR REMOVAL OF CARBON FROM THE SURFACE OCEAN

The increase in atmospheric CO_2 concentration over the last century, from about 290 ppm to 355 ppm, has been accompanied by an increase in total CO_2 in the ocean surface layer of about 40 μmols kg^{-1} (Brewer 1978, Chen and Millero 1979). Nevertheless, a net flux of carbon from the surface ocean to the ocean interior has maintained a partial pressure pCO_2 difference at the sea surface sufficient to drive a continued net flux of CO_2 from the atmosphere to the ocean, currently estimated to be about 2 GtC yr^{-1}. Clearly, understanding whether the ocean can continue to receive excess CO_2 from the atmosphere requires an understanding of the major pathways by which carbon is removed from the surface ocean to the ocean interior, whereby the net positive pCO_2 difference between the atmosphere and the ocean can be maintained.

The major pathways are two: the "biological pump" and what we may also refer to as a "physical pump." The biological pump can be described to any degree of complexity (see Longhurst 1991), but for our purposes it refers to that portion of marine planktonic photosynthetic (primary) production that eventually leaves the photic zone (light-lit surface ocean layer) as particulate organic matter—phytoplankton detritus, zooplankton faecal pellets, etc. Marine primary production is nutrient, not carbon, limited. A large fraction of primary production is supported by nutrients recycled by the microplankton food web within the photic zone, but a small fraction (4 of a total of 40 GtC yr^{-1} in Figure 3) either sinks out of the surface layer by means of gravity or (another small fraction) is released as particles at depth by zooplankton during their daily migration cycle between the photic zone (where they feed) and the deeper waters. This small fraction (4/40) of "export" production must be supported by an influx of "new" nitrogen to the photic zone, primarily from the ocean interior by upwelling and upward mixing, but also to a lesser extent from elemental atmospheric nitrogen entering the surface ocean, where it is fixed into biologically usable form by algae such as *Trichodesmium* (Carpenter and Romans 1991).

The physical pump refers to the removal of dissolved carbon by deep water formation, i.e., convective sinking due to extreme cooling of surface waters at high latitudes during winter, and by other processes causing large-scale subduction of water from the surface ocean. The 35 GtC yr^{-1} attributed to this process in Figure 3 sinks primarily as dissolved inorganic carbon (DIC). If recent estimates (obtained with a controversial new technique, Sugimura and Suzuki 1988) of surface to deep ocean differences in dissolved organic carbon concentrations (DOC) prove to be robust and widespread, then perhaps as much as 5 GtC yr^{-1} is transported as DOC by the physical pump. In the following discussion, let us assume that the 35 GtC yr^{-1} comprises 30 DIC plus 5 DOC.

A perusal of schematic diagrams of the global carbon cycle similar to Figure 3 published over the last decade would reveal most of the flux estimates to be ephemeral (e.g., Moore and Bolin 1986). In other words, we do not know them with much confidence or certainty (see JGOFS Science Plan, SCOR 1990 for a discussion of the uncertainties). For example, the 4 GtC yr^{-1} in Figure 3 that represents the flux of export particulate production may be as small as 1 (Moore and Bolin 1986) or as large as 20 (Packard et al. 1988). The 5 GtC yr^{-1} of DOC via deep water formation may be less than 1 if concentrations estimated from the new techniques turn out to be spurious. The 30 GtC yr^{-1} of DIC, also via deep water formation, translates into a downward volume flux of water of about 40 x 10^6 m^3 s^{-1} (40 Sverdrups). North Atlantic Deep Water formation accounts for about 15 - 20 Sverdrups (e.g., Gordon 1986). The rest is assumed to come from other sources, primarily formation of Antarctic Bottom Water (AABW) in the Weddell and Ross Seas, but recent studies (Poisson and Chen 1987) suggest that most AABW is formed under ice when the waters would not be in equilibrium with atmospheric CO_2 concentrations. Thus, an uncertainty in the flux via the physical pump of $\pm$ 10 GtC yr^{-1} would seem reasonable. Our estimates of the total downward flux of carbon from the surface ocean layer are therefore uncertain to tens of GtC yr^{-1}. Compare that uncertainty with the net flux of CO_2 from the atmosphere to the ocean of 2-3 GtC yr^{-1}.

THE ROLE OF THE "BIOLOGICAL PUMP" IN CLIMATE CHANGE

The total global oceanic primary production of 40 GtC yr^{-1} (Figure 3) may be as much as 20 times that entering the ocean from the atmosphere due to anthropogenic buildup. The question naturally arises: Has the biological pump responded to this buildup by increasing total and export production? The short answer, according to geochemists, is "No." They argue that the fluxes through the ocean biological pump are limited not by carbon, which is always in excess, but by the flux of nutrients from the ocean interior up into the photic zone. Increasing the carbon concentrations in the surface layer does not result in increasing export production, because the flux of nutrient (nitrate) entering the photic layer from below is closely controlled by physical advection and mixing processes. Unless the physical processes transporting nitrate into the photic layer from below change as the climate changes, the export production cannot change regardless of how much excess carbon enters the photic layer.

What are the major physical processes controlling the flux of nitrates up into the photic layer? It is a commonly held assumption that the bulk of the supply of nitrates to the photic layer is accomplished by the thermohaline flow pattern whereby the localized sinking of water at high latitudes in North Atlantic winter drives a distributed return flow to the surface oceans of waters rich in carbon and nutrients. Various indirect studies have led to estimates of a transport of 15 - 20 Sverdrups (Gordon 1986) associated with this thermohaline conveyer belt (Figure 6) and a timescale of centuries (Broecker and Peng 1982). If the nitrates supporting the "export" production are supplied primarily by this circulation pattern, then the time required for the biological pump to respond to circulation changes resulting from climate change would be centuries. For climate change projections over 10 and 60 years (the year 2000 and 2050 scenarios), the assumption of a constant biological pump would seem to be valid.

We can, however, estimate whether the thermohaline conveyer belt is capable of supplying sufficient nitrates to support contemporary estimates of "export" primary production. Consider an export production from Figure 3 of 4 GtC yr^{-1} = 4 x 10^{15} gC yr^{-1} = 1.3 x 10^8 gC s^{-1}. Carbon and nitrogen are taken up by phytoplankton and incorporated into their cellular bulk in an approximately constant ratio called the Redfield ratio. A recent estimate of the Redfield ratio for C:N of 6.44 (Takahashi et al. 1985) yields a value of 2 x 10^7 gN s^{-1} as the required flux of nitrate to the photic zone to support the export production of 4 GtC yr^{-1}. If we take a deep water nitrate concentration of 30 mmols m^{-3} = 0.4 gN m^{-3}, then the volume flow required to supply the necessary nitrate uptake is (2x10^7 gN s^{-1}) / (0.4 gN m^{-3}) = 50 x 10^6 m^3 s^{-1} = 50 Sverdrups. The thermohaline flow associated with North Atlantic Deep Water formation is estimated as 15 - 20 Sverdrups (Gordon, 1986), i.e. the thermohaline conveyer can supply perhaps only one-half the nitrate required for the global export ocean primary production.

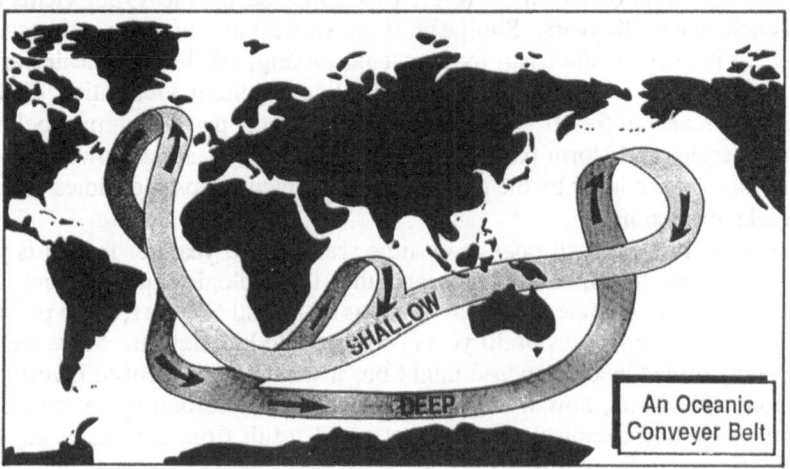

Figure 6. The long-term (centuries) thermohaline conveyer belt, whereby North Atlantic Deep Water formed near Iceland traverses the deep waters of the oceans and returns to the surface layer in a distributed fashion (from Steele 1986, after Gordon 1986).

NUTRIENT SUPPLY/CARBON TRANSPORT TIMESCALES

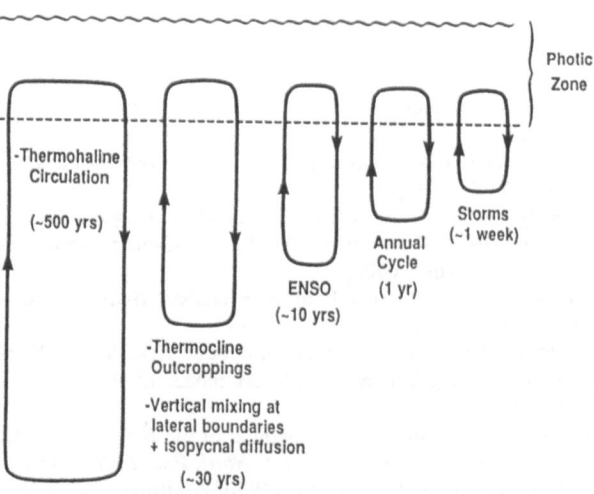

Figure 7. A schematic drawing of major vertical circulation processes with their associated time scales for nutrient supply and carbon transport.

The missing surface nitrate must be supplied by vertical exchange processes on shorter time scale over shallower depth ranges. Candidate processes are shown schematically in Figure 7, with time scales ranging from 30 years to 1 week. Thermocline outcroppings occur where thermal and density surfaces several hundreds of meters deep at low latitudes come to the surface at mid to high latitudes, perhaps 1000 km away. A conservative estimate of isopycnal eddy diffusion of 10^4 m^2 s^{-1} (e.g., Cox 1985, Sarmiento 1986) yields a time scale for vertical exchange of 30 years. Similarly, if the vertical mixing occurs at basin boundaries and reaches the interior of the basin by isopycnal mixing, the limiting scales are the same, yielding the same time scale of 30 years. El Niño-Southern Oscillation (ENSO) events reappear on time scales of 5 - 10 years, winter mixing to the permanent pycnocline occurs on an annual time scale, and storm passages have a time scale of about onw week. Mixing of nutrients into the surface layer by the passage of subsurface mesoscale eddies would also have scales of weeks to months.

Since these processes all operate on time scales of 30 years or less, it is possible that they could be altered by climate change and that the biological pump would respond on similar time scales. For climate change projections of 10 and 60 years, these possible changes to the biological pump on scales of 30 years or less could be relevant. Since we do not even know what sign any feedback response might be, it would be prudent to investigate how the biological pump functions, how it contributes to the ocean carbon cycle, and how it might respond to physical and chemical changes that could result from climate change.

ACKNOWLEDGMENTS

I have benefited from discussions with numerous colleagues on this topic. In particular, I wish to thank M. Fasham, C. Garrett, A. Longhurst, T. Platt, J. Sarmiento, U. Siegenthaler, and C.S. Wong for clarifying my thinking on many points.

REFERENCES

Barnola, J.M., D. Raynaud, Y.S. Korotkevich, and C. Lorius. 1987. Vostock ice core provides 160,000 year record of atmospheric CO_2. *Nature* 329:408-414.

Berger, W.H., V.S. Smetacek, and G. Wefer (eds.). 1989. *Productivity of the Ocean: Present and Past*. J. Wiley and Sons, New York, 471pp.

Brewer, P.G. 1978. Direct observation of the oceanic CO_2 increase. *Geophys. Res. Lett.* 5:997-1000.

Broecker, W.S., and T.-H. Peng. 1982. *Tracers in the Sea*. Lamont-Doherty Geological Observatory, Columbia University, Palisades, New York, 690pp.

Carpenter, E.J., and K. Romans. 1991. Major role of the cyanobacterium *Trichodesmium* in nutrient cycling in the North Atlantic Ocean. *Science* 254:1356-1358.

Chen, G.-T., and F.J. Millero. 1979. Gradual increase of oceanic CO_2. *Nature* 277:205-206.

Cox, M.D. 1985. An eddy-resolving general circulation model of the ventilated thermocline. *J. Phys. Oceanogr.* 15:1312-1324.

Dodimead, A.J., F. Favorite, and T. Hirano. 1963. Salmon of the North Pacific Ocean, Part II: Review of oceanography of the subarctic Pacific region, *Int. North Pac. Fish Comm. Bull.* 13, 195 pp.

Fasham. M.J.R., H.W. Ducklow, and S.M. McKelvie. 1990. A nitrogen-based model of plankton dynamics in the oceanic mixed layer. *J. Mar. Res.* 48:591-639.

Gordon, A.L. 1986. Interocean exchange of thermocline water. *J. Geophys. Res.* 91:5037-5046.

IPCC (Intergovernmental Panel on Climate Change). 1990. *Climate Change: The IPCC Scientific Assessment* (Houghton, J.T., G.J. Jenkins, and J.J. Ephraums, eds.). Cambridge University Press, Cambridge, U.K., 365pp.

Keeling, C.D., R.B. Bacastow, A.F. Carter, S.C. Piper, T.P. Whorf, M. Heimann, W.G. Mook, and H. Roeloffzen. 1989. A three-dimensional model of atmospheric CO_2 transport based on observed winds:

1. Analysis of observational data. In: D.H. Peterson (ed.). *Aspects of Climate Variability in the Pacific and Western Americas*, Geophysical Monographs of the American Geophysical Union 55, pp 165-236.

Longhurst, A.R. 1991. Role of the marine biosphere in the global carbon cycle. *Limnol. and Oceanogr.* 36: 1507-1526.

Moore, B., and B. Bolin. 1986. The oceans, carbon dioxide, and global change. *Oceanus* 29:9-15.

Packard, T.T., M. Denis, M. Rodier, and P. Garfield. 1988. Deep ocean metabolic CO_2 production: Calculations from ETS activity. *Deep-Sea Res.* 35:371-382.

Poisson, A., and C.-T. A. Chen. 1987. Why is there little anthropogenic CO_2 in the Antarctic Bottom Water? *Deep-Sea Res.* 34:1255-1275.

Ramanathan, V., R.J. Cicerone, H.B. Singh, and J.T. Kiehl. 1985. Trace gas trends and their potential role in climatic change. *J. Geophys. Res.* 90:5547-5566.

Rossby, H.T., S.C. Riser, and A.J. Mariano. 1983. The western North Atlantic - a Lagrangian viewpoint. In: A.R. Robinson (ed.). *Eddies in Marine Science*. Berlin, Springer-Verlag, pp. 66-91.

Sarmiento, J.L., 1986. Three-dimensional ocean models for predicting the distribution of CO_2 between the ocean and the atmosphere. In: J.R. Trabalka and D.E. Reichle (eds.). *The Changing Carbon Cycle*. Springer-Verlag, New York. pp. 279-294.

_____, M.J.R. Fasham, R. Slater, J.R. Toggweiler, and H.W. Ducklow. In press. The role of biology in the chemistry of CO_2 in the ocean. In: M. Farrell (ed.). *Chemistry of the Greenhouse Effect*. Lewis Publ., New York.

Scientific Committee on Oceanic Research. 1990. The Joint Global Ocean Flux StudyVm-JGOFS—Science Plan, JGOFS Rept. 5, SCOR, ICSU, Halifax, Canada, 61pp.

Semtner, A.J., Jr., and R. M. Chervin. 1988. A simulation of the global ocean circulation with resolved eddies. *J. Geophys. Res.* 93:15502-15522.

Steele, J.H. 1989. The message from the oceans. *Oceanus* 32(2):5-9.

Sugimura, Y., and Y. Suzuki. 1988. A high-temperature catalytic oxidation method for the determination of non-volatile dissolved orgainc carbon in seawater by direct injection of liquid samples, *Mar. Chem.* 24:105-131.

Tabata, S. 1975. The general circulation of the Pacific Ocean and a brief account of the oceanographic structure of the North Pacific Ocean, Part I - Circulation and volume transports. *Atmosphere* 13:133-168.

Takahashi, T., W.S. Broecker, and S. Langer. 1985. Redfield ratio based on chemical data from isopycnal surfaces. *J. Geophys. Res.* 90:6907-6924.

Tans, P.P., I.Y. Fung, and T. Takahashi. 1990. Observational constraints on the global atmospheric carbon dioxide budget. *Science* 247:1431-1438.

Thomson, R.E., P.H. LeBlond, and W.J. Emery. 1990. Analysis of deep-drogued satellite-tracked drifter measurements in the northeast Pacific. *Atmosphere-Ocean* 28:409-443.

Toggweiler, J.R. 1989. Is the downward dissolved organic matter (DOM) flux important in carbon transport? In: W.H. Berger, V.S. Smetacek and G. Wefer (eds.). *Productivity of the Ocean: Present and Past*. J. Wiley and Sons, New York, pp. 65-85.

Webb, D.J., et al. ("The FRAM Group"). 1991. An eddy-resolving model of the Southern Ocean, *EOS—Trans. Am. Geophys. Union* 72:169.

15
SHIFTING MOSAIC METAPOPULATION DYNAMICS

James S. Clark

INTRODUCTION

Many of the features long viewed as among the most important influences on tree population dynamics have only recently begun to be incorporated in analyzable models. These processes tend to operate at several spatial and temporal scales, and they represent factors that produce and/or depend on heterogeneity. Some of these considerations include:

- *Growth-dependent thinning:* Thinning rates at local scales (10^0 to 10^2 m^2) are determined by growth rates. There is no "carrying capacity" at such scales in the traditional sense, because plants are continually growing and, therefore, thinning.

- *Changing importance of density-dependent and density-independent mortality:* The relative importances of different mortality risks change with canopy coverage, and they influence recruitment. Thinning caused by crowding has different demographic consequences than do juvenile death and senescence.

- *Episodic recruitment:* Seedling establishment is locally episodic, being associated with "disturbance", i.e. specific types of mortality.

Numerical "gap" models (Botkin et al. 1972, Shugart 1984) assume and/or produce these features. They have served as a valuable tool for addressing many aspects of tree population dynamics (Huston and Smith 1987) and nutrient cycling (Pastor and Post 1986). Development of the complementary analyzable models needed for greater understanding of general relationships and for analysis of recruitment has lagged behind these efforts.

Here I discuss recent developments using metapopulation models specifically aimed at understanding dynamics of long-lived perennial plants. These models accommodate the above-mentioned features by treating the landscape as a collection of patches. The local (within-patch) dynamics consist of episodic recruitment events separated by periods dominated by thinning. Episodic recruitment is treated as a stationary renewal process (Cox 1962, Clark 1989). Intervening thinning phases are described by a model that builds on conceptual arguments of Yoda et al. (1963), Norberg (1988), Tait (1988), Valentine (1988), and Clark (1990), where thinning is driven by individual plant growth coupled with density-independent (DI) mortality (Clark 1992a). The metapopulation model assumes a landscape containing a collection of these patches, each of which is identical in all parameters except patch age (Clark 1991a, b). The metapopulation model is continuous in time and state, but discrete in space. By assuming that, except for dispersal, patches are independent, analytical versions of the model can be analyzed to address the consequences of local dynamics for metapopulation structure and life history evolution. The model is most closely related to those of Hastings and Wolin (1989), but it shares attributes with several others (e.g. Shmida and Ellner 1984, Warner and Chesson 1985, Metz and Diekmann 1986, Comins and Noble 1985, Armstrong 1989). The approach presented here differs from these related models primarily in the way local dynamics are incorporated into metapopulation structure to better reflect unique features of tree life history and interactions.

First, I use the model to show the dependency of local dynamics, including gap formation and net ecosystem production (NEP), on the growth of individual plants. These relationships establish the scale of autonomous heterogeneity. I then explore causes and consequences of patchiness at the metapopulation scale.

LOCAL DYNAMICS

Growth, Thinning, and Gap Formation

The local patch supports an even-aged monoculture of plants (a "stand") of density $x(a)$ that becomes established at age $a = 0$. Plants subsequently grow, the stand becomes crowded, and thinning ensues. Late in life growth rates slow, and senescence occurs. The model assumes that density-dependent (DD) thinning is driven by the degree of crowding in the stand, as measured by the exclusive crown-area projections of individual trees $A(a)$ summed over the number of trees in the stand. For simplicity, I assume that plant growth rate does not respond to crowding.

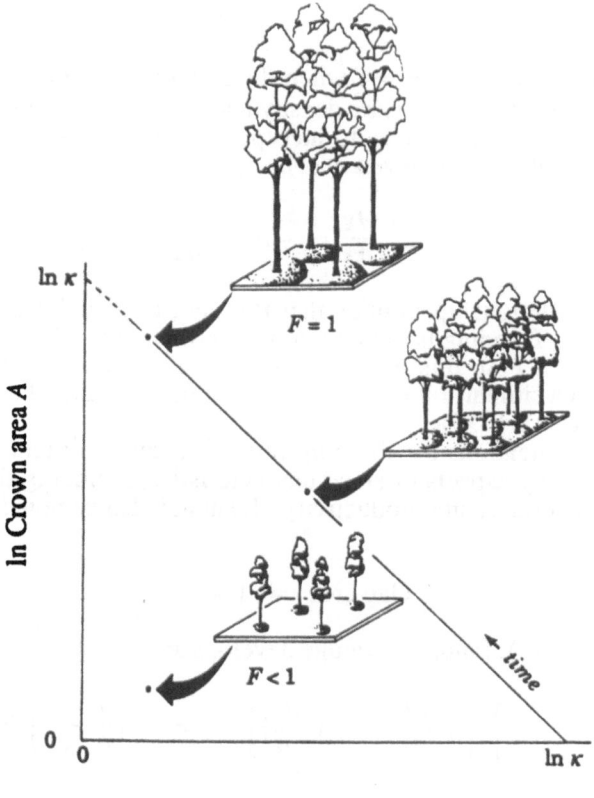

Figure 1. In a crowded even-aged stand, stand density x is proportional to the reciprocal of exclusive crown area A, described by $F = 1$ in equation (3). Open stands fall below the solid line described by $Ax = \kappa$. Redrawn from Clark (1990).

To begin with the simplest case, assume a crowded stand of growing trees. Because all of the projected area is filled, per capita mortality rate proceeds at a rate equal to the per area rate of crown area increase:

$$\frac{1}{x}\frac{dx}{da} = -\frac{1}{A}\frac{dA}{da} . \tag{1}$$

When this condition holds, density is proportional to the reciprocal of area per individual, $x \propto 1/A$. Then density can be expressed in terms of individual plant growth (Figure 1).

The relationship (1) is not realistic if the canopy is interrupted, which can occur if initial seed densities are low or at any time in the life of the stand if a disturbance results in mortality in excess of density-dependent thinning. Thinning rates should be lower than growth rates in an open stand, because growth does not immediately lead to mortality. Conversely, initial seed stocking may be extremely high, in which case we expect that thinning rates might exceed (1). Canopy closure of the stand can be defined as

$$F(x, A) = \frac{x A}{\kappa}, \tag{2}$$

where κ is a constant that makes $F = 1$ when the stand is fully occupied. κ depends on plant shape and the units used to measure density and crown area (Norberg 1988, Valentine 1988, Clark 1990). The simplest way to incorporate crowding effects into the thinning process is to use it to scale the rate given in (1),

$$\frac{1}{x}\frac{dx}{da} = -F \ \frac{1}{A}\frac{dA}{da} \qquad\qquad F > 0 \tag{3}$$

(Clark 1990). This relationship implies that thinning rate declines with the degree of crowding in the stand. Thinning is more rapid than growth rate when $F > 1$ and vice versa (Figures 2a, b). Use of F in this manner is motivated by the way in which density is observed to approach a similar thinning curve from a range of initial densities (Yoda et al. 1963, Harper 1977, Clark 1990).

Beyond its implications for thinning rates, F is an important quantity in its own right. It influences many aspects of stand characteristics, including thinning, understory conditions, recruitment rates, and productivity. It should also be noted that gap area is the complement of F,

$$\text{Gap area} = 1 - F$$

From equations (2) and (3), canopy coverage develops according to

$$\frac{dF}{da} = \frac{1}{\kappa}\left[x\,\frac{dA}{da} + A\,\frac{d\,x}{da}\right] = \left(\frac{1}{A}\frac{dA}{da}\right) F\,(1 - F). \tag{4}$$

From this relation we see that dF/da is logistic for a linear growth rate in A, and sigmoid in any case (Figure 2). F tends to unity over time, at which point (1) is approximately true. Thus we can express crowding in terms of growth, and our result shows that we cannot produce gaps late in stand development on the basis of factors thus far included in the model.

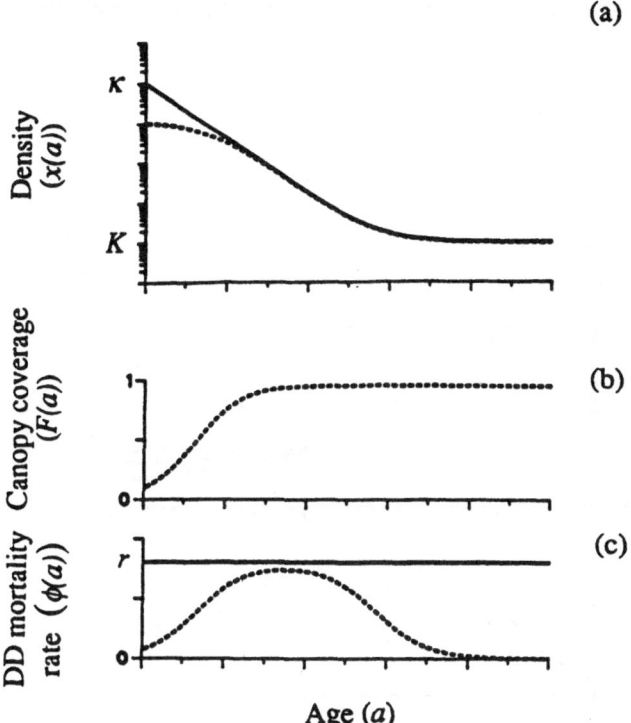

Figure 2. (a) Cohort thinning with complete canopy coverage (solid curve) and when the canopy is initially open (dashed curve). (b) Change in canopy coverage corresponding to the initially open stand in (a). (c) Density-dependent mortality rate for the open stand in (a) compared with the maximum rate of crown-area increase r. In this example, crown area growth rate is non-linear, slowing as plants approach maximum size. From Clark (in press, a).

How, then, do we produce heterogeneity? The answer is non-linear plant growth. Decreasing growth rate with age means that density-dependent mortality can never reduce density to zero. As $A(a)$ approaches a maximum crown-area projection A_m, density can only decline to $x(a) \to \kappa/A_m$, and F remains near unity. Even approaching this limit, canopy coverage does not decrease (equation (4), Figure 2b). The equilibrium case of complete canopy coverage is stable, however, only as long as growth is positive. As growth rate declines with age, complete canopy coverage tends toward neutral stability, and any small amount of DI mortality is sufficient to open the canopy. Canopy gaps result because canopy coverage becomes increasingly sensitive to DI mortality as growth rates slow. Mortality is no longer compensated by individual plant growth. A model that accommodates the interaction between DD and DI mortality factors and its consequences for local stand dynamics is discussed in Clark (in press, a).

In addition to demonstrating why canopy gaps occur, the analysis also illustrates the spatial and temporal scale of this autonomous heterogeneity. The temporal scale is governed by non-linear individual growth rates. Canopy gaps begin to form as growth rate slows. The spatial scale is determined by the size of trees at that age. Canopy gaps permit episodic recruitment on patches of a size determined by the crown areas of falling trees.

Figure 3 shows results of an analytical model that incorporates the interaction between density-dependent and density-independent mortality and their combined effects on total mortality rate and gap area.

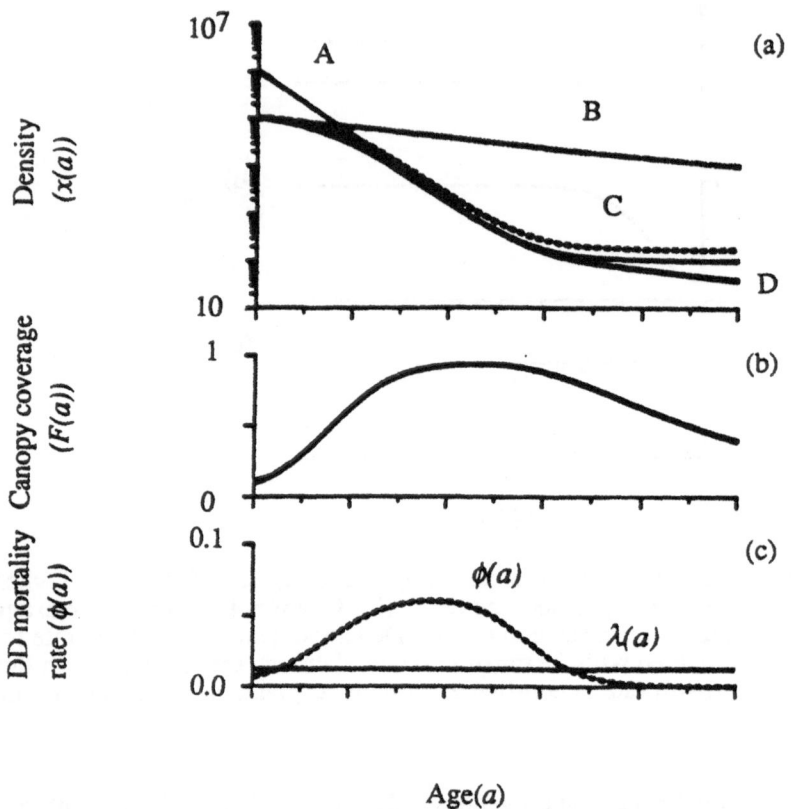

Figure. 3. Effects of different factors on the thinning process when DI mortality rate $\lambda(a)$ is constant. The thinning curves in (a) are: A–no effect of changing canopy coverage or DI mortality; B–only DI mortality, which is taken to be constant; C–effect of changing canopy coverage but not DI mortality; D–effect of changing canopy coverage and DI mortality. (b) shows change in canopy coverage that corresponds to the thinning curve D from (a). DD mortality rate $\phi(a)$ and DI mortality rate $\lambda(a)$ for thinning curve D are shown in (c). From Clark (in press, a).

Other Stand Characteristics

Individual plant growth can also be shown to serve as a basis for understanding net primary production (NPP), net ecosystem production (NEP), and the "self-thinning rule" or "3/2 power law". NPP is the annual rate at which new biomass y is added to the stand. NEP is NPP minus losses that result from mortality:

$$\text{NEP} = \frac{\text{gains due}}{\text{to growth}} + \frac{\text{biomass lost}}{\text{due to mortality}}$$

$$dy/dt = dy_g/dt + dy_m/dt$$

The first term on the right-hand side is NPP. The second term is negative. Both terms depend on exclusive plant area A and plant weight w. Because A increases in two dimensions over time, whereas increase in w follows volume (a three-dimensional quantity), Yoda et al. (1963) proposed that these relationships provide a basis for understanding thinning and plant size in crowded monocultures. The proportionate relationship between weight and area can be written as

$$\frac{dw}{w} \bigg/ \frac{dA}{A} = \frac{3}{2} \equiv \theta.$$

Now, in a crowded stand, where

$$\frac{dA}{A} = -\frac{dx}{x}$$

(equation 1), substitution gives

$$\frac{dw}{w} \bigg/ \frac{dx}{x} = -\frac{3}{2} = -\theta.$$

This relationship is the so-called "3/2 power law" or the "self-thinning rule", usually expressed as $w \propto x^{-\theta}$ and $y \propto x^{1-\theta}$. Although this relationship holds pretty well in the real world, it has sparked much debate. The important assumptions are that a) plants grow isometrically, and b) the stand remains crowded. Rather than assume the relationship between weight and density, we can use the relationship between weight and area to solve for it. In other words, we can retain the first assumption while relaxing the second. By "relaxing" the second assumption, I mean to say that we acknowledge that crowding and DI mortality can modify the relationship between plant weight and density. The relationship between plant weight and area is useful because it allows us to link a number of processes to individual plant growth.

Consider that the total biomass per unit area, call it standing crop y, is given by

$$y(a) = x(a)\, w(a).$$

We can exploit the relationships of the variables on the right-hand side to area A in order to derive expressions for factors related to standing crop. Some examples are given in Table 1. These relationships are extremely simple; even an "ecosystem" quantity like NEP is expressed only in terms of the parameters that govern individual plant growth. Indeed, the plant size at which NEP is maximized is well approximated by an expression that contains a single parameter (Figure 4). Obviously, other factors not included here affect NEP (Clark, (in press, a explores the effects of DI mortality). These simple relationships are nonetheless useful as a basis for comparison when NEP is affected by other complexities.

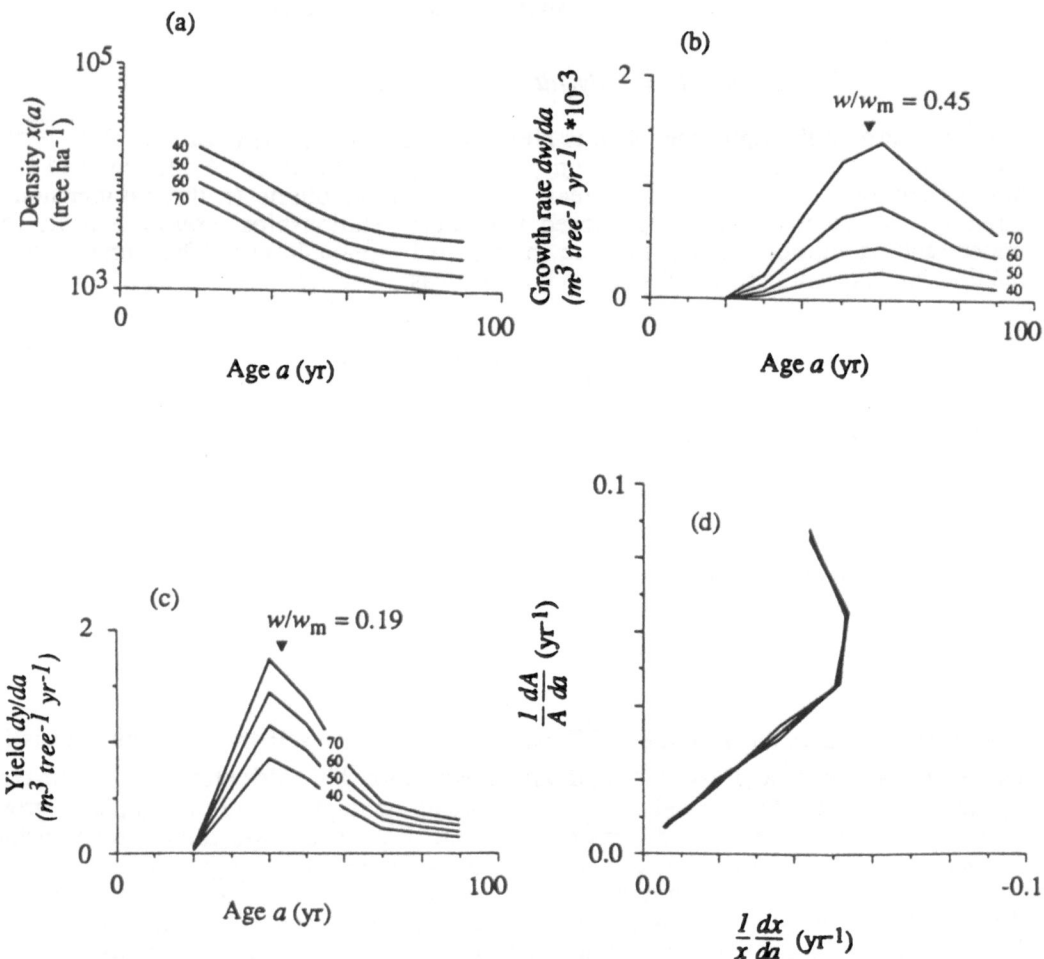

Figure. 4. Thinning, growth, and yield in *Abies balsamea* stands on four different site indices, which express the age at which the average tree attains a height of 50 ft (data from Meyer 1929). Observed constant per capita mortality rates (a) are predicted until individual plant growth rate begins to decline (see (d)). Growth and mortality can also be used to predict relationships between maximum growth increment (b) and maximum NEP (c), which closely match predicted values, w/w_m (see Table 1). If per capita mortality rate is equal to per-area rate of crown-area increase in crowded stands, then the slope of this plot should be -1, as observed once canopy closure obtains (d). From Clark (in press, b).

The self-thinning relationship is given by the slope of the log-log plot of x vs w and x vs y, call them b_w and b_y, respectively. Again, because w, x, and y each depend on crown area, the appropriate substitutions permit solutions using $w \propto x^{b_w}$ and $y \propto x^{b_y}$ to obtain

$$b_w = \frac{dw}{w} \bigg/ \frac{dx}{x}$$

and

$$b_y = \frac{dy}{y} \bigg/ \frac{dx}{x}$$

respectively, These relationships provide a basis for interpreting how individual plant growth influences the shape of the self-thinning curve and the basis for the stages in stand development traditionally identified by forest ecologists (Clark in press, a).

Table 1. NPP and NEP expressed solely on the basis of individual plant growth (from Clark 1990, in press, a).

Variable	Rate equation	Substitute for density
NPP (dy_g/da):	$= x \dfrac{dw}{da}$	$\propto \dfrac{F}{w^{1/\theta}} \dfrac{dw}{da}$
NEP* (dy/da):	$= \dfrac{dy_g}{da} + \dfrac{dy_m}{da}$	$\propto \dfrac{F}{w^{1/\theta}} \dfrac{dw}{da} \left(1 - \dfrac{1}{\theta}\right)$
	$\dfrac{dy_m}{da} = w \dfrac{dx}{da}$	

*Rates for NEP are additions due to growth, dy_g/da, and losses due to mortality, dy_m/da (negative).

Relationship between maximum growth rate and yield:

Rate	Proportionate size at maximum rate
dw/da	$\dfrac{w}{w_m} \approx \left(1 - \dfrac{1}{\theta}\right)^{\theta} = 0.45$
NEP:	$\dfrac{w}{w_m} \approx \left(\dfrac{\theta}{1 + \theta}\right)^{\theta} = 0.19$

Thus, many of the important features of local stand dynamics can be represented as simple functions of individual plant growth. Much realism is intentionally ignored to permit analysis of general relationships. Models based on these relationships can be used to track the changing roles of DD vs. DI mortality, the contribution of non-linear growth, and the time and spatial scales of gap formation (Clark 1990, in press, a). In the next section, we see how these local patterns permit simplification of metapopulation models based on a collection of local patches described by this process.

METAPOPULATION DYNAMICS

There are a number of questions concerning metapopulation behavior that depend on how these local dynamics "scale" to larger areas. What happens to metapopulation structure when local dynamics change? How do we describe the selective pressures on life history within this metapopulation? Local dynamics, particularly gap distributions (Watt 1947, Botkin et al. 1972, Shugart 1984), suggest spatial and temporal scales that can be used as a basis for simplification of metapopulation dynamics. The consequences of individual plant growth for metapopulation dynamics can be extended by analyzing a collection of small patches within each of which processes develop as discussed in the previous section.

These topics have been explored using an analytical model of the process (Clark 1991b). Numerical simulations that relax some of the more confining assumptions are in development. The general assumptions of the approach are:

• Local dynamics consist of an episodic recruitment process, described by a renewal function, separated by longer periods dominated by growth and thinning as per the previous section. "Disturbances" destroy adults and provide recruitment opportunities for new seedlings. Thus, a plant is the same age a as the patch that it occupies. The density within a patch $x(a)$ is constantly changing as plants increase in size.

• Seed production rate $b(a)$ is proportional to (β times as large as) crown-area projection.

• Longer lived species have longer maturation times. This relationship is roughly linear,

$$a_m = \alpha \cdot a_1 \qquad (5)$$

for a deterministic time of senescence a_m and maturation time a_1. α is about 5.5 for trees (Loehle 1988).

• The metapopulation is arranged as a grid of equal size patches linked by dispersal.

• Seeds are evenly dispersed over all patches within some finite dispersal area π.

I begin with some analytical methods applied to the single-species case followed by some simple comparisons with numerical results.

Life History and Demography

The growth rate of the metapopulation depends on the distribution of disturbances across the landscape, thinning and birth rates within local patches, and dispersal. Let π be the number of patches within a dispersal area; $m_\pi(a,t)$ be the renewal density for the number of disturbances to occur within the dispersal area of a patch of age a; $S_s(a)S_p(a)$ be the probability that a plant will not die (neither by disturbance s nor by thinning p) before age a; and $b(a)$ be the age-specific rate of seed production. Then a characteristic equation for metapopulation growth can be written as

$$1 = \int_{a_1}^{a_m} e^{-ra}\, m_\pi(a,t)\, b(a)\, S_s(a)\, S_p(a)\ da\,. \tag{6}$$

Root r is the Malthusian parameter that corresponds to the average metapopulation growth rate. Use of (6), of course, relies on the rather restrictive condition that growth rate r can be treated as a constant. Otherwise the functional form of $S_p(a)$, the survivor function that describes thinning, must vary in order to maintain the equality demanded by (6). One situation in which this condition might temporarily apply is that of a population expanding into a previously unoccupied landscape, where density is sufficiently low such that density-dependent effects are minimal. Under these circumstances, density-dependent survivorship described by $S_p(a)$ would be close to unity for all a. Then (6) can be used to solve for the life-history parameters that maximize fitness, an index of which is r (e.g. Hamilton 1966).

The second use of (6) corresponds to the case where the metapopulation is at equilibrium. At equilibrium, $r = 0$, and the net reproductive rate is on average unity and satisfies

$$R_0 = 1 = \int_{a_1}^{a_m} m_\pi(a,t)\, b(a)\, S_s(a)\, S_p(a)\, da. \tag{7}$$

Because the variance in stand density tends to decrease rapidly with stand age due to crowding effects on mortality rate (previous section), the thinning function $S_p(a)$ (which is derived directly from eqn (3)) can be approximated despite strong density-dependence (see below).

Maturation time/longevity. What kind of life history do we expect given this population structure and the constraint implied by equation (5)? A simple optimization approach can be used to explore the "best" maturation time for a particular combination of disturbance, thinning, and fecundity schedules, given the tradeoffs implicit in equation (5). Reasonable fitness criteria are r for a growing metapopulation and R_0 for a metapopulation at equilibrium. The optimal life history, given the assumed constraints, is given by the maturation time that results in the largest possible values for r or R_0, respectively. At first glance optimization of R_0 at equilibrium seems paradoxical, because equation (7) demands that $R_0 = 1$. It turns out that optimization of (7) results in the ESS life history, because all other strategies result in values of $R_0 < 1$ (Hastings 1978). Moreover, the optima for (6) and (7) can be solved in cases where the equations themselves cannot using the following approach. The optima occur where $r(a_1) = r(a_1 + \delta a_1)$ or $R_0(a_1) = R_0(a_1 + \delta a_1)$, respectively. We use the same method in either case. Let $I(a)$ be the integrand of (6) or (7). Then the optimum must satisfy

$$\int_{a_1}^{\alpha a_1} I(a)\ da = \int_{a_1+\delta a_1}^{\alpha(a_1+\delta a_1)} I(a)\ da.$$

With a little manipulation we obtain

$$\int_{a_1}^{a_1+\delta a_1} I(a)\ da = \int_{\alpha a_1}^{\alpha(a_1+\delta a_1)} I(a)\ da.$$

Dividing both sides by δa_1 and taking the limit $\delta a_1 \to 0$, we have the relation

$$I(a_1) = \alpha I(\alpha a_1) \tag{8}$$

The difference between applying (8) to optimization of r vs R_0 lies in how thinning survivorship $S_p(a)$ is treated. In the low-density case (optimization of r), this function remains near unity, and the solution is straightforward. At equilibrium (optimization of R_0), it is still possible to approximate the parameter combination that gives $R_0 = 1$ provided canopy coverage is close to complete ($F = 1$) by the time plants reach maturation age a_1 (Clark, in preparation). Seed production and dispersal will result in highly variable initial densities, which subsequently determine juvenile thinning rates. For reasons given in the foregoing section on growth, gap formation, and thinning, density becomes much more predictable once juvenile stages are passed (some numerical examples are given below). In this case

$$S_p(a) \approx \frac{x_0}{x(0)} e^{-\rho a}, \tag{9}$$

Box 1. Optimal maturation time

At equilibrium, the ESS life history (Hastings 1978, Charnov and Berrigan 1991) is that which optimizes (7). Assume a metapopulation of plants that increase in crown area at rate γ and seed production is proportional to canopy area. Then age-specific fecundity is

$$b(a) = \beta A_0 e^{\gamma a} \qquad\qquad a_1 \le a \le a_m$$

for initial crown area A_0 and per-area seed production rate β. Given the relationship among growth, crowding, and thinning that permits the approximation (9) together with (1) and (2), at equilibrium $\rho \approx \gamma$ and the product $S_p(a) b(a)$ is close to

$$\frac{\kappa \beta}{x(0)}$$

Now embed these local dynamics within a landscape where any given patch experiences a disturbance with constant probability λ. Then $S_s(a) = e^{-\lambda a}$ and $m_\pi(a) = \pi \lambda$ (Clark 1991a). Substitution in (7) gives an estimate for the expected density immediately following disturbance

$$x(0) \approx \kappa \beta \left[e^{-\lambda a_1} - e^{-\lambda \alpha a_1} \right]. \tag{1.1}$$

The integrand $I(a) = \dfrac{\kappa \beta}{x(0)} e^{-\lambda a}$ is substituted in (8) to solve for the optimal life history

$$a_1{}^* \approx \frac{\ln \alpha}{(\alpha - 1) \lambda}. \tag{1.2}$$

It is important to recall that this approximation is made possible by the assumptions that the effects of crowding on thinning rate are sufficient to cause convergence to the thinning line (i.e. equation (1) holds) by age a_1. Figure 6 compares the outcome of competition between this vs. alternative maturation times.

where $x_0 = \kappa/A_0$ (for $F = 1$ in equation (2)), and $x(0)$ is the density of seeds that fall on a patch of age $a = 0$.

Box 1 gives an example calculation of an optimal maturation time for linear growth rate, where thinning is driven by growth ($F = 1$, and equation (1) is approximately true), seed production rate is proportional to crown area, and disturbance occurs with constant probability λ. Clark (1991a) compares estimates from this approach with observed life histories of forest trees.

Dispersal. Several tradeoffs influence optimal dispersal distances in a heterogeneous metapopulation. Dispersal provides for avoidance of severe competition with siblings concentrated near the parent, but it involves the risk of landing in potentially unsuitable sites; the local site may be more suitable for the species than are those far away. This phenomenon has been explored from an ESS perspective by Hamilton and May (1977), Levin et al. (1984), Geritz et al. (1988), and Cohen and Levin (1991) among others.

This tradeoff is complicated by a difference between the investments in propagules that must disperse long distances vs those that remain nearby. For wind-dispersed seed, the carbohydrate reserves necessary for successful seedling establishment preclude long dispersal distances in most species (e.g. *Populus*, which pays the price of short seed viability in exchange for long-distance dispersal potential). Energy allocation to seed must balance the payoff of many small, well-dispersed seeds with limited potential for success against that of fewer well-supplied seeds, each of which has a better survival potential provided it finds a suitable site. Other considerations include investments in structures that increase drag and therefore dispersal distance and in structures needed to attract animal vectors.

Here I do not consider tradeoffs involving dispersal, but I do explore potential ways in which dispersal might bias analytical solutions outlined above and detailed in Clark (1991a, 1991b, and in press, a). In the numerical tests that follow I assess analytical predictions for populations having different dispersal distances.

Population Structure

The structure of the metapopulation can be viewed in any of several ways, depending on the perspective we adopt. I demonstrate three ways, but each is equivalent, emphasizing different aspects of the same metapopulation. They all use the same set of functions that are distilled down to three processes, i) fecundity at the local (within-patch) level, ii) thinning within patches, and iii) a disturbance process that can be viewed as a "local" (disturbance probability as a function of patch age) or a "regional" (patch age structure) process. The three perspectives are those of a metapopulation structured by i) age $X(a,t)$, ii) density $g(x,t)$, and iii) patch location $u(a,t)$. The perspective one adopts depends on the question at hand. To simplify the following discussion, I assume that all patches start are initiated at density x_0 at age $a = 0$. This assumption is subsequently relaxed in numerical simulations.

An age-structured population. This first perspective is one of a population structured by age. Because spatial structure is synonymous with age structure, it is possible to represent the effect of disturbance on metapopulation structure by its effect on plant age structure. Let $X(a,t)$ be the density of a species of age a at time t on this landscape. Then the age structure evolves according to

$$\frac{\partial X}{\partial t} + \frac{\partial X}{\partial a} = -(\rho(a) + \lambda(a)) X(a,t) . \tag{10}$$

This is simply the McKendrick-von Foerster equation (the derivation of which can be found in most demography texts) with mortality identified with either of two independent processes, thinning $\rho(a)$ vs disturbance $\lambda(a)$ ("senescence", described by (5), is ignored for simplicity). The right-hand side describes age-specific density losses. Because cohorts of different ages occupy different patches, (10) implies both age and spatial structure. The boundary condition for production of new individuals is

$$X(0,t) \;=\; \int_0^\infty m_\pi(a,t)\, b(a)\, X(a,t)\; da.$$

$m_\pi(a,t)$ is the rate at which sites come available within the dispersal area that contains π patches. Noting that $X(a,t) = X(0,t - a)\, S_p(a)\, S_s(a)$ and, at equilibrium, $X(0,t) = X(0,t - a)$, by substitution we recover equation (7) for the net reproductive rate.

This treatment differs from a typical age-structured model only to the extent that mortality is separated into two components, and only one of these components is associated with recruitment. As we have already seen this perspective is useful for exploring life history and demography of metapopulations.

Plant density distribution. We can use the same relationships to derive a distribution of densities. Because this approach is not so widely used, I discuss its derivation, an adaptation of the general bookkeeping methods described in Metz and Diekmann (1986). Let

$$G((x', x'+\delta x),t) \;=\; \int_{x'}^{x'+\delta x} g(x,t)\; dx$$

be the fraction of the metapopulation between density x' and $x'+\delta x$ (Figure 5). $g(x,t)$ is the p.d.f. of densities at time t. The rate of change in G is determined by the gradient in dx/dt with respect to x,

$$\frac{\partial G}{\partial t} \;=\; g(x,t)\, \frac{dx}{dt}\Big|_{x'+\delta x} \;-\; g(x,t)\, \frac{dx}{dt}\Big|_{x'} \tag{11}$$

Equation (11) represents the difference between the fluxes across the boundaries that define the integral G (Figure 5). Now expand the first term on the right-hand side,

$$g(x,t)\, \frac{dx}{dt}\Big|_{x'+\delta x} \;=\; g(x,t)\, \frac{dx}{dt}\Big|_{x'} \;+\; \frac{\partial}{\partial x}\left[g(x,t)\, \frac{dx}{dt} \right]_{x'} \delta x \;+\; o(\delta x^2),$$

and do the same for the left side,

$$G((x', x'+\delta x),t) \;=\; g(x',t)\, \delta x \;+\; o(\delta x^2).$$

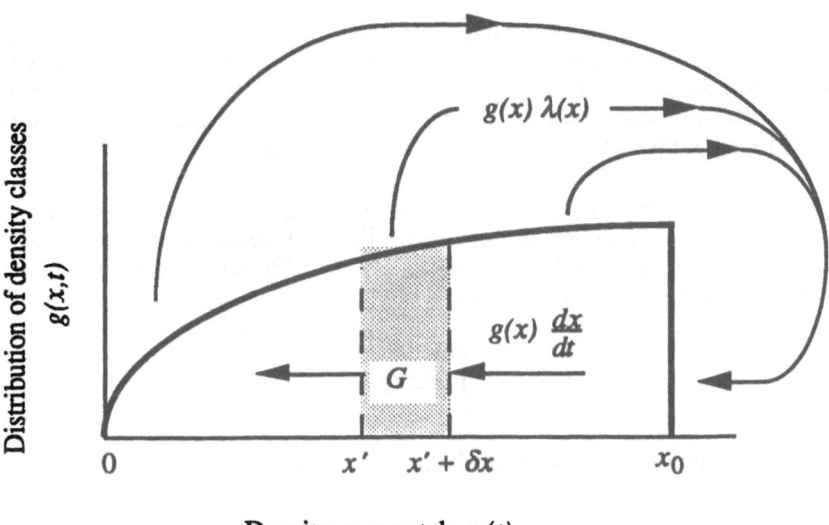

$g(x) \lambda(x)$

$g(x) \dfrac{dx}{dt}$

G

Distribution of density classes $g(x,t)$

0 x' $x' + \delta x$ x_0

Density on a patch $x(t)$

Figure 5. Metapopulation dynamics as a consequence of thinning, represented by advection from x_0 toward the left at rate $g\ dx/dt$, and disturbance, which acts as a sink $g\lambda B$ that returns patches to density x_0. At the boundary x_0, the flux in as a result of disturbance on patches of all ages balances flux away from the boundary due to thinning as reflected in boundary condition (13).

G is approximated as a rectangle g high by δx wide. Substitution into (11) and division by δx gives the relation

$$\frac{\partial g}{\partial t} = -\frac{\partial}{\partial x}\left[g(x,t)\,\frac{dx}{dt}\right].$$

(The minus sign reflects the fact that we calculated density flow to the left in Figure 5, because it is decreasing.) So far we have described how the distribution changes as density "moves through" the rectangle as a consequence of thinning (Figure 5). Rather than move density along the horizontal axis, disturbance returns patches to the x_0 state at a rate $g(x)\ \lambda(x(t))$ (Figure 5). Taken together, these two processes are used to describe the density distribution by advection along the density dimension (thinning) with a sink term reflecting the role of disturbance,

$$\frac{\partial g}{\partial t} = -\frac{\partial}{\partial x}\left[g(x,t)\,\frac{dx}{dt}\right] - \lambda(x)\,g(x,t). \tag{12}$$

The boundary condition is determined by setting the rate at which density passes out of the x_0 class (as a result of thinning) equal to the rate at which it enters (disturbance),

$$-g(x_0,t)\,\frac{dx}{dt}\bigg|_{x_0} = \int_0^{x_0} g(x)\,\lambda(x)\,dx \tag{13}$$

(Figure 5). If (13) were not true, then density would either "pile up" or disappear from the boundary x_0, and the distribution would not be in equilibrium. This perspective is useful, because it permits analysis of how parameter values affect the distribution of density classes across this landscape (Clark 1991b, Hastings 1991). Equation (13) would require modification to accommodate variability in initial seed densities.

Patch age distribution. Because density develops with patch age, we might choose to focus on the ages of patches themselves. This approach was used by Hastings and Wolin (1989), Armstrong (1989), and Clark (1991b). The disturbance process determines the fraction of patches between ages a and $a+\delta a$,

$$\int_{a}^{a+\delta a} u(a,t)\ da.$$

The distribution of patch ages $u(a,t)$ evolves according to

$$\frac{\partial u}{\partial t} + \frac{\partial u}{\partial a} = -\lambda(a)\ u(a,t).$$

The rate of new patch production is

$$u(0,t) = \int_{0}^{\infty} u(a,t)\ \lambda(a)\ da.$$

The distribution of patches at time t is $u(a,t) = u(0,t-a)\ S_s(a)$. At equilibrium, $u(0,t-a) = u(0,t)$, so the equilibrium patch-age distribution is given by

$$u(a) = u_0\ S_s(a)$$

$$= \frac{S_s(a)}{E[a]}, \tag{14}$$

(e.g., Cox 1962) where $1/E[a] \equiv u_0$ is the average disturbance rate and can be derived from the distribution of disturbance events (e.g., Clark 1989).

This patch structure provides a useful perspective on metapopulation structure. Combined with information on local (within-patch) dynamics, it can be used to determine age or density structure. For example, the equilibrium distribution of plant age classes,

$$f_p(a) = \frac{u(a)\ x(a)}{\displaystyle\int_{0}^{\infty} u(a)\ x(a)\ da} \tag{15}$$

and densities,

$$g(x) = u(a)\left|\frac{da}{dx}\right|, \tag{16}$$

can be determined from knowledge of the patch age structure $u(a)$ and the local dynamics within patches summarized by $x(a)$ (Clark 1991b). The advantage of this perspective is the explicit treatment of metapopulation structure as the result of processes at two scales, the local dynamics and the patch-age structure. Nonetheless, these different perspectives are equivalent ways of examining the same process. Box 2 gives an example.

Box 2. Equilibrium distributions of metapopulation densities and ages

The equivalence of the different perspectives can be demonstrated in any of a number of ways. Consider plant ages and densities given by dynamic equations (10) and (12) compared with the equilibrium distributions (15) and (16). At equilibrium, $\partial X/\partial t = 0$ and $\partial g/\partial t = 0$ in equations (10) and (12). For constant thinning rate ρ and disturbance rate λ, integration of (10) gives the average survivorship $X(a) = X_0 e^{-(\lambda + \rho)a}$. Normalization over total density $\int_0^\infty X(a)\, da$ gives the age-class distribution

$$f_p(a) = (\lambda + \rho)\, e^{-(\lambda + \rho)a}. \tag{2.1}$$

Differentiation of (12) at $\partial g/\partial t = 0$ for constant thinning and disturbance rates ρ and λ, gives

$$\rho \left(g + x\, \frac{\partial g}{\partial x} \right) = \lambda g.$$

Rearranging terms,

$$\frac{\partial g}{g} \Big/ \frac{\partial x}{x} = \frac{\lambda}{\rho} - 1,$$

implying

$$g(x) = C\, x^{(\lambda/\rho) - 1}. \tag{2.2}$$

To obtain C, we use (2.2) to obtain $g(x_0) = C\, x_0^{(\lambda/\rho) - 1}$ and boundary condition (13), which results in $g(x_0) = \lambda/(\rho x_0)$. Solving for C from these two expressions and substitution into (2.2) gives the distribution

$$g(x) = \frac{\lambda}{\rho x_0} \left[\frac{x}{x_0} \right]^{(\lambda/\rho) - 1} \qquad 0 \le x \le x_0$$

(Clark 1991b).
    Now consider the patch-age perspective of equations (15) and (16). For the same assumptions, $u(a) = \lambda e^{-\lambda a}$, and $dx/da = -\rho x$. $x(a) = x_0 e^{-\rho a}$ can be solved for a and substituted in $u(a)$. Then (15) is equal to (2.1) and (16) is equal to (2.2).

In order to solve these distributions, thinning must be treated deterministically, and we must decide on a fixed initial density $x_0 = x(0)$, a limitation this approach shares with that of Hastings and Wolin (1989). Fortunately, the relationship between growth and thinning discussed in the previous section on life history permits some reasonable approximations of these quantities. They nonetheless introduce biases that should have their greatest impacts on younger age/higher density classes (Clark 1991b). Below, some examples of these distributions are compared with results from a numerical patch model.

Numerical Results

A number of simplifying assumptions are made in order to permit analytical treatment. These assumptions each contain biases that might importantly influence interpretation. Here I show that analytical results provide good approximations even when the more restrictive assumptions are relaxed in a numerical model. The principal assumptions that are relaxed in this model are:

• *Thinning rates*: The deterministic thinning survivorship and fixed boundary condition for $x(0)$ in the analytical models are relaxed. In the numerical model, thinning rates respond to initial seed densities, changing local (within-patch) densities, and canopy closure (i.e. crowding).

• *Fecundity:* In the numerical model seed production varies according to local dynamics, depending on densities and sizes of plants within individual patches. Patch density affects, in turn, the establishment success of seeds dispersed in from other patches.

• *Interspecific competition:* The numerical model permits differential thinning responses to crowding to allow exploration of the effects of competition during the thinning phase.

The numerical model consists of a grid of local patches linked by dispersal. The grid contains 100×100 cells with the outer four cells serving as a buffer (they are not used in calculations). Disturbances affect single patches. Disturbances occur stochastically with constant probability λ. All patches are identical with the exception of the particular realization of the disturbance process (they are out of phase). Like the analytical model, the numerical model assumes that all individuals of a species within a patch behave identically. Recruitment occurs immediately following (in the same time step as) disturbance at a density $x(0)$ determined by the densities and sizes of plants on all patches within dispersal distance of that patch. Seed dispersal is "local"; seed production is divided evenly among the nearest π patches. The edge of the grid is an absorbing boundary for seeds. Global dispersal is the special case as π tends to the total number of patches on the grid. Thinning depends on the degree of crowding in the patch and on the rate of plant growth. Crown area growth rate is linear, i.e. $(1/A)dA/da$ is a constant γ following initial establishment at size A_0. A plant's seed production is proportional to (β times as great as) its crown area. Seed production begins at age a_1 and continues until the plant dies from thinning, disturbance, or senescence ($a_m = \alpha\, a_1$). The response of thinning rate to growth and crowding is equivalent to a two-species case of equation (3):

$$\frac{dx_1}{da} + \frac{dx_2}{da} = -F\,(\gamma_1 x_1 + \gamma_2 x_2)$$

where crowding is now given by

$$F = \frac{x_1 A_1 + x_2 A_2}{\kappa} \tag{17}$$

In order to solve to thinning rates, we partition mortality according to abundance, under the assumption that thinning is random with respect to species identity. Solving for the thinning rates of individual species, we have

$$\frac{\delta x_1}{\delta a} = - F (\gamma_1 x_1 + \gamma_2 x_2)\, p^c \qquad\qquad (18a)$$

$$\frac{\delta x_2}{\delta a} = - F (\gamma_1 x_1 + \gamma_2 x_2)\, (1 - p^c) \qquad\qquad (18b)$$

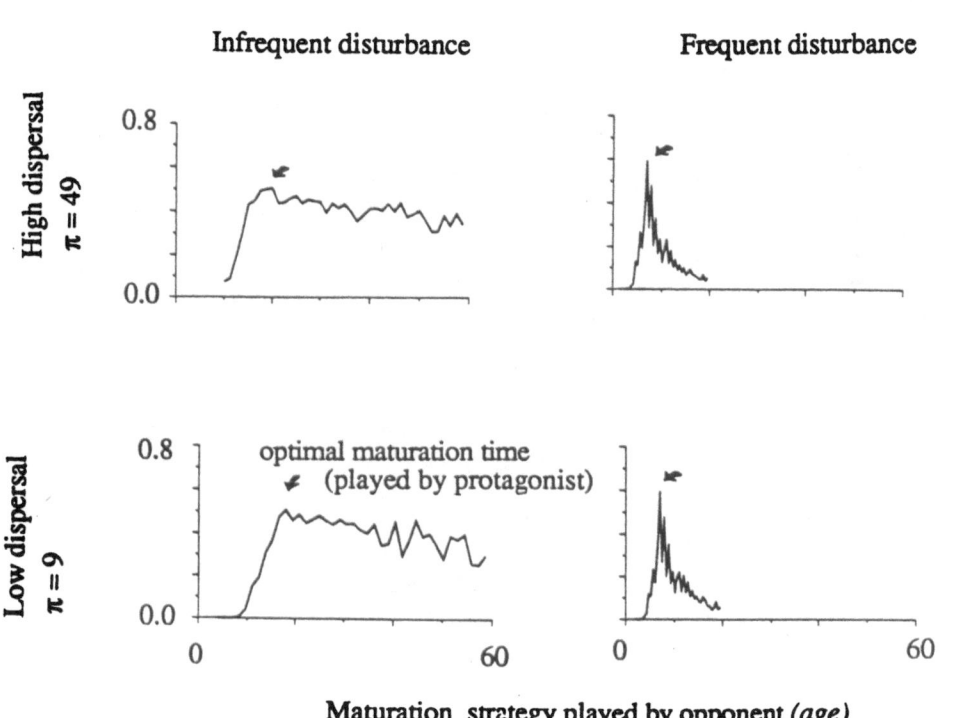

Maturation strategy played by opponent *(age)*

Figure 6. Optimal life histories (maturation times) calculated analytically using equation 1.2 (indicated by arrows) compared with numerical results (curve). The dependent axis shows the relative abundance achieved by the opponent when playing a range of life histories against the "optimal" solution for different dispersal rates and disturbance frequencies. The opponent does as well as the optimal strategy (i.e. it achieves a density of 50%) only when it too plays the optimal strategy indicated by the arrow.

where $p = \dfrac{x_1}{x_1 + x_2}$ is the fraction of local individuals that are of species 1, δx is the change in density that occurs over an age increment of duration δa, and c is a competition coefficient describing the extent to which thinning preferentially falls on species 1 vs species 2. If $c \neq 1$, then thinning is not random with respect to species

Simulation results provide support for aspects of the simpler models, and they demonstrate consequences of factors not contained in those models. Below are some examples.

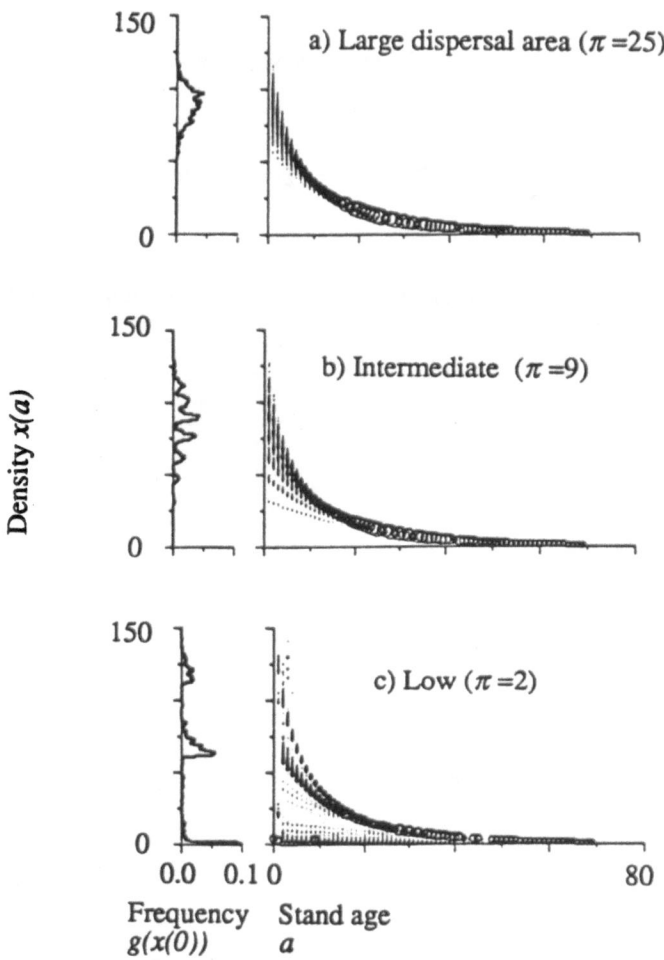

Figure 7. Effect of dispersal area on initial densities (left) and subsequent convergence of densities to the common thinning equation (18) with increasing stand age (right). High variances and multiple modes on initial seed densities ($g(x(0))$) are characteristic of populations having low dispersal area (bottom). The different modes of the initial densities define separate thinning curves, each of which converge to equation (18) with increasing patch age. Parameter values are $\kappa = 50$, $A_0 = 1$, $\gamma = 0.06$, $\lambda = 0.03$, $a_1 = 12.6$ (the optimal maturation time for this parameter combination), and $\beta = 3$. Sizes of symbols are proportional to density at a given age.

Life history. The optimization approach used above ignores potentially important consequences of frequency dependence (e.g., Hamilton and May 1977, Levin et al. 1984, Cohen and Levin 1991). The optimum life history (that satisfying (8)) was competed against nearby strategies for different dispersal rates and disturbance frequencies. I compared the densities of species playing the two strategies after changes in their average densities were too small to detect. The opponent strategy did as well as the protagonist (playing (8)) only when it also played the optimal strategy (Figure 6). It therefore appears that the added complexity of the numerical model does not importantly bias the result obtained by simple optimization methods.

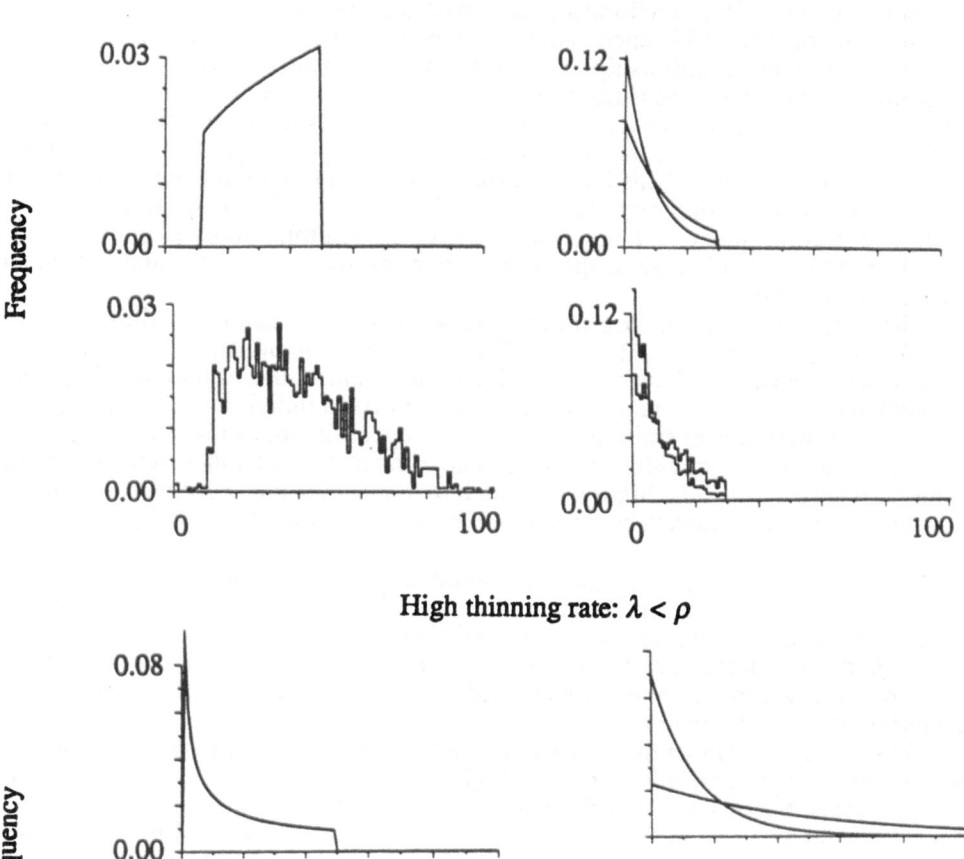

Figure 8. Population structure simulated by the numerical model (below) and analytical approximations (above) at high ($\lambda = 0.07$) and low ($\lambda = 0.02$) disturbance rates. $\kappa = 50$, $A_0 = 1$, $\gamma = 0.05$, $\beta = 2$, $\pi = 9$. Maturation times are the "optimal" ones: 18.9 and 5.4 for low and high disturbance frequencies respectively.

Population structure. Metapopulation structure can be thought of as consisting of two components, one associated with the phase relationships among patches, and a second resulting from differences in the way density changes with age on different patches. We have already addressed both of these topics, although we did not identify them in this way. Whenever a fixed boundary condition x_0 is assumed (or, equivalently, crowding F remains fixed at unity), potential differences among patches are ignored; the resulting structure reflects only the phase relationships among patches. If $x(0)$ is permitted to vary as a consequence of changing conditions on patches within dispersal distance, then a second source of variability must be considered. Because different crowding levels are associated with different initial densities, the thinning process on any given patch depends on the initial density it experiences. This initial density depends in turn on the numbers and sizes of plants on all patches within dispersal distance. This source of variance diminishes with stand age as a consequence of crowding effects on thinning rates (Figure 7). The numerical model provides an opportunity to explore their combined effects on metapopulation structure.

The analytical solutions regarding population structure can be used together with numerical results to identify the effects of differences in how thinning procedes on different patches. The analytical solutions show the distributions that would result if local dynamics were completely deterministic under the assumption of fixed initial density $x(0)$ (Figure 8). The principle result of this exercise is that analytical approximations are extremely close to the numerical solutions provided that disturbance rates are not too much greater than thinning rates (Figure 8). The highly variable initial densities under frequent disturbance explain the poor correspondence between distributions in the upper half of Figure 8.

Thus, numerical models suggest that optimization results hold up rather well when the more obvious assumptions are relaxed. Optimal life histories achieve higher densities in competition with competing strategies, and there is some justification for equating maximum density with the ESS life history (Hastings 1978). In the absence of tradeoffs between fecundity and/or dispersal distance and competitive ability, dispersal area does not influence the winning maturation time.

Analytical predictions of population structure are best when results are not overly influenced by recruitment, which can be highly variable. Results are poorest at high disturbance rates, because the recruitment phase represents an increasing fraction of the total metapopulation dynamics. Analytical approximations are best at lower disturbance rates, because then dynamics are dominated by the more predictable thinning phase.

FUTURE DIRECTIONS

Populations of sessile organisms characterized by locally episodic recruitment followed by thinning continue to pose a challenge to ecologists. A range of processes contribute heterogeneity to metapopulation structure. Simple models are needed to identify contributions of individual plants to composite landscape structure, together with more complex models that allow for their combined influences. Much work is needed in the areas of integrating and simplifying the information from numerical models so that it can be interpreted in terms of processes that operate at local and landscape scales. Shifting mosaic metapopulations display a rich spectrum of behaviors, even in these extremely simplified models that abstract only a few of the most important processes. Nonetheless, it is likely that further development will elucidate much of the underlying cause for observed patterns.

ACKNOWLEDGEMENTS

For their helpful discussions, I thank O. Diekmann, R. Nisbet, S. Levin, S. Pacala, and J. Steele. Preparation of the manuscript was supported by NSF BSR-9011661.

245

REFERENCES

Armstrong, R. A. 1989. Fugitive coexistence in sessile species: Models with continuous recruitment and determinant growth. *Ecology* 70:674-680.

Bartlett, M. S. 1960. *Stochastic Population Models in Ecology and Epidemiology*. Methuen, London, England.

Botkin, D. F., J. F. Janak, and J. R. Wallis. 1972. Some ecological consequences of a computer model of forest growth. *Journal of Ecology* 60:849-872.

Charnov, E. L., and D. Berrigan. 1991. Dimensionless numbers and the assembly rules for life histories. *Philosophical Transactions of the Royal Society of London, Series B* 332:41-48.

Clark, J. S. 1989. Ecological disturbance as a renewal process: theory and application to fire history. *Oikos* 56:17-30.

Clark, J. S. 1990. Integration of ecological levels: Individual plant growth, population mortality, and ecosystem processes. *Journal of Ecology* 78:275-299.

Clark, J. S. 1991a. Disturbance and tree life history on the shifting mosaic landscape. *Ecology* 72:1102-1118.

Clark, J. S. 1991b. Disturbance and population structure on the shifting mosaic landscape. *Ecology* 72:1119-1137.

Clark, J. S. in press, a. Density-independent mortality, density compensation, gap formation, and self-thinning in plant populations. *Theoretical Population Biology*.

Clark, J.S. in press, b. Functional groups and ecological consistencies: population perspectives on regional forest dynamics. In J. Ehleringer and C. Field (eds.). *Scaling processes Between Leaf and Landscape Levels*, Academic Press, New York, NY.

Clark, J. S. in press, c. Relationships between individual plant growth and the dynamics of populations and ecosystems. In D. DeAngelis and L. Gross, (eds.). *Populations, Communities, and Ecosystems: an Individual Perspective*. Chapman and Hall, New York, NY.

Cohen, D., and S. A. Levin. 1991. Dispersal in patchy environments: The effects of temporal and spatial structure. *Theoretical Population Biology* 39:63-99.

Comins, H. N., and I. R. Noble. 1985. Dispersal, variability, and transient niches: Species coexistence in a uniformly variable environment. *The American Naturalist* 126:706-723.

Cox, D. R. 1962. *Renewal Theory*. Chapman and Hall, London.

Geritz, S.A.H., J.A.J. Metz, P.G.L. Klinkhamer, and T.J. DeJong. 1988. Competition in safe-sites. *Theoretical Population Biology* 33:161-180.

Hamilton, W.D. 1966. The moulding of senescence by natural selection. *Journal of Theoretical Biology* 12:12-45.

Hamilton, W. D., and R. M. May. 1977. Dispersal in stable habitats. *Nature* 269:578-581.

Harper, J. L. 1977. *Population Biology of Plants*. Academic Press, New York, NY.

Hastings, A. 1978. Evolutionary stable strategies and the evolution of life history strategies: I. Density dependent models. *Journal of Theoretical Biology* 75: 527-536.

Hastings, A. 1991. Structured models of metapopulation dynamics. *Biological Journal of the Linnean Society* 42:57-71.

Hastings, A., and C. L. Wolin. 1989. Within-patch dynamics in a metapopulation. *Ecology* 70:1261-1266.

Huston, M., and T. Smith. 1987. Plant succession: Life history and competition. *The American Naturalist* 130:168-198.

Levin, S. A., D. Cohen, and A. Hastings. 1984. Dispersal in patchy environments. *Theoretical Population Biology* 26:165-191.

Loehle, C. 1988. Tree life history strategies: the role of defenses. *Canadian Journal of Forest Research* 18: 209-222.

Metz, J. A. J., and O. Diekmann. 1986. *The Dynamics of Physiologically Structured Populations*. Springer-Verlag, Berlin.

Meyer, W. H. 1929. Yields of Second-Growth Spruce and Fir in the Northeast. United States Department of Agriculture Technical Bulletin Number 142.

Norberg, Å. 1988. Theory of growth geometry of plants and self-thinning of plant populations: Geometric similarity, elastic similarity, and different growth modes of plant parts. *The American Naturalist* 131: 220-256.

Pacala, S. W., and J. A. Silander. 1985. Neighborhood models of plant population dynamics: I. Single-species models of annuals. *The American Naturalist* 125:385-411.

246

Pastor, J., and W. M. Post. 1986. Influence of climate, soil moisture, and succession on forest carbon and nitrogen cycles. *Biogeochemistry* 2:3-27.

Shmida, A., and S. Ellner. 1984. Coexistence of plant species with similar niches. *Vegetatio* 58:29-55.

Shugart, H. H. 1984. *A Theory of Forest Dynamics: the Ecological Implications of Forest Succession Models*. Springer-Verlag, New York, NY.

Tait, D. E. 1988. The dynamics of stand development: A general stand model applied to Douglas-fir. *Canadian Journal of Forest Research* 18:696-702.

Valentine, H. T. 1988. A carbon-balance model of stand growth: A derivation employing pipe-model theory and the self-thinning rule. *Annals of Botany* 62:389-396.

Warner, R. R., and P. L. Chesson. 1985. Coexistence mediated by recruitment fluctuations: A field guide to the storage effect. *The American Naturalist* 125: 769-787.

Watt, A. S. 1947. Pattern and process in the plant community. *Journal of Ecology* 35:1-22.

Yoda, K., T. Kira, H. Ogawa, and K. Hozumi. 1963. Self-thinning in overcrowded pure stands under cultivated and natural conditions (intraspecific competition among higher plants). *Journal of Biology* (Osaka City University) 14:107-129.

16
MODELING FIRE REGIME IN MEDITERRANEAN LANDSCAPES

Frank W. Davis and David A. Burrows

INTRODUCTION

This chapter summarizes some current approaches to modeling fire spread and fire regime over heterogeneous landscapes. Applications of satellite remote sensing and Geographic Information Systems (GIS) are emphasized. A regional fire regime simulation model (REFIRES) is described that has been designed to analyze the relationships between fire history and vegetation pattern in chaparral landscapes. For additional details the reader should consult Davis and Burrows (In Press) and Burrows (1987).

FIRE REGIME

The kind of fire history that characterizes an area can be termed its "fire regime," the elements of which include fire type and intensity, size, return interval, and spatial pattern. Fire regime plays a major role in determining regional patterns of species distributions, vegetation patterns, and fluxes of matter and energy (Kilgore 1981).

The fire regime of Mediterranean shrublands has received considerable attention because of fire's dominant role in determining ecosystem structure and function. In the chaparral of California, for example, fire recurs every 20 - 100 years, its spread promoted by the continuous shrub canopy, high proportion of fine fuel, and low fuel moisture during the prolonged summer drought period. Wildfires are extremely hot, removing all but the largest stems. Chaparral stands regenerate rapidly after fire from vegetative sprouts and from buried refractory seed (Horton and Kraebel 1955). The general successional pattern can be summarized as:

Years 1-2: Dominance by annuals; shrub resprouting and seedling establishment;
Years 3-7: Maximal cover by sub-shrubs and perennial herbs;
Years 7-15: Closure of the shrub canopy;
Years 15-50+: Gradual changes in shrub layer, increasing dead fuel and stand flammability.

Stand-replacing fires produce a mosaic of even-aged patches in chaparral landscapes. In theory, the spatial pattern of the mosaic influences other elements of fire regime such as ignition timing, fire type and intensity, fire management, and post-fire vegetation recovery and land use patterns (Parsons 1976, Minnich 1983). Some of these relationships are illustrated in Figure 1. For example:

In landscapes with infrequent ignitions or active fire suppression, fuel accumulation fosters infrequent, large, hot burns.

Areas with frequent ignitions and where fire is not suppressed tend to develop mosaics with smaller patches, smaller and less intense burns, and a greater diversity of stand ages due to the tendency for fires to extinguish at the boundaries of recent burns.

In areas where vegetation has been extensively fragmented by land-use conversion, fire spread is retarded, patch size decreases, and patch age increases.

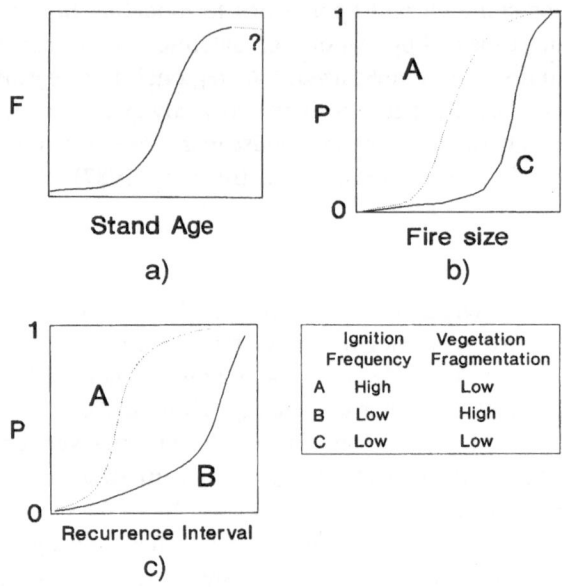

Figure 1. Some hypothesized relationships of fire regimes in chaparral ecosystems: a) the relationship between stand flammability (F) and stand age; b) cumulative frequency (P) of fire sizes for different ignition regimes and landscapes; c) cumulative frequency (P) of fire recurrence interval for different ignition regimes and landscapes.

DOCUMENTING CHAPARRAL FIRE REGIME

Knowledge of the relationship among fire history, modern fire regime, and vegetation pattern is of theoretical and practical significance for understanding chaparral dynamics on time scales of decades to centuries or millennia. Unfortunately, it is difficult to assemble long fire histories for shrublands. Approaches include:

    1) Analysis of charcoal stratigraphy in sediments. For example, Byrne et al. (1977) analyzed varved sediments in the Santa Barbara channel to reconstruct the chaparral fire recurrence interval in the Santa Ynez Mountains. They concluded that the interval between large fires has remained relatively constant at 60 - 70 years despite changes in management policy.

2) Reconstruction of fire history from historical narratives. Minnich (1987) used this approach to study uncontrolled burns in southern California. One important conclusion was that fires could continue to burn over weeks or months, smoldering as embers during cooler or more humid periods, and flaring and running during hotter, windier conditions.

3) Fire mapping from historical aerial photography and satellite imagery (Minnich 1983, Davis et al. 1989). For example, Minnich (1983) used Landsat MSS imagery to map all fires greater than 5 ha in a large area of coastal southern California and northern Baja Peninsula. His analysis indicated that fires were smaller in Baja, where fires were not suppressed, than in adjacent areas of California, where suppression resulted in less frequent, larger burns.

SIMULATING FIRE BEHAVIOR AND FIRE REGIME

Physical Models

Most physical models of fire behavior are based on the fire spread equations of Rothermel (1972), which predict fire reaction intensity (heat energy/fuel bed area/time) and rate of spread (distance/time) in homogeneous, continuous fuel beds as a function of fuel properties and meteorological conditions. Meteorological variables include air temperature, humidity, wind speed, and wind direction. Fuel parameters include,

a) fuel per land area,
b) fuel size distributions (diameter classes),
c) height of the fuel bed,
d) fuel heat content,
e) particle density,
f) fuel moisture content.

Natural vegetation rarely meets the pre-conditions of a continuous, uniform fuel bed, and efforts have been made to account for heterogeneous fuels (Frandsen and Andrews 1979). The Rothermel model has been elaborated to describe fire spread in large fuel burnouts (Albini 1976). Other companion fire spread models have been developed to predict spotting distance from fire fronts (Albini 1979), wind speed at midflame levels (Albini and Baughman 1979), wind-driven fire size, and fire shape (Anderson 1983, Baker 1983).

Malanson (1984) used the Rothermel equations to simulate stand composition as a function of fire intensity and fire regime in California coastal sage scrub.

Probabilistic Models

Spatial stochastic models of fire behavior have been developed in an attempt to study interactions among fuel accumulation, fire behavior, vegetation recovery and community composition (e.g., Van Wagtendonk 1985, Keane et al. 1990). Turner et al. (1989) used cellular automata to model the spread of a disturbance such as fire as a percolation process on a binary grid (flammable vs. non-flammable). Green (1989) also used a cellular formulation to simulate fire spread, seed dispersal, and interspecific competition in forest ecosystems. Fires were described by randomly located ellipses whose frequency and size were described by the poisson and negative exponential distributions, respectively.

Applications of Remote Sensing and Geographic Information Systems

Many spatial variables influence fire regime, notably ignition sources, topography, soils, vegetation, and land-use patterns. A number of studies have demonstrated that digital satellite and terrain data can be used to map general fuel types and to model fire spread using Geographic Information System (GIS) technology for spatial data manipulation and display (Burrough 1986).

Cosentino et al. (1981) described an approach to fire hazard mapping in southern California that combined classified Landsat MSS data, digital topographic data, fire history maps, and site quality maps derived from soils, climate, and topographic data. Using a similar approach, Burgan and Shasby (1984) combined classified Landsat MSS data with digital topographic data and meteorological data to maintain fire hazard maps based on the U.S. National Fire Danger Rating System (NFDRS) (Deeming et al. 1977).

Chuvieco and Congalton (1989) incorporated vegetation data derived from Thematic Mapper satellite imagery, topographic data, and road maps to map fire hazard on the Mediterranean coast of Spain based on vegetation, slope, aspect, roads, and elevation. The map was not a good predictor of the behavior of a documented fire, but was effective in predicting the probability of an area burning.

Other recent applications of GIS to fire regime analysis include analysis of vegetation recovery from fires of varying intensity (Jakubauskas et al. 1990) and statistical analysis of vegetation succession under different management scenarios (Lowell and Asroth 1989).

THE REFIRES (REGIONAL FIRE REGIME SIMULATION) MODEL

The goal of much ongoing work in fire modeling is to link biologically based stand succession models, physically based fire spread models, and some features of spatially and temporally stochastic cellular automata. When parameterized with local physical and biological data, such models could be extremely useful for understanding and managing fire dynamics in specific regions. Fire spread equations have been used to simulate individual fires over gridded surfaces with varying fuel characteristics (Frandsen and Andrews 1979), but not to model longer-term fire regime. The REFIRES model (Davis and Burrows 1991) is one attempt at fire regime simulation that couples physical and stochastic elements of fire modeling. We adapted the following modeling objectives:

1) Fire behavior is modeled using the physically based fire spread equations of Rothermel (1972) and Albini (1976);
2) The model of fire spread must account for the ability of unsuppressed fires to burn for weeks or longer, through diurnal and nocturnal meteorological conditions;
3) Spatial variation in topography, vegetation pattern, and fire history must be

modeled explicitly;

4) REFIRES should be operable over a relatively large grid to realistically represent variation in topography and fuels;

5) The model should be modular to the degree possible to facilitate modifications and additions, and should not depend on specific spatial data handling software.

The model is written in the C programming language and operates on a surface of hexagonal cells (see below). Each cell is described by fuel and topographic variables. Spread of fires through the grid depends on cell characteristics and meteorological conditions. Fire history can be simulated through hundreds of fire seasons, as specified by the user. Output includes a map of simulated fire history, as well as summary statistics on fire recurrence interval, fire size distribution, final patch size, and age distributions.

Spatial Data Handling

Like many spatial fire models, REFIRES represents the landscape as a set of co-registered digital maps. Most fire models utilize a *raster* data structure, gridding space into cells and storing information for each cell. The most common raster structure is the square lattice whose values are stored as two-dimensional arrays. This structure is employed for imaging systems, and the uniform cell size and shape are well suited to spatial analysis. However, the diagonal neighbors in the grid raster representation pose a problem in modeling contagious spread by fire, which is better handled by triangular or hexagonal rasters.

REFIRES utilizes a hexagon-based sept-tree data structure using a spatial addressing system known as a Generalized Balanced Ternary (GBT) (Gibson and Lenzmeier 1981). GBT addressing is based on a hierarchy of hexagonal cells. Each cell possesses a unique address number whose digits correspond to its position at nested levels of aggregation. Movement across the grid is accomplished by address addition, subtraction, or multiplication. This addressing system is especially well suited to modeling contagious diffusion processes. One disadvantage is that output maps must be converted to a square grid for display.

Model Operation

Model flow is depicted in Figure 2. Input data for the model consist of:

Digital maps of elevation, slope angle, slope aspect, potential vegetation (which may include non-flammable land use classes such as agricultural or residential), and actual vegetation (a function of potential vegetation and time since burning).

Meteorological data, including daily maximum and minimum temperatures and humidities and accompanying live fuel moisture (by fuel size class) for a set of fire season dates (we have used large random samples of fire season days from the study region). A frequency distribution of wind speeds and directions is also required.

Fuels data for age classes of each vegetation type including dead fuel surface area to volume ratio (by fuel size class), live fuel surface area to volume ratio (also by size class), dead fuel load, fuel bed height, and dead fuel moisture of extinction.

User-supplied parameters include:

a) grid size,
b) ignition frequency (F_i, average potential ignitions per year),
c) a threshold fire spread rate (R_t, in ft/minute) below which fire spread does not occur,
d) duration of model run (in years),
e) seed for the random number generator.

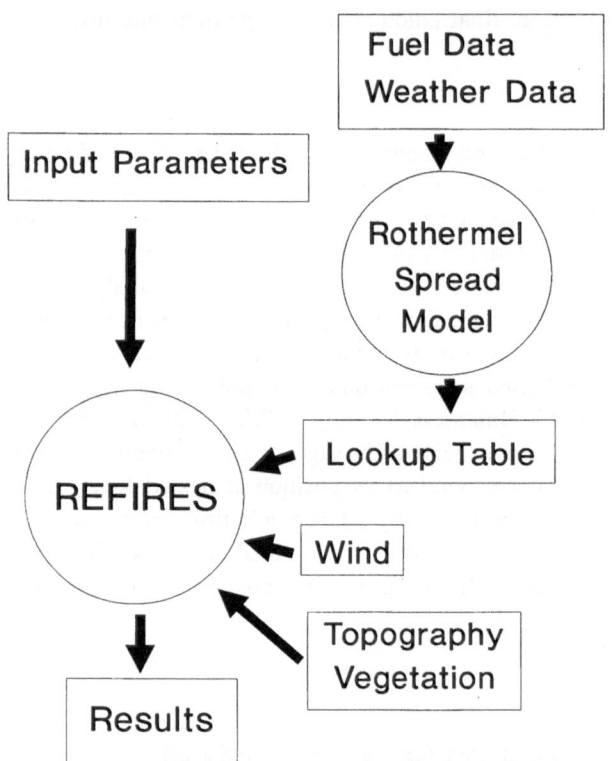

Figure 2. General formulation and data flow in the REFIRES model.

To reduce computation time, at the outset of a model run a lookup table is produced that contains rates of fire spread as calculated by the Rothermel equations for each unique combination of vegetation, temperature, humidity, and fuel moisture conditions. These spread rates are modified during the run to account for wind and topographic effects.

For each year in a run, the number of possible ignitions is selected from a poisson distribution with parameter F_i, as defined previously. The location, date, time, and weather conditions for a possible ignition are selected at random (Figure 3). The target cell has a probability of ignition P_i that is calculated based on its dead fuel moisture and the weather conditions.

If the cell is ignited, the fire may spread throughout that cell based on the spread rate R_s as calculated from fuel, topography, temperature, humidity, and wind conditions. If $R_s > R_t$, where R_t is the user-defined threshold rate of spread, then the fire spreads with probability $P_s = 1$. If $R_s < R_t$, fire may spread with P_s proportional to R_s/R_t. If fire spreads through the cell, the cell's age is set to zero and the fire then continues to spread to neighboring cells with probabilities and rates calculated on a cell-by-cell basis as determined by local fuel conditions, topography, and wind.

Based on calculated fire spread rates and distances, the program keeps track of the time elapsed between ignition and when the burn reaches a cell. Daily maximum temperatures and minimum humidities are applied to cells that burn during daylight hours, and minimum temperatures and humidities are applied during nighttime hours. Wind speed and velocity are kept constant during the day or night.

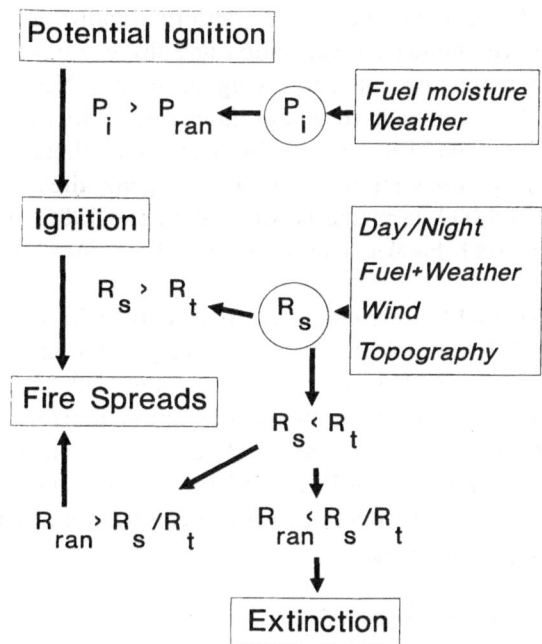

Figure 3. Inputs and operations for a fire event in REFIRES. See text for parameter definitions.

Fires spread only by contagious diffusion, and not by spotting. A crude approximation to spotting can be implemented by setting minimum flammability to some value greater than 0, which allows fires to spread across fire breaks with low probability.

Fires cease to spread when they reach the edge of the grid or encounter conditions in which the predicted spread rate falls below R_t. After the model is run for a specified number of years, the program returns fire regime statistics and fire history maps. Statistics include average ignitions per year, number of possible ignitions, number of actual ignitions, number of fires (by cell), fire size distribution, stand age frequency distribution, patch size distribution, and fire recurrence interval distribution.

MODEL APPLICATION

The Rothermel model has been tested, refined, and validated over the past 15 years. However, we have not yet validated other components of REFIRES (such validation data must be obtained from a number of unsuppressed burns). Our use of the model thus far has been to test its sensitivity to several key parameters and to ensure that model output appears reasonable and consistent with existing theory.

Study Site

The model has been applied to a region of coastal California known as Burton Mesa. For the past several years we have studied the vegetation and fire ecology of this area to provide better information for managing and conserving its unique biota (e.g., Davis et al. 1988, Davis et al. 1989). The dominant vegetation is chamise (*Adenostoma fasciculatum*) chaparral, which has been extensively fragmented by roads, agricultural fields, and residential development. Lightning ignitions are rare, but anthropogenic fires occur frequently. At least 27 fires occurred over a 4,000 ha region between 1938 and 1986. These fires are typically arrested at roads or fuel breaks, and thus give little indication of natural fire behavior.

Preliminary tests of REFIRES were conducted over a 1865 ha region represented by 20,720 0.09 ha cells. Two different potential vegetation maps were used: 1) the entire area was modeled as potentially chamise chaparral; 2) the area was modeled as chamise chaparral, but modern land uses were included and mapped as non-flammable. We refer to these two vegetation models as *Prehistoric* and *Modern*. Other details concerning the study area and model parameterization are provided in Davis and Burrows (1991).

Ninety 500-year model runs (5 runs per parameter set) were conducted to test 18 combinations of the following parameters:

a) Prehistoric versus Modern vegetation,
b) F_i = 0.1, 1.0 or 3.0 average possible ignitions per year,
c) R_t = 1.0, 2.0 or 8.0 feet per minute.

Model Results

A statistical summary of model runs is provided in Table 1. To summarize some of the results: [1]

1) Ignition frequency and fire frequency increase linearly with increasing F_i. The ratio of ignitions to fires decreases with increasing ignition frequency, because with increasing ignition and fire frequency the average stand age decreases, reducing

average flammability over the landscape.

2) Average fire size decreases non-linearly with increasing ignition frequency under all scenarios except on the Modern grid with R_t set to 8 ft/minute, when average fire size remains relatively constant (Figures 4a,b). Fire size distribution on Prehistoric landscapes does not differ much between simulations using an R_t of 1.0 and 2.0 ft/min. At an ignition frequency of 0.1/yr, nearly half of the fires burned the entire landscape. The range of fire sizes is much smaller on the Modern landscapes, and no fires spread over more than 50% of vegetated areas.

3) Frequency distributions of fire recurrence are sensitive to changes in Fi, less so to R_t (Figures 4c,d). For example, on the Prehistoric landscape with $F_i = 0.1$ and $R_t = 1.0$, 55% of burns spread across the entire landscape, and 80% of cells burned at ages greater than 95 years. With Fi = 1.0, roughly 75% of burns did not exceed 400 hectares in size, and 80% of the cells burned before reaching 45 years of age.

4) Fire sizes are much smaller on the Modern landscape, and size distributions are less sensitive to F_i because of the isolation of old, highly flammable stands. The fire recurrence distribution on the Modern grid is shifted strongly toward longer intervals. With $F_i = 0.1$, a cell has only 50% probability of burning during 200 years. Vegetation fragmentation isolates small chaparral patches that behave as "fire refugia." These areas are conspicuous in the simulated fire history maps (Figures 5a, 5b).

Table 1. Statistical summary of simulated fire regime under varying vegetation fragmentation, ignition frequencies and threshold rates of spread. Values are means of five runs for each combination of modeled conditions.

	Vegetation Model					
$R_t = 1.0$	Prehistoric			Modern		
Ignition Frequency (F_i)	0.1	1.0	3.0	0.1	1.0	3.0
Possible Ignitions	49	495	1493	50	497	1496
Actual Ignitions	14	155	464	10	90	307
Fires	11	83	226	9	64	176
Mean fire size	1052	285	123	268	137	70
Patch Number	9	18	30	138	184	216
$R_t = 8.0$						
Ignition Frequency (F_i)	0.1	1.0	3.0	0.1	1.0	3.0
Possible Ignitions	46	473	1471	49	90	303
Actual Ignitions	13	135	452	9	90	303
Fires	9	55	176	6	56	157
Mean fire size	489	232	101	33	45	35
Patch Number	16	28	47	28	211	314

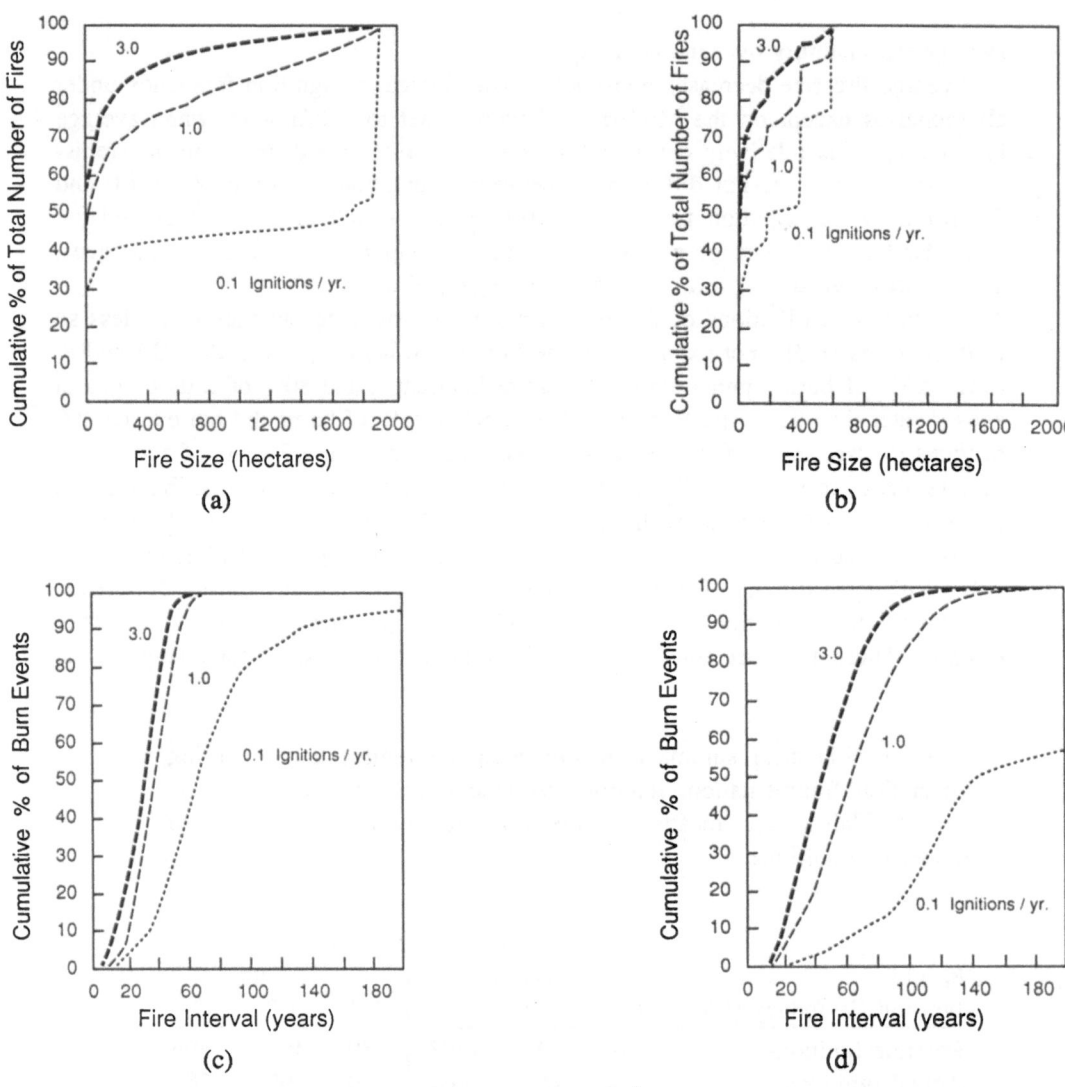

Figure 4. Cumulative fire size distributions obtained for (a) Prehistoric landscapes and (b) Modern landscapes with F_i = 0.1, 1.0 and 3.0 ignitions/year, and for R_t = 2.0 ft/minute. Cumulative frequency distribution of fire recurrence intervals obtained for (c) Prehistoric landscapes and (d) Modern landscapes with F_i = 0.1, 1.0 and 3.0 ignitions/year, and for R_t = 2.0 ft/minute. Each parameter set was run five times to produce the curves.

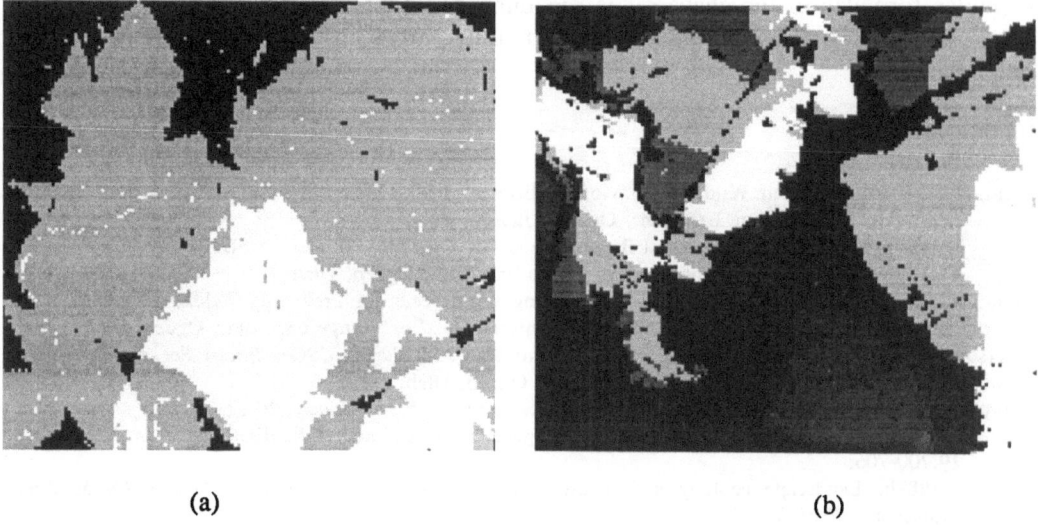

(a)   (b)

Figure 5. Fire histories produced by single 500-year runs of REFIRES. Image brightness is proportional to the number of fires. $R_t = 1.0$, $F_i = 3$ ignitions/year. (a) Prehistoric grid, number of fires per location ranges from 8 to 19. (b) Modern grid, number of fires ranges from 0 (developed sites) to 13. The unburned swath running from the southwest to the northeast corner of the image is residential areas and cropland. The large, dark "island" of chaparral in the southeast quadrant is surrounded by developed sites and behaves as a fire refugium.

SUMMARY

Fire mosaic properties such as patch size and age distributions are closely coupled to ignition frequency and are quite sensitive to vegetation fragmentation. GIS modeling of fire behavior based on a combination of physical and stochastic processes provides a means of studying relationships among landscape heterogeneity and other components of fire regime.

In simulations of fire regime in coastal California, the spatial distribution of fire recurrence is very uneven across the fragmented modern landscape, with fire refugia located, as expected, in areas generally downwind from large fuel breaks. Despite the many stochastic elements to the model, there is only slight variation among runs made using the same parameter settings.

There are a number of improvements that can and should be made to REFIRES. The current implementation locates ignitions at random in the region, but it is often the case that ignitions are strongly clustered in landscapes. For example, lightning strikes are more likely at higher elevations, and anthropogenic fires begin near roads and trails. Including a probability surface for ignition should be straightforward. The meteorological model could be improved by introducing higher frequency variation in wind, temperature, and humidity. A model to simulate fire spotting should be incorporated. Equations have been developed to predict the probability, direction, and distance of spotting from burning trees (Albini 1979).

A similar formulation for chaparral is recommended. This introduces new complexity, however, because of the need to simulate multiple fires burning simultaneously on the landscape.

REFERENCES

Albini, F. A. 1976. Estimating Wildfire Behavior and Effects. USDA Forest Service Gen. Tech. Rep. INT-30, Intermount. For. Range Exp. Stn., Ogden, Utah.
_____. 1979. Spotfire Distances from Burning Trees—A Predictive Model. USDA Forest Service Gen. Tech. Rep. INT-268, Intermount. For. Range Exp. Stn., Ogden, Utah.
Albini, F. A., and R. G. Baughman. 1979. Estimating Windspeeds for Predicting Wildland Fire Behavior. USDA Forest Service Res. Paper INT-221, Intermount. For. Range Exp. Stn., Ogden, Utah.
Anderson, H. E. 1983. Predicting Wind-driven Fire Size and Shape. USDA Forest Service Res. Paper INT-305, Intermount. For. Range Exp. Stn., Ogden, Utah.
Baker, D. G. 1983. Shapes of simulated fires in discrete fuels. *Ecol. Model.* 20:21-32.
Baker, W. L. 1989a. Effect of scale and spatial heterogeneity on fire-interval distributions. *Can. J. For. Res.* 19:700-706.
_____. 1989b. Landscape ecology and nature reserve design in the Boundary Waters Canoe Area, Minnesota. *Ecol.* 70:23-35.
Bradshaw, L. S., J. E. Deeming, R. E. Burgan, and J. D. Cohen. 1983. The 1978 National Fire Danger Rating System: Technical Documentation. USDA Forest Service Gen. Tech. Rep. INT-169, Intermount. For. Range Exp. Stn., Ogden, Utah.
Burgan, R. E., and M. B. Shasby. 1984. Mapping broad-area fire potential from digital fuel, terrain, and weather data. *J. Forest.* 82:228-231.
Burrough, P. A. 1986. *Principles of Geographic Information Systems for Land Resources Assessment.* Clarendon Press, Oxford.
Burrows, D. A. 1987. The REFIRES (Regional FIre REgime Simulation) Model: A C program for Regional Fire Regime Simulation. M. A. Thesis, University of California, Santa Barbara.
Byrne, R., J. Michaelsen, and A. Soutar. 1977. Fossil charcoal as a measure of wildfire frequency in southern California: A preliminary analysis. In: H. A. Mooney and C. E. Conrad (eds.). Proceedings of the Symposium on the Consequences of Fire and Fuel Management in Mediterranean Ecosystems, Palo Alto, California. U. S. Forest Service General Technical Report WO-3. Washington, D.C., pp. 361-367.
Chuvieco, E., and R. G. Congalton. 1989. Application of remote sensing and geographic information systems to forest hazard mapping. *Remote Sens. of Environ.* 29:147-159.
Cosentino, M. J., C. E. Woodcock, and J. Franklin. 1981. Scene analysis for wildland fire. In: Proceedings of the Fifteenth International Symposium on Remote Sensing of Environment, Environmental Research Institute of Michigan, Anne Arbor, pp. 635-646.
Davis, F. W., and D. A. Burrows. In Press. Spatial Simulation of Fire Regime in Mediterranean-Climate Landscapes. In: M. C. Talens, W. C. Oechel, and J. M. Moreno. (eds.). *The Role of Fire in Mediterranean-Type Ecosystems.* Springer-Verlag, New York.
Davis, F. W., D. E. Hickson, and D. C. Odion. 1988. Composition of maritime chaparral related to fire history and soil, Burton Mesa, Santa Barbara County, California. *Madroño* 35: 169-195.
Davis, F. W., M. I. Borchert, and D. C. Odion. 1989. Establishment of microscale pattern in maritime chaparral after fire. *Vegetatio* 84: 53-67.
Deeming, J. E., R. E. Burgan and J. D. Cohen. 1977. The National Fire Danger Rating System—1978. USDA Forest Service Gen. Tech. Rep. INT-82, Intermount. For. Range Exp. Stn., Ogden, Utah.
Frandsen, W. H., and P. L. Andrews. 1979. Fire Behavior in Non-Uniform Fuels. USDA Forest Service Research Paper INT-232. Intermount. For. Range Exp. Stn., Ogden, Utah.
Gibson, L. D., and C. Lenzmeier. 1981. A Hierarchical Pattern Extraction System for Hexagonally Sampled Images. U. S. Air Force Office of Scientific Research Report # AFOSR-TR-81-0845, Interactive Systems Corporation.
Green, D. G. 1989. Simulated effects of fire, dispersal, and spatial pattern on competition within forest mosaics. *Vegetatio* 82:139-153.

Horton, J. S., and C. J. Kraebel. 1955. Development of vegetation after fire in the chamise chaparral of southern California. *Ecol.* 36:244-262.

Jakubauskis, M. E., K. P. Lulla, and P. W. Mausel. 1990. Assessment of vegetation change in a fire-altered forest landscape. *Photogramm. Eng. Remote Sens.* 56:371-377.

Keane, R. E., S. K. Arno, and J. K. Brown. 1990. Simulating cumulative effects in Ponderosa pine/Douglas fir forests. *Ecol.* 71:189-203.

Kilgore, B. M. 1981. Fire in ecosystem distribution and structure. In: H. A. Mooney et al. (eds.). Proceedings of the Conference on Fire Regimes and Ecosystem Properties. USDA Forest Service Gen. Tech. Rep. WO-26, Washington, D.C., pp. 58-59.

Lowell, K. E., and J. H. Asroth. 1989. Vegetative succession and controlled fire in a glades ecosystem. *Int. J. Geogr. Info. Sys.* 3:69-81.

Malanson, G. P. 1984. Fire history and patterns of California coastal sage scrub. *Vegetatio* 57:121-128.

Minnich, R. A. 1983. Fire mosaics in southern California and northern Baja California. *Science* 219: 1287-1294.

_____. 1987. Fire behavior in southern California chaparral before fire control: The Mount Wilson burns at the turn of the century. *Ann. Assoc. Amer. Geogr.* 77:599-618.

Parsons, D. J. 1976. The role of fire in natural communities: An example from the southern Sierra Nevada, California. *Environ. Conserv.* 3:91-99.

Rothermel, R. C. 1972. A Mathematical Model for Predicting Fire Spread in Wildland Fuels. USDA Forest Service Res. Paper INT-115, Intermount. For. Range Exp. Stn., Ogden, Utah.

Shasby, M. B., R. E. Burgan, and G. R. Johnson. 1981. Broad area forest and topography mapping using digital Landsat and terrain data. In: Proceedings of the Seventh Symposium on Machine Processing of Remotely Sensed Data. Purdue University, pp. 529-537.

Turner, M. G., R. H. Gardiner, V. H. Dale, and R. V. O'Neill. 1989. Predicting the spread of disturbance across heterogeneous landscapes. *Oikos* 55:121-129.

Van Wagtendtonk, J. W. 1985. Fire suppression effects on fuels and succession in short-fire-interval wilderness ecosystems. In: J. E. Lotan, B. M. Kilgore, W. C. Fischer, and R. W. Mutch (technical coordinators). Proceedings of the Symposium and Workshop on Wilderness Fire. USDA Forest Service Gen. Tech. Rep. INT-182, Intermount. For. Range Exp. Stn., Ogden, Utah, pp. 119-126.

17

THE INFLUENCE OF REGIONAL PROCESSES ON LOCAL COMMUNITIES: EXAMPLES FROM AN EXPERIMENTALLY FRAGMENTED LANDSCAPE

Robert D. Holt and Michael S. Gaines

INTRODUCTION

Until recently, most empirical research in terrestrial community ecology—and, in particular, experimental studies (Hairston 1989)—concentrated on phenomena at small spatial scales (Kareiva and Anderson 1988). Yet local communities are embedded in a spatially heterogeneous world and may be influenced by processes operating at a multiplicity of spatial and temporal scales (Ricklefs 1987, Roughgarden et al. 1988, Levin 1988, Wiens 1989, Hastings 1990). There is increasing urgency in understanding the role of spatial processes in community ecology, given that a pervasive effect of humans on the earth is the destruction and fragmentation of natural habitats. Habitat fragmentation potentially influences a multitude of ecological phenomena, ranging from individual behavior to population persistence, to the strength and predictability of interspecific interactions, to ecosystem fluxes (Saunders et al. 1991). Ameliorating the effects of habitat fragmentation will require a deep understanding of the role of spatial processes in population and community dynamics.

A burgeoning body of observational (e.g., Dickman 1987, Quinn and Harrison 1987) and theoretical (e.g., Wilcove et. al 1986; Fahrig and Paloheimo 1988) analyses of habitat fragmentation now exists, but experimental studies of habitat fragmentation are still relatively infrequent (e.g., Robinson and Quinn 1988, Kareiva 1987, Lovejoy et al. 1984). For the past seven years we have been examining community dynamics on an experimentally created archipelago of patches undergoing secondary succession (Figure 1), surrounded by a "sea" of low turf maintained by regular mowing. When we designed this patch array in 1984, we were guided by theoretical notions about how the spatial context of a local community—the size of the area it occupies, and its position relative to source pools for colonization—might influence successional dynamics, either by direct effects on the plant community (Holt et al., ms.) or by indirect effects via small mammal herbivores (Louda et al. 1990). We now believe our system provides a more general opportunity for studying the population and community processes involved in habitat fragmentation, and that is the focus of our current research. In this chapter, we present case studies from this system to illustrate a conceptual framework for analyzing the influence of regional processes on local populations and communities (Holt, in press, a).

AN EXPERIMENT IN LANDSCAPE ECOLOGY

The study site is a 12 ha field on the University of Kansas' Nelson Environmental Study Area

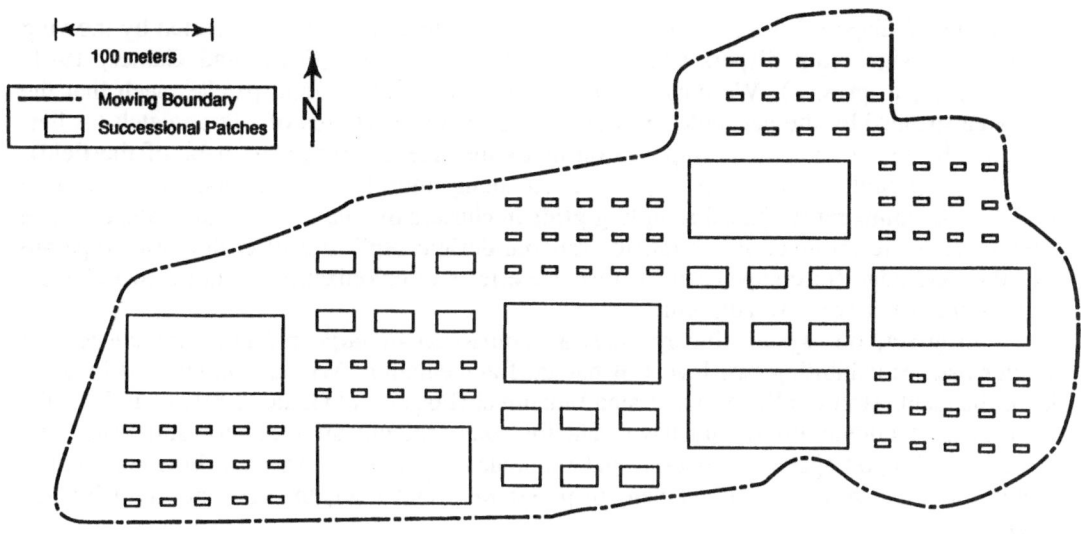

Figure 1. The Biotic Succession Facility at the Nelson Environmental Study Area (NESA), University of Kansas, Leavenworth County, Kansas. The patch sizes are 50 m x 100 m (large), 12 m x 24 (medium), and 4 m x 8 m (small. The medium and small patches are arranged in clusters spanning 50 m x 100 m. Because of the irregular shape of the field, two of the clusters of small patches had to be smaller than this. The low turf plant community on the interstitial habitat (maintained by mowing) substantially differs in species composition from the successional old-field community on the patches (Holt, Robins, and Gaines ms).

(NESA), 16 km north of Lawrence, Kansas. The archipelago of habitat islands shown in Figure 1 was created from an agricultural field in 1984 by intensively mowing interstitial areas between patches at bi-weekly intervals and allowing succession to proceed unhindered within the patches. Old-field succession in our system has followed the familiar script documented throughout the central U.S. in habitats free of fire (e.g., Bazzaz 1968). Substantial changes occurred during the first three years, but the rate of change in the plant community thereafter greatly slowed as perennial forbs and then invading woody plants began to replace annual species (Robinson et al., in press, Holt et al., ms.).

Our choices of patch sizes, their spatial separation, and the spatial arrangement of patches were guided by a knowledge of the dispersal biology of plants and small mammals, by a consideration of local landscape patterns, and by logistical concerns. The largest patches (0.5 ha) are comparable to the sizes of enclosures typically used in experimental studies of small mammal communities (e.g., Johnson and Gaines 1987). The smallest patch size (0.0032 ha), though small, can potentially harbor many hundred individuals of herbaceous plant species. The minimum distance between any two plots was determined from data on plant and small mammal dispersal (e.g., Gaines et al. 1979); the interstitial habitat created by mowing constitutes a significant dispersal barrier for many organisms (Foster and Gaines, 1991; Gaines et al., in press a). We clustered groups of small and medium patches such that the total area spanned by the perimeter of a cluster equals the total area of a large patch (as best as we could, given constraints imposed on us by the size and irregular shape of the field). This design permits us to compare grids of sampling units (e.g., quadrats, traps) in large patches with comparably spaced sampling grids in clusters of small or medium patches. The patch clusters are arranged in a stratified random design; replicates of each cluster type are near the field edge (to compensate for distance effects in colonization), and no two clusters of the same patch type are adjacent.

In effect, our system incorporates a comparison of large patches with clusters of smaller patches exhibiting two levels of habitat fragmentation. We have monitored in some detail the plant (Holt et al., ms.) and small mammal (Foster and Gaines 1991, Gaines et al., in press, a,b) communities, and have data for some but not all years for soil water and nitrogen, arthropod species richness, snake abundances, and the breeding bird community (Robinson et al., in press; Teravainen, in press; Roth and Holt, in press; and unpublished data).

Local Community Responses

As Levin has remarked (Levin 1988, in press), the spatial boundaries of what we call a "population" or "community" usually cannot be specified unambiguously, and so ecological patterns and processes ideally should be examined across a range of spatial scales. In practice, in experimental community ecology there usually is a maximal spatial scale that is logistically feasible. At or near these logistic limits, one way to conceptualize the spatial processes that have been necessarily excluded from the experimental system is to carry out a thought experiment, schematically illustrated in Figure 2 (after Brown and Gibson 1983; Holt in press 6). Imagine that the area defined to be the "local community" is walled off by an invisible force-field that cuts off dispersal. The magnitude and rate of change in population size and species composition in the patch following isolation measures the importance of dispersal in the dynamics of the initial, non-isolated community. Habitat fragmentation creates an archipelago of patches varying in size and degree of isolation, and so is a complex realization of this thought experiment.

A variety of responses on different time scales may occur in a freshly isolated local community (see Holt, in press (a) for more detail). Four such responses, with case studies from our fragmentation project that seem to fit them, are described below.

Historical Source-pool Effects

There is always some temporal scale over which dispersal linking a local community to a larger source pool is relevant to understanding the species composition of that community. The real question for terrestrial community ecologists (as opposed to historical biogeographers) is whether or not dispersal episodes subsequent to the initial colonization event that "seeded" the local community with a given species must be considered to explain its persistence or abundance there.

Certain species in a local community might well show no discernible effect of cutting off dispersal with the surrounding landscape, over the time scale with which one is concerned.

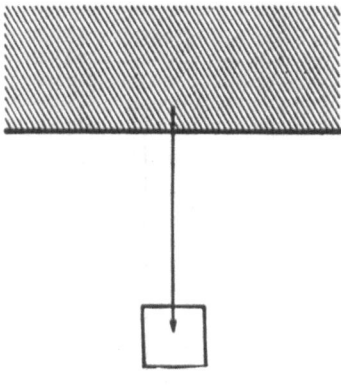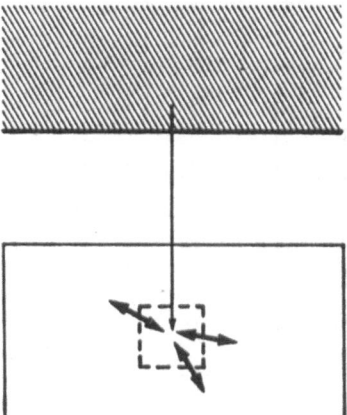

Figure 2. A schematic representation of a thought experiment. To assess regional influences on a local community, indicated by dotted lines (right), one should imagine that dispersal is cut off, in effect creating an isolated small patch (left). In empirical fragmentation studies (e.g., the NESA patch study), where habitat fragments are surrounded by partial or complete barriers to dispersal, one can compare small patches to samples of the same size taken at random from large patches. (Reprinted by permission from Holt, in press, h).

As a case in point, a classical theory of plant succession is the "initial floristic composition" hypothesis. Egler (1954) proposed that chance determines the collection of starting propagules available at the time a site becomes available for succession; subsequent dynamics, he surmised, reflect a sorting out of this initial array of species, in accord with their different life histories and interspecific interactions. Spatial processes may determine the initial condition of the community (e.g., the relative abundance of seeds in a seed bank may reflect the rain of seeds over a broader landscape), but may not matter much once succession starts.

Some patterns in our plant community data are consistent with this idea. Our system displays the expected total species vs. total area relationship (MacArthur and Wilson 1967, see Figure 3a). But if one examines the relationship between species richness per unit area (e.g., sampling quadrat) and the total area of a patch, there is no effect whatsoever of patch area on local species richness (Figure 3b). Local species richness, in fact, was higher during the first year of our study than it has been in any subsequent year, presumably because a diverse preexisting seed bank was present, ready to spring up after the final wheat harvest. Most of the species now present were in that initial community, and succession has mainly consisted of changes in their relative abundances, rather than the addition of species by long-distance colonization. Application of standard ordination techniques used in plant community ecology (e.g., reciprocal averaging, canonical correlation analysis) fails to generate significant

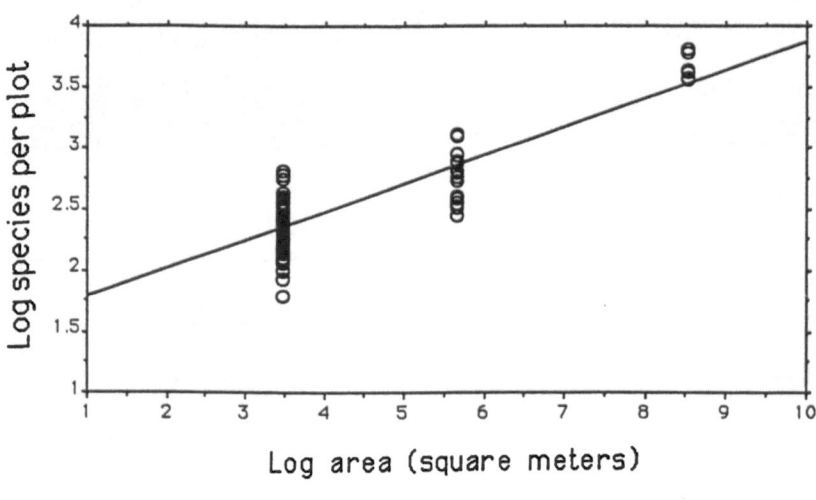

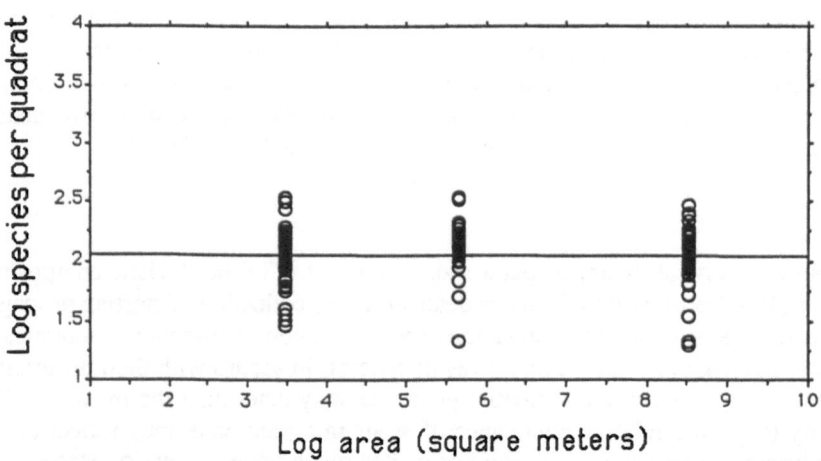

Figure 3. Two kinds of species-area relations for the plant community of the NESA patch study. The top figure shows that total species richness per patch increases with total patch size (as is typical in island biogeographic surveys). The bottom figure shows that local species richness (= the number of species present in permanent, 1 m² quadrats) is independent of patch size. (Reprinted by permission from Holt, in press, a).

aggregate axes separating the plant communities occupying different-sized patches. This suggests that there is no systematic impact of patch size on the temporal development of the vegetation. Moreover, there is a temporal autocorrelation for permanent quadrats in species richness, which has decayed with increasing sampling interval but persists even between 1985 and the present (Holt et al., ms.). In other words, the spatial and temporal patterning of species richness during succession on a given patch seems largely to involve an unfolding of a structure latent in the initial conditions of that patch, with little indication of spatial dynamics such as colonization from outside the system. Dispersal from the surrounding landscape was doubtless important as part of the historical explanation of the initial condition of our site, and the significance of dispersal is of course beginning to increase as woody species progressively invade. To a first approximation, however, such dispersal appears relatively unimportant in determining the broad pattern or rate of local succession in our system following site initiation.

The Spatial Implications of Species' Autecological Requirements

Species differ greatly in the spatial scale required to successfully complete their life cycles (Wiens 1989). Most communities contain a few (and sometimes a great many) species that could not persist as resident members for even one generation, were that community isolated (as in our thought experiment). For instance, large-bodied species with large home range requirements (Peters 1983) should quickly disappear from isolated communities smaller than a single home range. Likewise, many species exploit predictable temporal variation in the environment by migrating among distinct habitats; such species should vanish following isolation. Because these vulnerable species are a biased subset of a local community, fragmentation should lead to a "nesting" in community composition, with fragments containing predictable subsets of the original community. Evidence from biogeographic surveys supports this hypothesis (e.g., Patterson and Atmar 1990).

Taking our system as a whole, certain species (e.g., raptorial birds) would rapidly disappear following isolation from the surrounding landscape matrix. We believe that the spatial expression of autecology requirements also helps to explain the distribution of one member of the small mammal community on our site (Figure 4). The three small mammal species regularly found on the study site, and their respective average body sizes (adult females) are as follows: hispid cotton rats (*Sigmodon hispidus*, 135 g), prairie voles, (*Microtus ochrogaster*, 43 g), and deer mice (*Peromyscus maniculatus*, 20 g). Resident populations of the large-bodied cotton rat are present on the large patches, but on small patches individuals are rarely captured and are virtually never recaptured there (Foster and Gaines 1991) suggesting that small patches are actively selected against by dispersing individuals. The overall distribution of this species in our fragmented landscape almost surely reflects its autecology, and in particular, its home range requirements.

We initially predicted that all three mammal species would be less abundant on smaller patches. This holds for *S. hispidus*, the largest species, but, to our surprise, the other two species achieved their highest densities on smaller patch sizes. In fact, the modal abundances of these three species neatly sort out by body size on different patch sizes (Figure 4)! This intriguing pattern has been present throughout our study, though changing in sharpness with fluctuations in density. Our current hypothesis to explain this distributional pattern involves the interplay of source-sink dynamics and interspecific competition. Before describing this, it is useful to deal with source-sink dynamics at a more general level.

Source-Sink Population Structures in Heterogeneous Landscapes

For mobile organisms, the effect of fragmentation should reflect individual spatial behavior. Vertebrate community ecologists have explored the consequences of one kind of dispersal behavior—optimal habitat selection—for population dynamics and community structure (Holt 1984, 1985, 1987, in press, a; Sih 1987; Morris 1987; Rosenzweig 1991; Pulliam and Danielson 1991). A principal result of habitat selection theory (Milinski and Parker 1991) is that if individuals choose to reside in habitats where they have the highest fitness, then the distribution of individuals among habitats is adjusted toward the "ideal free distribution" (Fretwell 1972) in which individual fitness is equalized across habitats. The prediction of an ideal free distribution rests on several assumptions (Milinski and Parker 1991): 1) individuals should be able to move freely among habitats; 2) individuals can assess habitat-specific fitness accurately and without cost; and 3) individuals are not excluded from some habitats by intraspecific interference competition.

Relaxing these assumptions leads to source-sink population structures (Holt 1985, Shmida and Wilson 1985, Pulliam 1988, Wiens 1989, Pulliam and Danielson 1991). If in some high-quality habitats ("sources") a species' rate of reproduction exceeds its rate of mortality, whereas in other low-quality habitats ("sinks") its reproduction is less than its mortality, a net flux of dispersers away from sources may sustain populations in sinks. Several distinct mechanisms can produce source-sink population structures in heterogeneous environments, including: 1) density-independent dispersal (Holt 1985), 2) time or energy costs of dispersal (Rosenzweig 1974, 1991), and 3) intraspecific preemptive or interference competition (Pulliam 1988).

Source-sink dynamics is beginning to receive explicit attention from theoretical ecologists (Holt 1985, 1987, in press a; Pulliam 1988; Pulliam and Danielson 1991), and is indeed implicit in many earlier studies (e.g., Levin 1974). Source-sink effects should be most noticeable over moderate spatial scales (e.g., a small multiple of the root-mean-square dispersal distance of individuals), and where there is a slow rate of population decline in the sink. A simple continuous-generation model illustrates the latter point. Assume that density source saturates at N_{source}, that these individuals continue to reproduce at a per capita rate r_{source}, and that these excess individuals are forced into a sink where the expected per capita growth rate is r_{sink} (by assumption, r_{sink} is negative in a sink). The rate of change of the sink population is described by

$$dN_{sink} = [\text{immigration}] + [\text{local recruitment}] = N_{source}r_{source} + N_{sink}r_{sink}$$

and the equilibrial abundance of the sink population is

$$N^*_{sink} = N_{source}r_{source}/|r_{sink}|.$$

A large sink population may be maintained if the rate of decline in the sink is small and the rate of recruitment in the source is large. Analogous predictions emerge from a wide range of models (Holt in press (a)).

Source-sink effects potentially have important implications for community structure and dynamics (Holt 1984, 1985, in press, a; Pulliam 1988; Rosenzweig 1989; Bowers and Dooley 1991). Optimal habitat selection (assuming the conditions for an ideal free distribution) can magnify the effects of spatial heterogeneity in the environment, sharpening habitat partitioning patterns and promoting the coexistence of competitors (Chesson and Rosenzweig 1991). By contrast, given the kinds of dispersal that lead to source-sink dynamics, 1) habitat partitioning is blurred, for immigration permits species to persist in habitats from which they would otherwise be excluded (Levin 1974); and 2) a species may be individually superior in a particular habitat, yet competitively suppressed there because of a "spillover" of competitors from other habitats—in extreme cases even leading to competitive exclusion (Christiansen and Fenchel 1977, Holt, in press, a). Indeed, whether or not a habitat is a source or a sink for a species may depend upon the presence and abundance of competitors or predators (Holt in press, a).

A Threefold Example of Source-Sink Dynamics?

Convincing examples of sink populations maintained by density-independent dispersal have been found in several plant species (Keddy 1981, Kadmon and Shmida 1990). However, no direct experimental demonstrations of source-sink population structures exist for vertebrates, though considerable indirect evidence suggests that source-sink structures may be important in vertebrate communities. Many small mammal species display among-habitat variance in fitness and intraspecific aggression (e.g., white-footed mice, Morris 1991; Pulliam and Danielson 1991), a pattern consistent with source-sink population structures. Previous field experiments have shown that preventing dispersal in small mammals can lead to increases in local density (the "fence effect," Krebs et al. 1969), but it is not known whether these local increases are accompanied by population declines elsewhere in the landscape.

We believe that source-sink dynamics are a significant aspect of the small mammal distributions shown in Figure 4. Our reasoning and preliminary supporting evidence, species by species, are as follows.

Cotton rats. As noted above, this species is resident on large patches but essentially absent on small patches. It is present in low abundance on medium patches. Individuals there are recaptured at a low rate, but do not persist for long (Gaines et al., in press, a). Moreover, the presence of individuals on the medium patches is correlated with population highs on the large patches. It is reasonable to surmise that cotton rats on the medium patches are a transient sink population maintained by emigration from large patches.

S. hispidus is the likely competitive dominant in the system. In the lab, it aggressively dominates *M. ochrogaster* (Terman 1974). Several previous studies have shown that *M. ochrogaster* densities decline when *S. hispidus* is abundant (Glass and Slade 1980, Fleharty and Olson 1969, Frydenhall 1969). We have also observed a significant negative temporal correlation between *S. hispidus* and *P. maniculatus* densities ($r = -0.26$, $p < 0.05$), indicating that *S. hispidus* may also competitively suppress *P. maniculatus*. Because *S. hispidus* is relatively uncommon in small and medium patches (we believe by virtue of its large spatial requirements), both *M. ochrogaster* and *P. maniculatus* may have a refuge from competition in these smaller patches.

Prairie voles. Microtus ochrogaster is most abundant on medium patches, which we think are sources for it. We believe this species tends to be excluded from the large patches by *S. hispidus*, and that it persists there because of immigration. Our trapping data indicate that a greater proportion of between-patch movements by *M. ochrogaster* occurred from smaller to large patches than in the reverse direction. These results suggest that large patches

268

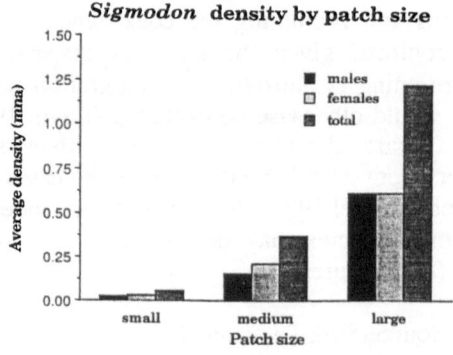

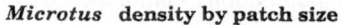

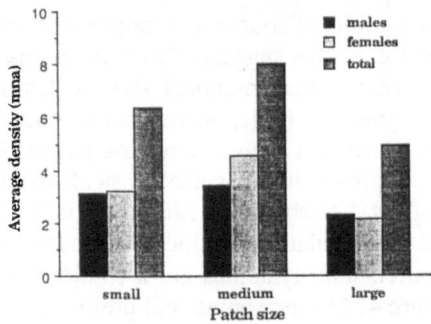

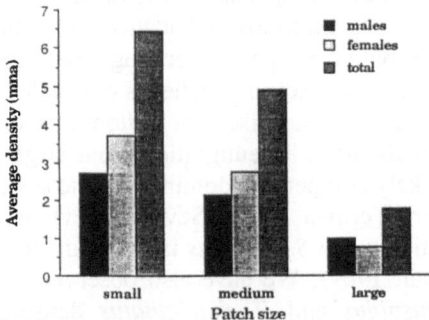

Figure 4. The distribution of three small mammal species on an experimental gradient in patch size. Each species is respectively most abundant on a different patch size. (From data in Foster and Gaines 1991, Gaines et al., in press).

are indeed sinks. *M. ochrogaster* is relatively common in the small patches. But small patches may be sinks for this species, too, because individual persistence rates are much lower on the small patch clusters than on larger patches (Gaines et al., in press, a). This suggestion is consistent with the observed age structure, which is biased toward nonreproductive subadults on small patches (Gaines et al., in press, b).

M. ochrogaster could be behaviorally dominant to *P. maniculatus*. We have no direct evidence of interference between these two particular species, but other studies have shown that members of the genus *Microtus* aggressively dominate smaller rodents (Linzey 1984, Blaustein 1980), including *Peromyscus* (Grant 1971).

Deer mice. *Peromyscus maniculatus* may have a refuge from interspecific competition in the small patches and in the interstitial area. The high density of *P. maniculatus* on the small patches could be due to this species' ability to exploit, to a moderate degree, the interstitial mowed areas (Foster and Gaines 1991), which in turn allows it to use small patches. Low-density *P. maniculatus* populations on the larger patches could be sink populations maintained by dispersal from small patches and the interstitial area; most movement occurs from small to larger patches, and age structure is biased toward nonreproductive younger individuals in larger patches.

Working hypothesis. The following table summarizes our working hypothesis for the three species' respective source-sink structures in our system:

PATCH SIZE	SPECIES BY BODY SIZE		
	Sigmodon	Microtus	Peromyscus
Large	**Source**	Sink	Sink
Medium	Sink	**Source**	Sink
Small	Sink	Sink	**Source**

In the entries below the diagonal, a patch is a sink because it is too small given that species' autecology; in the entries above the diagonal, a patch is a sink because it contains one or more dominant competing species. We intend to test this hypothesis by using semipermeable exclosures to manipulate dispersal, with and without competitors. For instance, removing the two larger species should convert the two larger patches from sinks into sources for deer mice.

Metapopulation Dynamics

In its most general sense, a "metapopulation" is defined as a system of local populations coupled by dispersal (Hanski 1991). Usually the term is used more narrowly to describe systems in which local populations go extinct and are then recolonized. In the original metapopulation models of Levins and others (see Hanski 1991), the populations in the metapopulation are on average assumed to be statistically homogeneous, with localized disturbances leading to a probabilistic distribution in extinctions. Dispersal allows a species to exploit this variance in phase among sites, thereby forestalling regional extinction even though every local population eventually goes extinct.

A recent review by Harrison (1991) suggests that this classical metapopulation structure is rather rare. Instead, one more commonly observes mainland-island and source-sink structures, in which a few local populations resist extinction and permit a species to persist despite frequent extinctions in other local populations. The "size" of a metapopulation is given by the number of constituent local populations making it up. Large metapopulations inevitably occupy a larger total area than do small metapopulations. Given that, in the real world, there is usually a relationship between area and habitat heterogeneity, large metapopulations almost inevitably will subsume substantial variation in local demographic parameters, colonization, and extinction rates. This spatial heterogeneity permits a number of important processes to operate, including source-sink dynamics within single species, the sorting out of species in communities along spatial gradients, and the regional balancing of colonizations and extinctions.

Many plant species in our experimentally fragmented landscape have short-distance dispersal (e.g., vegetative growth or large seeds) and can have large populations even within our smallest patches. For such species, our system may usefully be viewed as a hierarchically structured metapopulation. Each successional patch is a metapopulation of contiguous local populations, potentially coupled more strongly by dispersal among themselves than to populations on other patches; patch size scales the "size" of a metapopulation. The classic "area effects" of island biogeography may in part result from the internal spatial dynamics of such contiguously coupled metapopulations (Holt, in press, b).

There are intriguing hints in our data to suggest the operation of within-patch metapopulation dynamics (Holt et al., ms.). Because our plant samples are taken within grids of permanent quadrats, we can examine temporal trends in the spatial structure of the community. Figure 5 shows the spatial scaling of community dissimilarity at two points in succession (1985, the summer following the fall 1984 site initiation; and 1989, four years into succession). The data leading to this figure consist of all quadrat pairs within a given patch or patch cluster spanning .5 ha. For each quadrat pair, we computed a measure of dissimilarity in community composition (percentage remoteness, see figure legend), which ranges from 0 when two community samples have identical species lists and relative abundances of those species, to 1 when the species lists of the two samples are nonoverlapping (Pielou 1984).

Early in succession, the further apart two points are, the more dissimilar are the plant communities on those points. Interestingly, paired quadrats in the large patches separated by > 50 m in the large patches are more heterogeneous than paired quadrats the same distance apart in the medium-sized or smaller patches. It is likely, we suspect, that this reflects a more homogeneous physical environment in the smaller patches in the very first year of succession, but unfortunately we do not have the data to support or refute this suggestion.

As succession proceeded, community dissimilarity grew at all spatial scales, but particularly at small scales. The initial spatial scaling of between-community heterogeneity dissipates; so that the three patch size classes are essentially indistinguishable at scales $> =$ 10m. The exception to this trend is at the smallest spatial scale, where the rate of divergence between nearby quadrats on the large patches has been much less than on the smaller patches. This is consistent with the idea that there is some kind of short-range spatial coupling on large patches that tends to break down on smaller patches.

An additional line of evidence pointing to such coupling comes from analyses of local population persistence, defined here as the presence of a given species in a permanent quadrat in successive years. When we compute persistence for each vascular plant species in our system, and then combine the species into two broad functional groups—clonal species, which

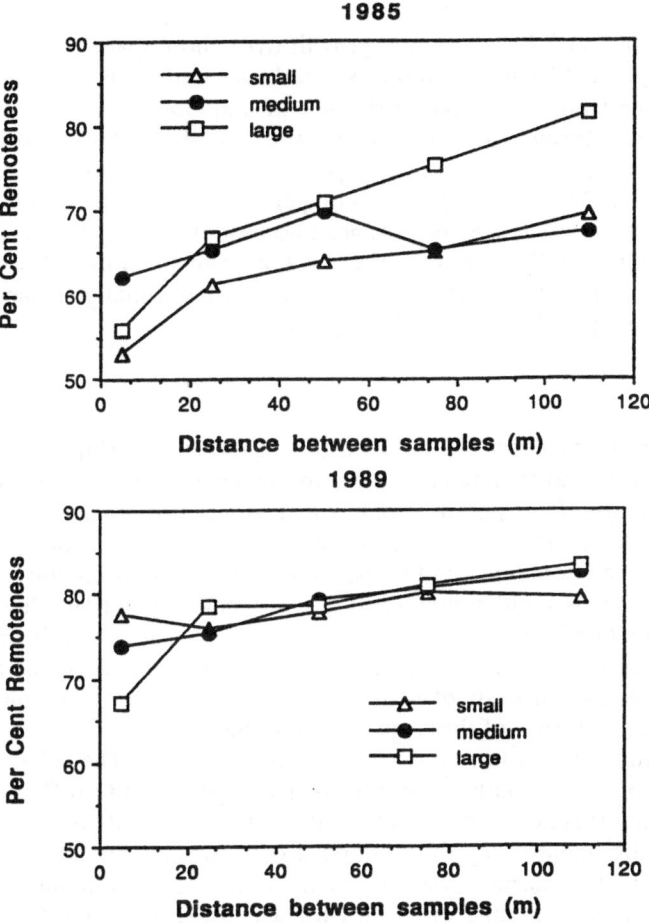

Figure 5. The spatial scaling of community dissimilarity with distance at two points in succession. The plant data includes point cover estimates in a grid of permanent quadrats, sampled from 1984 until the present. Let $N_{i1(t)}$ and $N_{i2(t)}$ be the amounts of plant species i in quadrats 1 and 2 at sampling period t, and s(t) the total number of species in both quadrats combined. A metric measure of dissimilarity between pairs of community samples (Pielou 1984) is <u>percentage remoteness</u>, PR, defined as

$$PR = 100 \ X \ (1 - \frac{\sum_{i=1}^{s} \min(N_{i1}(t), N_{i2}(t))}{\sum_{i=1}^{s} \max(N_{i1}(t), N_{i2}(t))})$$

We computed PR for each quadrat pair within the .5 ha large patches, and within the .5 ha clusters of small and medium patches. The figure plots PR (for a given patch size class) as a function of the distance separating the quadrat pair. By 1989, the initial strong spatial scaling of community dissimilarity had largely disappeared, except at small spatial scales within the large patches. (Holt et al., ms.)

reproduce in large measure by vegetative growth over short spatial scales; and non-clonal species reproducing by seed, which can be dispersed over longer distances—we find that over the three years of the study, the local persistence of non-clonal species was independent of patch size, but clonal species persisted significantly better on larger patches (Figure 6). Individuals or ramet populations of clonal species are akin to amoebae moving over the landscape. The probability that a species will persist on a given sample area (as in the embedded quadrat of Figure 2) may be enhanced because there may be a greater opportunity for rapid re-colonization following a local extinction, and possibly a lower probability of extinction in the first place (Holt, in press, b). These are essentially metapopulation effects reflecting the internal spatial structure of the patch.

CODA

The four spatial processes described above differ in the relationship assumed between the probability of extinction and the amount of time following isolation, relative to a comparable, non-isolated community. If dispersal is only of historical importance, there should be no increase in local extinction rates because of isolation. If, by contrast, individuals in a species require more space than contained within the local community to complete their life cycle, extinction should be very rapid following isolation. Sink populations deterministically go extinct following isolation, at a rate dependent upon initial densities and the rate of population decline. In a classical metapopulation, some isolated populations may last a long period of time. As in our experimental study of habitat fragmentation, any given community is likely to include species that fit each of these spatial syndromes.

The above ruminations have a decidedly terrestrial cast. The relative importance of the various spatial processes outlined above are likely to be quite different in marine vs. terrestrial ecosystems. It is not obvious that our thought experiment--isolating a patch of land and watching the subsequent decay of the trapped community therein--has much relevance, even as a metaphor, for marine systems. Many terrestrial communities, we tentatively suggest, have a substantial number of species for which dispersal is relatively unimportant in determining their persistence and abundance (except at very small spatial scales, or if one is considering
large biogeographic time scales). Over evolutionary time, this cadre of low-dispersal species finds expression in endemic species with highly restricted geographic ranges, and in spatially reticulated patterns of genetic differentiation within widespread species. Such species are essentially absent in marine communities. Source-sink dynamics, by contrast, are likely to be at least as important in marine systems as in terrestrial systems, if not more so.

A largely unexplored dimension of community ecology is to ascertain how local community structure reflects the interplay of the different spatial processes different species in the ensemble. For instance, the vast literature on the structure and associated with dynamics of food webs (Cohen et al. 1990) has until very recently ignored the potential importance of spatial dynamics in governing food webs (Holt, in press, b). Yet there is often a strong positive correlation between the trophic rank of a species and the spatial scale relevant to its dynamics. Grasping the full implications of spatial dynamics in determining the structure of local communities is, we believe, one of the most important challenges facing community ecologists today.

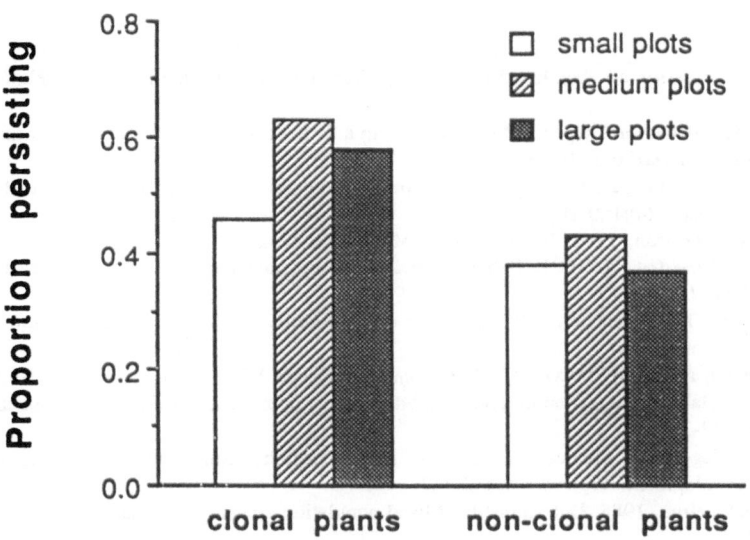

Figure 6. Local population persistence as a function of patch size and life history. If $N_{ij}(t)$ is the abundance of plant species i on quadrat j in sampling year t, our operational measure P_{ij} of local population persistence (for samples one year apart) is that a population persists (P_{ij} = 1) if $N_{ij}(t + 1) > 0$, given that $N_{ij}(t) > 0$, and that a population does not persist (P_{ij} = 0) if $N_{ij}(t + 1) = 0$, given that $N_{ij}(t) > 0$. For each species on each patch size, we computed its mean persistence during the years 1987-1990. We then categorized species by basic life history into those with clonal growth (runners, rhizomes, etc.), and those reproducing solely by seed. Clonal species collectively show a significantly higher rate of local persistence on medium and large patches, compared to small patches. Non-clonal species show no effect of patch size, and overall persist less well than do clonal species (Robinson et al., in press).

ACKNOWLEDGEMENTS

We would like to thank a long list of faculty, students, and research associates who have assisted in one way or another in this project, all of whom are colleagues in the best sense of the word: George Robinson, Jay Diffendorfer, Jo Foster, Mike Johnson, Wendy Sera, Kevin Parker, Kevin O'Brien, Jim Roth, Esa Teravainen, Terry Shistar, Cathy Gorman, Steve Hamburg, Ken Armitage, Dean Kettle, Henry Fitch, and Ed Martinko. We also thank the General Research Fund of the University of Kansas, as well as the National Science Foundation (BSR-8718088), for their continued support.

REFERENCES

Bazzaz, F.A. 1968. Succession on abandoned fields in the Shawnee Hills, southern Illinois. *Ecology* 49:924-936.

Blaustein, A. R. 1980. Behavioral aspects of competition in a three-species rodent guild of coastal California. *Behav. Ecol. Sociobiol* 6:247-255.

Bowers, M.A., and J.L. Dooley, Jr. 1991. Landscape composition and the intensity and outcome of two-species competition. *Oikos* 60:180-186.

Brown, J.H., and A.C. Gibson. 1983. *Biogeography*. Mosby, St. Louis.

Chesson, P., and M. Rosenzweig. 1991. Behavior, heterogeneity, and the dynamics of interacting species. *Ecology* 72:1187-1195.

Christiansen, F.B., and T.M. Fenchel. 1977. *Theories of Populations in Biological Communities*. Springer-Verlag: New York.

Cohen, J.E., F. Briand, and C.M. Newman. 1990. *Community Food Webs*. Springer-Verlag, New York.

Dickman, C.R. 1987. Habitat fragmentation and vertebrate specie richness in an urban environment. *J. Appl. Ecology* 24:337-351.

Egler, F.E. 1954. Vegetation science concepts. I. Initial floristic composition, a factor in old-field vegetation development. *Vegetatio* 4:412-417.

Fahrig, L., and J. Paloheimo. 1988. Determinants of local population size in patchy habitats. *Theor. Pop. Biol.* 34:194-213.

Fleharty, E.D., and L.E. Olson. 1969. Summer food habitats of Microtus ochrogaster and Sigmodon hispidus. *J. Mamm.* 50: 475-486.

Foster, J., and M.S. Gaines. 1991. The effects of a successional habitat mosaic on a small mammal community. *Ecology* 72:1358-1373.

Fretwell, S.D. 1972. *Populations in Seasonal Environments*. Princeton University Press, Princeton.

Frydenhall, M.J. 1969. Rodent populations on four habitats in central Kansas. *Trans. Kansas Acad. Sci.* 72:213-222.

Gaines, M.S., G.R. Robinson, J.E. Diffendorfer, R.D. Holt, and M.L. Johnson. In press (a). The effects of habitat fragmentation on small mammal populations. In: *Wildlife 2001: Populations*, D.R. McCullough and R.H. Barret (eds.).

Gaines, M.S., J. Foster, W. Sera, J.E. Diffendorfer, and R.D. Holt. In press (b). Population processes and biodiversity. *Trans. N. Amer. Wildlife and Natural Resources Conference* 57

Gaines, M.S., A.M. Vivas, and C.L. Baker. 1979. An experimental analysis of dispersal in fluctuating vole populations: Demographic parameters. *Ecology* 60:814-828.

Glass, G.E., and N.A. Slade. 1980. Population structure as a predictor of spatial association between *Sigmodon hispidus* and *Microtus ochrogaster*. *J. Mammal.* 61:473-485.

Grant, P.R. 1971. Experimental studies of competitive interaction in two-species systems III. *Microtus* and *Peromyscus* species in enclosures. *J. Anim. Ecol.* 40:323-350.

Hairston, N.G., Sr. 1989. *Ecological Experiments: Purpose, Design, and Execution*. Cambridge University Press, Cambridge.

Hanski, I. 1991. Single-species metapopulation dynamics: Concepts, models, and observations. *Biol. J. Linn. Soc.* 42:17-38.

Harrison, S. 1991. Local extinction in a metapopulation context: An empirical evaluation. *Biol. J. Linn. Soc.* 42:73-88.

Hastings, A. 1990. Spatial heterogeneity and ecological models. *Ecology* 71:426-428.

Holt, R.D. 1984. Spatial heterogeneity, indirect interactions, and the coexistence of prey species. *Am. Nat.* 124:377-406.

_____. 1985. Population dynamics in two-patch environments: Some anomalous consequences of an optimal habitat distribution. *Theor. Pop. Biol.* 28:181-208.

_____. 1987. Prey communities in patchy environments. *Oikos* 50:276-290.

_____. In press (a). Ecology at the mesoscale: The influence of regional processes on local communities. In: R. Ricklefs and D. Schluter (eds.). *Community Diversity: Historical and Geographical Perspectives*, University of Chicago Press, Chicago.

_____. In press (b). Internal spatial dynamics: A neglected facet of island biogeography. *Theor. Pop.*

Biol.

Holt, R.D., G.R. Robinson, and M.S. Gaines. Manuscript. Vascular plant diversity and succession in experimental habitat fragments.

Johnson, M.L., and M.S. Gaines. 1987. The selective basis for dispersal of the prairie vole, *Microtus ochrogaster. Ecology* 68:684-694.

Kareiva, P. 1987. Habitat fragmentation and the stability of predator-prey interactions. *Nature* 326:388-390.

Kareiva, P., and M. Anderson. 1988. Spatial aspects of species interactions: The wedding of models and experiments. In: A. Hasting (ed.). *Community Ecology.* Springer-Verlag, New York.

Kadmon, R., and A. Shmida. 1990. Spatiotemporal demographic process in plant populations: An approach and a case study. *Am. Nat.* 135: 382-397.

Keddy, P.A. 1981. Experimental demography of the sand-dune annual, *Cakile eduntula*, growing along an environmental gradient in Nova Scotia. *J. Ecol.* 69: 615-630.

Krebs, C.J., B. Keller, and R. Tamarin. 1969. *Microtus* population biology. *Ecology* 50:587-607.

Levin, S.A. 1974. Dispersion and population interactions. *Am. Nat.* 108:207-228.

_____. 1988. Pattern, scale, and variability: An ecological perspective. In: A. Hastings (ed.). *Community Ecology.* Springer-Verlag, New York, pp. 1-24.

_____. In press. Concepts of scale at the local level.

Linzey, A.V. 1984. Patterns of coexistence in *Synaptomys cooperi* and *Microtus pennsylvanicus. Ecology* 56:382-393.

Louda, S.M., K.H. Keeler, and R.D. Holt. 1990. Herbivore influences on plant performance and competition interactions. In: J.B. Grace and D. Tilman (eds.). *Perspectives on Plant Competition.* Academic Press, New York, pp. 414-444.

Lovejoy, T.E., J.M. Rankin, R.O. Bierregard Jr., K.S. Brown Jr., L.A. Emmons, and M.E. Van der Voort. 1984. Ecosystem decay of Amazon forest remnants. In: M.H. Nitecki (ed.). *Extinctions.* University of Chicago Press, Chicago, pp 295-326.

MacArthur, R.H., and E.O. Wilson. 1967. *The Theory of Island Biogeography.* Princeton Univ. Press, Princeton.

Milinski, M., and G.A. Parker. 1991. Competition for resources. In: J.R. Krebs and N.B. Davies (eds.). *Behavioural Ecology*: Sunderland, MA, pp. 137-169.

Morris, D.W. 1987. Habitat-dependent population regulation and community structure. *Evol. Ecol.* 2:232-252.

_____. 1991. Fitness and patch selection in white-footed mice. *Am. Nat.* 138:702-716.

Patterson, B.D., and W. Atmar. 1990. Nested subsets and the structure of insular mammalian faunas and archipelagos. *Biol. J. Linn. Soc.* 28:65-82.

Peters, R.H. 1983. *The Ecological Implications of Body Size.* Cambridge University Press, Cambridge.

Pielou, E.C. 1984. *The Interpretation of Ecological Data.* Wiley, New York.

Pulliam, H.R. 1988. Sources, sinks, and population regulation. *Am. Nat.* 110:107-119.

Pulliam, H.R., and B.J. Danielson. 1991. Sources, sinks, and habitat selection: A landscape perspective on population dynamics. *Am. Nat.* 137:50-60 (supplement).

Quinn, J.F., and S.P. Harrison. 1987. Effects of habitat fragmentation on species richness: Evidence from biogeographic patterns. *Oecologica* 75:132-140.

Ricklefs, R.E. 1987. Community diversity: Relative roles of local and regional processes. *Science* 235:167-171.

Robinson, G.R., and J.F. Quinn. 1988. Extinction, turnover and species diversity in an experimentally fragmented California annual grassland community. *Oecologia* 76:71-82.

Robinson, G.R., R.D. Holt, M.S. Gaines, S.P. Hamburg, M.L. Johnson, H.S. Fitch, and E.A. Martinko. In press. Consequences of habitat fragmentation vary within and among ecosystem components. *Science.*

Rosenzweig, M.L. 1974. On the evolution of habitat selection. Proceedings of the First International Congress of Ecology. Wageningern, Centre for Agricultural Publishing and Documentation, The Hague, Netherlands, pp. 401-404.

_____. 1989. Habitat selection, community organization and small mammal studies. In: D.W. Morris, Z. Abramsky, B.J. Fox, and M.R. Willig (eds.). *Patterns in the Structure of Mammalian Communities.* Texas Tech University Press, Lubbock, pp. 5-23.

Abramsky, B.J. Fox, and M.R. Willig (eds.). *Patterns in the Structure of Mammalian Communities.*Texas Tech University Press, Lubbock, pp. 5-23.

_____. 1991. Habitat selection and population interactions: The search for mechanism. *Am. Nat.* 137:5-28 (supplement).

Rosenzweig, M.L., Z. Abramsky, B. Kotler, and W. Mitchell. 1985. Can interaction coefficients be determined from census data? *Oecologia* 66: 194-198.

Roth, J.D., and R.D. Holt. Manuscript. Habitat selection of birds in a fragmented landscape: Effects of habitat structure, patch size, and edge on a bird community in northeastern Kansas.

Roughgarden, J., S. Gaines, and H. Possingham. 1988. Recruitment dynamics in complex life cycles. *Science* 241:1460-1466.

Saunders, D.A., R.J. Hobbs, and C.R. Margules. 1991. Biological consequences of ecosystem fragmentation: A review. *Conservation Biology* 5:18-32.

Shmida, A., and M. V. Wilson. 1985. Biological determinants of species diversity. *J. Biogeography* 12:1-20.

Sih, A. 1987. Prey refuges and predator-prey stability. *Theor. Pop. Biol.* 31:1-12.

Teravainen, E. In press. Distribution of the soldier beetler Chauliognathus pennsylvanicus in a spatially fragmented environment. *Oecologia.*

Terman, M.R., 1974. Behavioral interactions between *Microtus ochrogaster* and *Sigmodon*: A model for competitive exclusion. *J. Mamm.* 55: 705-719.

Wiens, J.A. 1989. Spatial scaling in ecology. *Functional Ecology* 3:385-397.

Wilcove, D.S., C.H. McLellan, and A.P. Dobson. 1986. Habitat fragmentation in the temperate zone. In: M.E. Soule (ed.). *Conservation Biology.* Sinauer, Sunderland, MA, pp 237-256.

18
ECOLOGICAL AND EVOLUTIONARY CONSEQUENCES OF PATCHINESS: A MARINE-TERRESTRIAL PERSPECTIVE

Pablo A. Marquet, Marie-Josee Fortin, Jesus Pineda, David O. Wallin, James Clark, Yegang Wu, Steve Bollens,Claudia M. Jacobi and Robert D. Holt

INTRODUCTION

A quantitative description of patchiness and the assessment of its effects on ecological and evolutionary processes represents a major research focus as well as a challenge for ecologists and evolutionary biologists (e.g., Pickett and White 1985, Shorrocks and Swingland 1990, Kolasa and Pickett 1991). Patchiness is neither unique in origin nor characteristic of particular temporal or spatial scales; rather, patchiness emerges from the interactions between physical and biotic processes (Levin 1976, 1978) and is apparent at any scale of resolution. The scale dependency of patchiness and the complexity it generates calls attention to the need for new modeling approaches where spatial and temporal heterogeneity is explicitly incorporated (e.g., Hassell et al. 1991, Deutschman et al., this volume) and for new methodological tools to deal with problems of scale (e.g., Milne 1992, Garcia-Moliner et al., this volume).

In this chapter we address some of the ecological and evolutionary consequences of patchiness in both terrestrial and marine ecosystems. To do this we focus on the interactions across space and time scales. We illustrate particular phenomena by means of simulation models and by pointing out new approaches and conceptual frameworks for the analysis of patchiness and its consequences. We emphasize that the ecological and evolutionary consequences of patchiness vary according to the scales at which different organisms operate within a particular environment (intrinsic component), as well as on the scales of abiotic variability that characterize each environment (extrinsic component). First we compare terrestrial and marine ecosystems with respect to their characteristic spatial and temporal scales of variability. Next we address some of the ecological consequences of patchiness at two levels: population and landscape. Finally, we discuss the consequences of patchiness for microevolutionary change and macroevolutionary patterns and demonstrate the connections with the ecological scales. Throughout the chapter, the word *patchiness* is used to refer to both spatial and temporal heterogeneity or environmental variability.

THE TEMPORAL AND SPATIAL SCALES OF SOME ECOLOGICAL PHENOMENA: A MARINE-TERRESTRIAL COMPARISON

There are many ways to contrast marine and terrestrial systems (e.g., Fuentes and Jaksic 1988). Here we will compare temporal scales of some ecological variables in both systems.

This approach has been used by Steele (1989, 1991a) and by Dayton and Tegner (1984). We start by exploring some consequences of marine and terrestrial physical variability for life-history attributes. Later, we focus on a particular methodology that can help in contrasting marine and terrestrial systems and explore some of the resulting insights.

Types and Scales of Variability

Variability in the physical environment has several aspects that are important for understanding biological heterogeneity and how it differs between terrestrial and marine ecosystems. Environments of low abiotic variability may still be highly variable as a consequence of the behavior of the organisms themselves (e.g., treefall gaps in forests, foraging activities of predators). However, under highly variable abiotic conditions, the processes that produce biotic variability will tend to be disrupted. The limit cycles associated with simple predator-prey interactions might not be expected in nature if they are readily disrupted by other sources of variability.

The scales of variability also govern whether a species with a given life history perceives a particular source of variance as a general feature of its environment or as a rare event. In the former case, an organism might adapt to the heterogeneity it is sure to face on a regular basis. In the latter case, the cost of adaptation to a perceived rare event might outweigh the benefit that is realized only occasionally.

Predictable events are perceived much differently than are unpredictable ones. Highly predictable variability in time includes periodic "events" (e.g., diel, fortnightly, seasonal) and temporally autocorrelated processes, such as dissipative structures in ocean currents. Organisms can adapt to variability that is highly predictable in time. Zooplankton migrate vertically with diel variation in light. Many temperate angiosperm plants construct relatively inexpensive leaves each year, shedding them in step with the highly predictable climate variation represented by seasonal changes. The predictability of events in this sense appears to be reflected in the life-history strategies of the dominant organisms present in an environment. Predictable variability can be the focus of much behavioral and physiological adaptation. For population persistence, the critical aspects are environmental changes where variance is large, unpredictable, and of a time scale that demands adaptation. There is a great diversity of responses or bet-hedging strategies (Stearns and Crandall 1981). These include iteroparity of a relatively resistant mature stage, dispersal to spatially restricted sites of recruitment, and dormancy (Cohen 1966, Gadgil 1971, Roff 1975, Levin et al. 1984, Levin 1985, Cohen and Levin 1985, 1991). In this case the organism must integrate over areas and/or times of unfavorable conditions in order to realize comparatively large reproductive gains at suitable times and places. In general, the gains realized during rare favorable periods must be sufficiently large to offset the losses that accumulate during unfavorable times. Marine/terrestrial comparisons are interesting here, because of some general patterns in costs associated with bet-hedging strategies in the two environments. Consider the role of dispersal. Seeds possessing built-in carbohydrate reserves have a greater chance of survival. However, such reserves are not conducive to dispersal by wind. Due to the physical characteristics of the atmosphere, adaptations that result in high atmospheric residence times (e.g., structures that decrease drag) are almost universally achieved at the cost of low seed reserves. Because of the high juvenile mortality associated with low seed reserves, terrestrial plants are not expected to engage in a bet-hedging strategy of high dispersal rates. It is likely that the costs of high dispersal are a strong force regulating many aspects of terrestrial plant assemblages.

On the other hand, the physical characteristics of marine environments (gravity is not a problem) renders increased stored reserves less costly in terms of dispersal and subsequent juvenile survival, setting a different scenario for the evolution of bet-hedging strategies. As pointed out by Strathmann (1990), differences in the physical characteristics of terrestrial and marine habitats have resulted in the prevalence of different life histories in these environments. These relationships between life history and the physical environment may provide some insights for understanding the regularities observed in Stommel diagrams.

The Stommel Diagram

Stommel diagrams summarize variability in natural systems. They are plots that identify which scales contribute most to the total variance of the system (Figure 1). The characteristic scale of a phenomenon can be defined as the length or time interval necessary for the variable of interest to show a significant degree of variation (Powell 1989). Stommel diagrams (or their heuristic equivalents) have been used as descriptors of the most relevant spatial and temporal scales of physical phenomena (Stommel 1963), zooplankton biomass (Haury et al. 1978), or as an aid in defining the relevant scales in paleobiological phenomena (Schopf 1972). Other diagrams (e.g., Steele 1978, 1989, 1991a; Delcourt et al. 1983; Harris 1986; Wiens 1989) are also used to portray the spatial and temporal scales of the most relevant phenomena. Haury et al.'s (1978) Stommel diagram is, to our knowledge, the only one that has been published for an ecological variable. An important feature of Haury et al.'s diagram is that it shows that biomass variability peaks at the largest spatial and temporal scales, implying that only phenomena characterized by large temporal scales produce changes at the large spatial scale.

Technically, the Stommel diagram is a three-dimensional representation of the power spectrum, Sq, of a relevant record q (i.e., q represents some variable, phenomenon, process, etc.) as a function of frequency (W) and wave number (K) (see Figure 1). Since period (T) $\alpha\ W^{1}$ and wave length (λ) $\alpha\ K^{-1}$, the analyst commonly plots S_q vs T (a measure of temporal scale) and λ (a measure of spatial scale). The volume under the S_q surface represents the total variance associated with the record (Figure 1).

Some ecological variables may be poorly suited for spectral analysis, violating some necessary assumptions (for examples see Haury et al. 1978). One case is represented by variables that show secular trends in space or time (i.e., are non-stationary), such as species richness. Further, large variability may be related to single (i.e., non-stationary) catastrophic events. Moreover, one can also envision situations in which there are no apparently characteristic time or spatial scale associations. Another difficulty is that in marine and terrestrial systems (e.g., Harris 1986, Pagel et al. 1991) longitudinal, latitudinal, and vertical (altitudinal) transects are likely to produce different dominant scales. In this case, one might plot different contours for the different axes. Finally, the existence of a maximum spatial scale value, given by the size of the earth, implies that at some point the matching between temporal and spatial scales is no longer possible.

Because of the difficulty in collecting the appropriate data for the largest time and space scales, Stommel diagrams are likely to be the result of educated guesses in these spatial and temporal domains. In our case (Figure 2), we followed Haury et al. (1978) in completing the diagrams with our best knowledge and best guesses (in some other cases inferring the patterns from the processes). Therefore, the diagrams are schematic representations; plots with real data might be restricted to smaller scales. Nonetheless, we believe the general

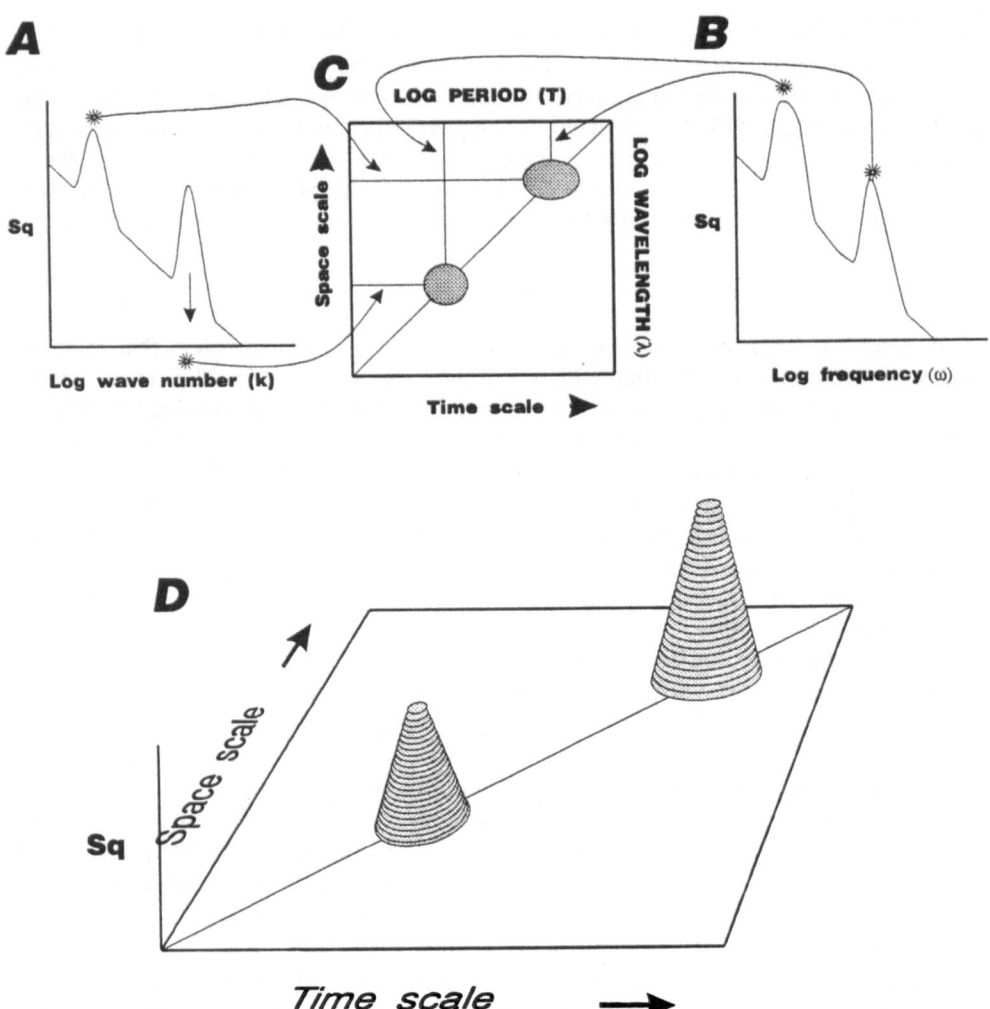

Figure 1. Schematic representation of the procedure to construct a Stommel diagram. Panel A represents the power spectrum, S_q, of a spatially distributed variable, while panel B represents the spectrum of a time series. Panel C represents the resulting Stommel diagram. Notice that the peaks in the spectral diagrams represent the most energetic frequencies or wave numbers (i.e., the important scales). The peaks in time and space are matched in the Stommel diagram. Panel D is a three-dimensional depiction of panel C.

281

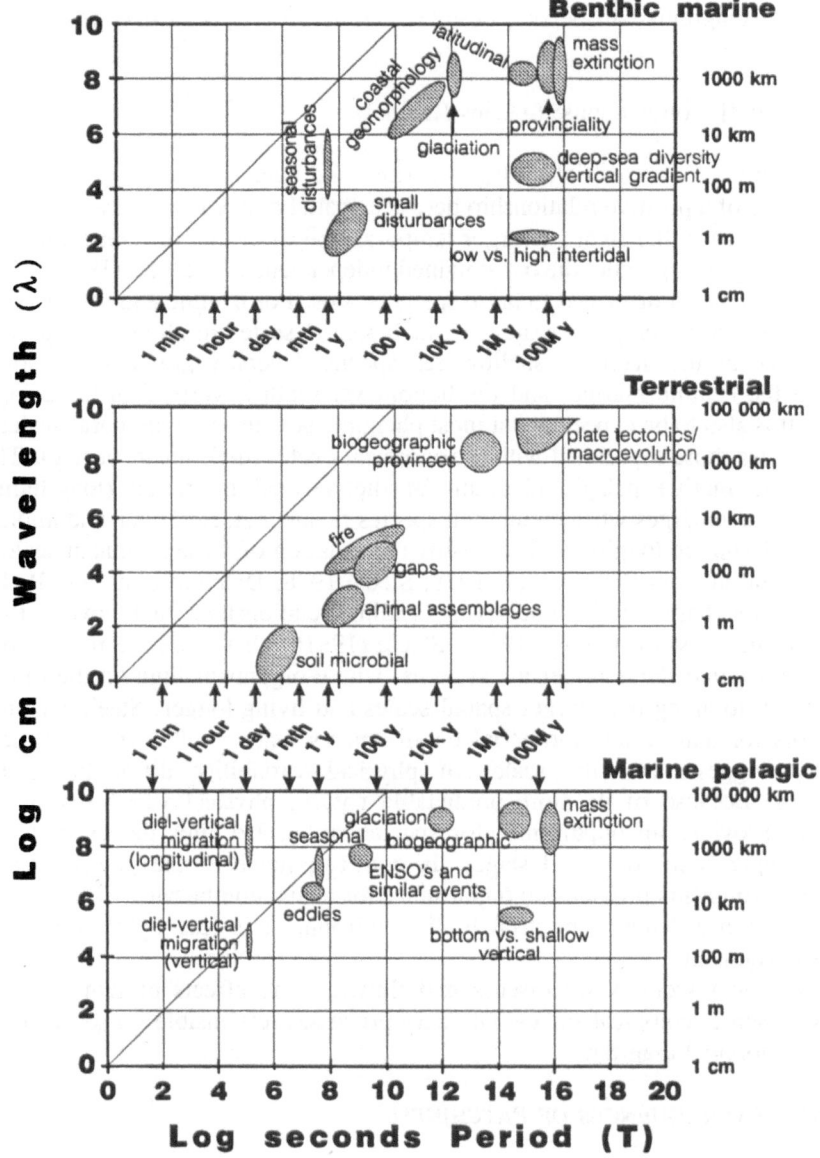

Figure 2. Stommel diagrams for species richness. Marine benthos, marine pelagic, and terrestrial systems. Marine benthos does not include coral communities. (ENSO = El Niño-Southern Oscillation). Stippled figures represent contours of maximum variance (see text for an explanation).

features of our diagrams will persist, despite some fine-tuning that may occur as more data accumulates.

Marine-Terrestrial Comparisons: Species Richness

Stommel diagrams for terrestrial, pelagic marine, and benthic systems (Figure 2) show the general features of a positive relationship between spatial and temporal scales. The terrestrial pattern is displaced to the right (to larger temporal scales) of that for pelagic marine; in this aspect, these diagrams parallel the one obtained independently by Steele (1991a). Further, the pelagic diagram is also displaced to large spatial scales. For benthic and pelagic systems, the phenomena with large temporal scale and small spatial scale are all associated with vertical gradients, such as the deep vs. shallow sea species diversity gradient, the low vs. high intertidal in the benthic marine, and the bottom vs. shallow vertical in the pelagic marine (Figure 2). It is also to be expected that most phenomena with small temporal scales and large spatial scales are those associated with astronomical cycles (diel, lunar, yearly). This is most obvious in the marine pelagic diagram for diel-vertical migration (longitudinal). This particular pattern emerges when comparing species richness at, say, noon and midnight at two sites separated 180° in longitude. The positive association of these dominant scales has been noted by several authors (Haury et al. 1978, Steele 1978, Delcourt et al. 1983). It has been explained on the basis of the underlying scaling relationships that govern the physical heterogeneity in the sea (Stommel 1963). Steele (1991b) also notes the regular pattern that emerges with trophic status in pelagic systems, where organisms high on the food chain are larger, thereby foraging over larger spatial scales and living longer. Steele's comparison of such patterns for marine and terrestrial environments featured the offset between the two curves. He suggests that the scales of physical variability differ between the two environments. Because of its more predictable nature, physical variability in the marine environment provides an opportunity for specialization. He cites the example of a near-universal pelagic habit for larval stages. In contrast, much of the physical variability in terrestrial environments provides no opportunity for adaptation because of its high frequency and unpredictability. Thus, terrestrial biotic environment is less tightly coupled with the physical environment.

In the next sections we discuss and illustrate the effects of temporal and spatial variability on some ecological and evolutionary processes responsible for some of the patterns observed in Stommel diagrams.

ECOLOGICAL CONSEQUENCES OF PATCHINESS

At small spatial and temporal scales, biotic and physical variability in the environment is of paramount importance in affecting ecological patterns and processes. In this section, we discuss some specific implications of patchiness on ecological processes, starting with the characterization of patch properties with an emphasis on boundary effects. We then consider some of the population and landscape-level ecological consequences.

Patch Boundary Properties and Ecological Processes

The bulk of the literature on patchiness emphasizes patch size and shape as salient features of patches, while comparatively little attention has been directed to the functional characteristics of boundaries and their potential influence on ecological processes. In part this is because, in order to characterize and understand ecosystem processes, terrestrial researchers have focused primarily on homogeneous regions, avoiding the heterogeneous areas linking them. As a result, properties of boundaries have been overlooked, and their spatial representation often reduced to a line on a map. In marine systems, however, boundaries associated with water masses such as fronts, thermoclines, or pycnoclines are well-studied features (e.g., Le Fèvre 1986; Steele 1989, 1991b; Barry and Dayton 1991) and are generally regarded as having, sometimes, characteristic communities (Steele 1991b, Ray 1991).

The boundary of a patch can be viewed as the location in space at which the rate of change of a given variable or assemblage of variables is the highest (Burrough 1986). This definition encompasses both structural and functional boundaries (Allen and Starr 1982). While some boundaries may completely surround a patch, others simply denote a discontinuity in rate of change, structure, or composition between adjacent systems (O'Neill et al. 1986). From a dynamic point of view, boundaries and ecotones are transition zones, at which flux of material or energy and other ecological processes take place (Margalef 1963, Naiman and D'Camps 1990).

The terms *boundary* and *ecotone* were primarily used to define the limits between terrestrial biomes based on vegetation types. Recently they have been extended to account for discontinuities that occur at all spatial scales. At a given scale, boundary edges might be perceived in a series of forms, ranging from a sharp to a fuzzy transition zone (Ferson 1988). The detection of boundaries depends on the scale of observation; boundaries that at some scale are recognized as a whole ecosystem, such as intertidal systems, can be viewed as an ecotone or even a very sharp edge at larger scales. Boundary properties have recently started to be the focus of an increasing number of studies in terrestrial environments (e.g., Forman and Godron 1986). This reflects the relative ease in visualizing them (e.g., forest-prairie, hemlock-maple groves) compared to pelagic realms, where patches and their boundaries are much more dynamic. Indeed, at small scale in aquatic ecosystems, patches due to biological behavior grow and decay at a speed that does not allow the observer to make accurate assessments of the patch size and its boundary. For these ecosystems, a statistical definition of patchiness using variance/mean indices may be more appropriate (Downing 1991). On the other hand, boundaries in large-scale pelagic communities are generally determined by mesoscale physical dynamics such as fronts and pycnoclines (Brinton 1962, McGowan 1972, Brandt and Wadley 1981) and are typically more persistent in time than small scale boundaries. This is also apparent in terrestrial ecosystems, where physical variables such as temperature may impose limits on the geographical distribution and survival of organisms such as birds (Root 1988) or trees (Davis et al. 1986, Davis 1987). Although the importance of large-scale boundaries as a general feature of pelagic ecosystems is well established, our perceptions may be biased to some degree by the limited capabilities of sampling pelagic habitats on smaller scales. The further development and use of new methodologies such as remote sensing and continuous recording systems—including fluorometry, flow cytometry, hydroacoustics, and optical plankton counters—may change our view of the dominant sizes, shapes, and boundary characteristics of patches in the pelagic realm.

Patch dynamics refers to the temporal response of patches to either autonomous (intrinsic) or induced (extrinsic) processes or forces. A patch may react to these forces by changing its internal state, thus reinforcing or reducing the difference between the patch and its surroundings, which in turn leads to changes in boundary properties. In terrestrial landscapes, the shape of boundaries has implications for ecological processes such as colonization: forest recovery is faster when forest patches present concave boundaries rather than convex ones (Hardt and Forman 1989), due to the combined effect of microclimate and larger seed pool. In the same way that boundary shape may affect the dynamics of some ecological processes, other processes such as spatial competition, disturbance, or organismal response to environmental changes may influence the shape of boundaries (Forman and Godron 1986, Paine and Levin 1981, Navarrete and Castilla 1990). Thus, there is a two-way interactive process where the ecological phenomena influence the properties of the boundary and where the boundary characteristics may determine specific biological and ecological responses.

Boundaries affect processes occurring at disparate spatial scales, from local movements of an individual up to processes such as long-term species migration or dispersal. However, whatever the scale, a boundary acts either as a barrier or as a permeable membrane (Wiens et al. 1985). The degree of permeability might be different for fluxes leaving or entering the patch and could depend on the process or species involved. The interaction among boundary shape, permeability, and patch size is particularly important in the outcome of migration processes. Stamps et al. (1987) explored the impact of geometry and permeability of boundary on emigration in a patch using computer simulations. Patch shape and size were not found to be relevant as long as the degree of permeability was low. These results were supported by field studies of some terrestrial insects (Turchin 1986) and vertebrates (Stamps et al. 1987).

Permeability influences population interactions such as competition and predator-prey dynamics by varying the rate of dispersal of different species across heterogeneous systems. For example, a refuge is a patch where the boundary is less permeable to the predator than to its prey. Taylor and Pekins (1991) found such a functional boundary in the case of wolf-deer interactions. A refuge for the deer was a patch that the wolf cannot occupy due to behavioral constraints. Similarly, Menge et al. (1985) found that the survival of many intertidal organisms, and thus diversity and community structure, depend on the availability of refuges from both vertebrate and invertebrate predators.

Boundaries are a salient feature of patches. Their functional characteristic can strongly affect the outcome of species interactions and species persistence on heterogeneous landscapes. Their characterization is important for a more complete understanding of the ecological consequences of patchiness, and for a more precise assessment of the effect that alteration of landscape heterogeneity (mainly by humans) can have on species survival. One of the effects of human use of the landscape is to modify the shape of patches by straightening the boundaries. In this way, the edge-to-area ratio is altered, and this in turn may affect the level of permeability. Besides acting at the boundary level, humankind also fragments the landscape in such a way that suitable patches for a given species become discontinuous and usually smaller. This discontinuity and size alteration can be critical depending on the dispersal capabilities of each species, and on its minimum territorial requirements. Thus, some aspects of landscape management, such as reserve planning, should take into account not only boundary characteristics but also the spatial requirements of the species involved. This is a difficult task to accomplish in terrestrial landscapes, and this kind of space management may not be applicable at all in seascapes.

Some Population-Level Consequences of Patchiness

Much of the appreciation for the importance of patchiness on population dynamics was stimulated by the classic work of Huffaker (1958). Unfortunately, logistical constraints have severely limited the amount of experimental work conducted in natural systems (Kareiva 1990, Kareiva and Anderson 1988). Thus much of the work in this field has been conducted using mathematical models (see reviews by Taylor 1990, Hastings 1990, Reeve 1990, Harrison, in press). These models explore the relationship between patchiness and population persistence and demonstrate the importance of species attributes (e.g., life history and dispersal behavior).

The importance of both patchiness and species dispersal capabilities is illustrated by a predator-prey model recently developed by Schumaker and Wallin (in prep). This model (Table 1) tracks predator and prey population dynamics on a 50 x 50 gridded landscape. Populations dynamics for each grid cell are calculated independently using standard Lotka-Volterra difference equations (Table 1). For all simulations, a limited proportion of the landscape is defined as suitable habitat. The remainder of the grid cells are defined as unsuitable for both predator and prey. The grain specifies the size distribution of suitable patches. For a grain size of one, suitable patches are randomly placed onto the landscape one grid cell at a time. For a grain size of two, suitable patches of two-grid cells by two-grid cells are randomly placed on the landscape. As grain size increases, the suitable habitat occurs in a smaller number of larger blocks and the inter-patch distances increase (Figure 3). At the maximum grain size all of the suitable habitat is present in a single large block.

When predator or prey populations in a grid cell exceed a predefined carrying capacity (see Table 1), the excess individuals disperse. Individuals can move through unsuitable grid cells, but they must continue to walk until they die from natural causes (Table 1) or find a suitable grid cell with a population size below the carrying capacity.
The edges of the landscape are wrapped into a torus so that animals experience a continuous landscape. The dispersal track for an individual animal consists of one or more segments or "steps," and the characteristics of the track are defined by three movement parameters (Table 1). For both predator and prey, the direction of the first step is selected at random and the direction of all subsequent steps is constrained by the species directional autocorrelation. The prey uses an uncorrelated walk; that is, the direction of each step is independent of the previous step direction. The predator uses a self-avoiding walk; after random selection of the initial direction, the animal continues in the same direction. The length of each step is variable; an integer value is selected at random between the range of zero and the species' maximum step size. During dispersal, animals are assumed to experience higher mortality rates. The more steps involved in a dispersal track, the lower the survival rate. The predator and prey each have a specified survival probability per step.

Two sets of simulations were conducted (Table 1). In the first set, the predator's maximum step size (a measure of dispersal distance) was set to 1.0; in the second to 2.0. Each set of simulations included thirty model runs, five for each grain size. For each run, summary statistics for the predator population were computed without considering the first 100 time steps, in order to eliminate transient responses due to initial conditions. For both sets of simulations, there is a highly significant, positive relationship between mean predator population size and habitat grain (Figures 4a, 4b; $r^2 = 0.89$ and 0.50, respectively, $P < 0.05$). For the less mobile predator (i.e., small step size), population size is considerably reduced in the fragmented landscapes (i.e., small grain size, Figure 3a). For the more mobile predator, population size is not as strongly affected by grain size. Figures 4c and 4d present the coefficient of variation in population size for these two predators ($r^2 = 0.46$ and 0.77, respectively, $P < 0.05$). For the less mobile predator, there is a significant, inverse relationship between the coefficient of variation and habitat grain (Figure 4c). For the more

Table 1. Model parameters and initial conditions.

Simulation Length: 250 time steps
Landscape in suitable habitat: 22%
Habitat grain sizes:1, 2, 4, 8, 16, 23
Five replicate simulations run for each grain size
Lotka-Volterra Equations:

$$H_{t+1} = c_1 H_t - c_2 P_t H_t$$

$$P_{t+1} = -d_1 P_t + d_2 P_t H_t$$

$$c_1 = 2.0 \qquad d_1 = 0.1$$

$$c_2 = 5.0 \qquad d_2 = 0.2$$

	Prey (H)	Predator (P)
Initial Population Size	5000	500
Maximum Density/Cell	12	5
Maximum Step Size	2	1 or 2
Directional Auto-Correlation	0	1.00
Survival Probability/Step	0.8	0.70

mobile predator, the relationship between the coefficient of variation and habitat grain is positive (Figure 4d).

Given that a high coefficient of variation in population size implies a high probability of extinction (Pimm et al. 1988), our results suggest that the less mobile predator is at a higher risk in fragmented landscapes. Extinctions occurred in three of five simulations with the less mobile predator in the most highly fragmented landscape. In these cases, it appears that following local extinctions of prey, the fragmented landscape and the predator's limited dispersal capabilities make it difficult to reach new concentrations of prey. Conversely, the more mobile predator is at a higher risk of extinction in large intact blocks of habitat. With the more mobile predator, extinctions occurred in eight of fifteen simulations in landscapes with the three largest grain sizes. In these large blocks of habitat, it appears that the prey are unable to find refuge from these highly mobile predators. The predators quickly build up to very large population sizes and drive the prey and themselves to extinction. Although additional simulations need to be conducted over a wider range of conditions, these results demonstrate that conclusions about the effect of habitat heterogeneity on population persistence cannot be made without knowledge of species dispersal capabilities and life history attributes

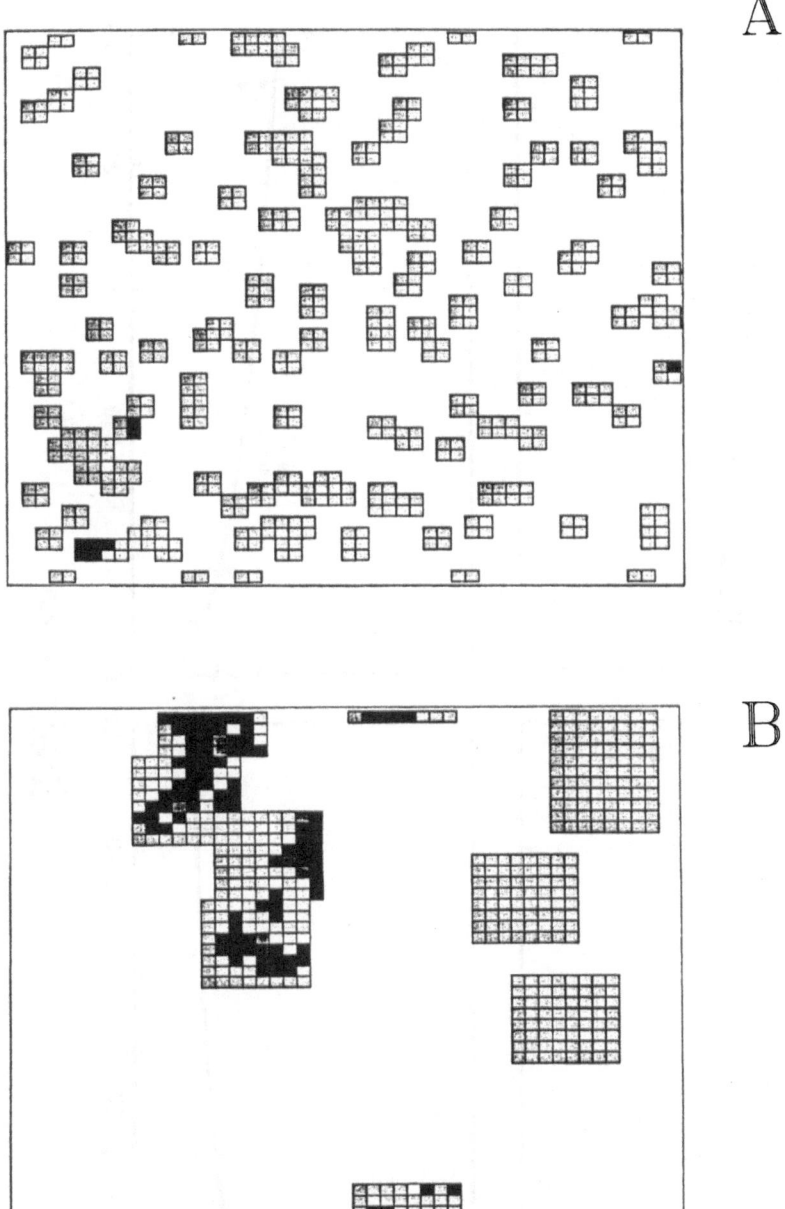

Figure 3. Sample landscapes used in the simulations. Total landscape size is 50 x 50 cells. Cells outlined in black represent suitable habitat; remaining white background is unsuitable habitat. The solid black cells contain predators and prey; dark grey cells—predators only; light grey—prey only; open cells are unoccupied. (a) habitat grain = 2; (b) habitat grain = 8.

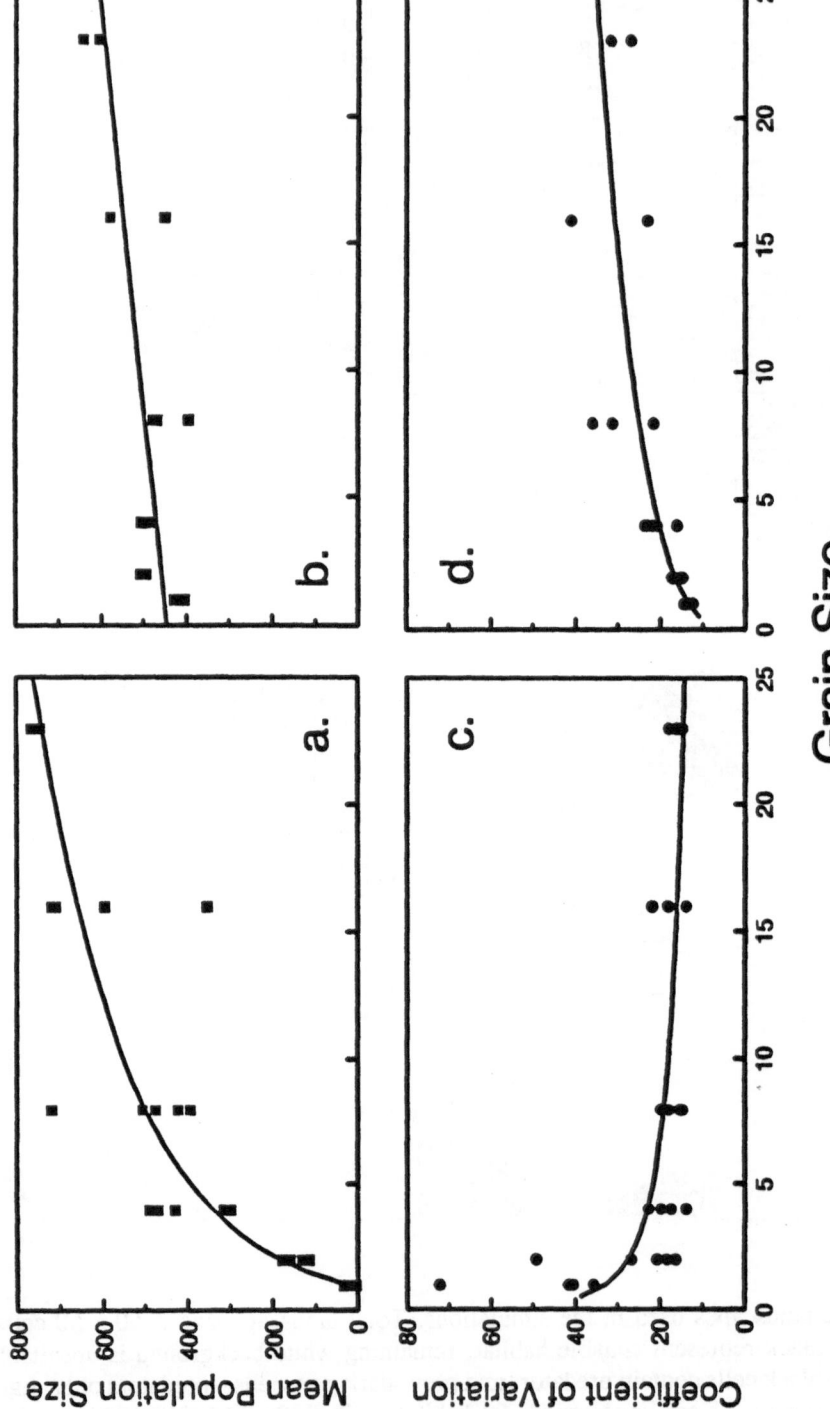

Figure 4. Predator population size and corresponding coefficient of variation as function of habitat grain. Approximately 22% of the landscape in suitable habitat. Each point based on the last 150 time steps of a 250-step simulation. See text for additional details. (a,c) Maximum predator step size = 1; (b,d) Predator step size = 2.

related to habitat requirement. As pointed out by Kareiva (1987, p. 389), "...instead of making robust generalization about habitat fragmentation (such as "patchiness is stabilizing") we should seek predictions that are based on the details of an organism's dispersal behavior and demography." For some species, population persistence may be highest in fragmented landscapes; other species may do best in large blocks of habitat. Unfortunately, the life history and dispersal characteristics for many species are not well known. Nevertheless, an important conclusion is that major changes in habitat grain—e.g., due to anthropogenic factors—are likely to have a major impact on population dynamics. For species that have evolved in environments with a large habitat grain, a substantial reduction in grain is likely to increase the probability of extinction. Such changes are also likely to increase the likelihood of invasions by species that have evolved in fine-grained habitats. This could ultimately result in substantial shifts in species composition and biodiversity.

Ecological Consequences of Landscape-Level Patchiness

The importance of landscape patchiness has been widely recognized in the study of both terrestrial (Wiens et al. 1985) and marine ecosystems (Paine and Levin 1981, Steele 1991b). Here, we focus on two aspects: the effect of landscape patchiness on the spread of disturbances, and landscape patchiness as it affects the redistribution of available resources and species survival.

Patchiness and the spread of disturbance. One of the challenges for landscape ecologists is to understand how disturbance, as an ecological process, spreads across heterogeneous landscapes (Turner et al. 1989, Green 1989, Costanza et al. 1990). A process-based landscape model, developed by Wu (1991), simulates fire spread across a Rocky Mountain foothill landscape and suggests that different interactive overlays of landscape patchiness such as vegetation, fuel complex, stand age, topography, and weather conditions affect fire spread differentially. Wu found that some vegetation patches in the study area, such as juniper woodlands, may play an important role in fire spread under conditions of high wind speed but not at low wind speed because of the distance between individuals trees. Usually, fuel complex correlates to stand age, as older stands accumulate higher fuel loadings and have a higher fire spread rate. However, steep slopes reduce fuel accumulation rate and therefore alter the effect of stand age on fire spread. Also, stand age composition may be a consequence of the effect of topography on fire. Thus, the interaction of abiotic and biotic patchiness plays an important role in fire spread. In other words, all the modeled components of landscape patchiness interact to determine the spread rate of fire. It is also suggested that under different conditions there may be different components of patchiness controlling the process.

Disturbance regimes may significantly affect landscape stability. For fire spread in woodland ecosystems, a set of disturbance regimes creates a set of landscape patterns; and, by the same logic, a set of landscape patterns makes a set of disturbance regimes more prevalent than others. This dynamic property suggests the existence of metastability in landscape patchiness. This means that landscape patchiness will remain relatively unchanged under a disturbance regime of a certain frequency and intensity, but could undergo a radical change at others (O'Neill et al. 1989). By way of example, disturbances of low frequency and high intensity may create a landscape with low patchiness. In contrast, disturbances of high frequency and low intensity may prevail in highly patchy landscapes. Different recurrence times for fire in different landscapes have been reported (Clark 1989, Wu 1991), suggesting that variable fire regimes may control different landscape patchiness (even-aged stand mosaics). For example, as Turner and Romme (1991) found out, a fire cycle of 26 - 113 years in interior Alaska creates landscape patches of less that 200-year-old stand age (Yarie 1981). But with a 434-year cycle, the patch mosaic of the landscape may include many stands with age greater that 1000 years old.

Patchiness and the effect of resource distribution on species survival. Landscape patchiness affects the spatial and temporal availability and distribution of resources and the way species use those resources. At the landscape level, several questions on patchiness and resource utilization may be asked:

(1) How do the abundance and spatial distribution of resources affect species survival?
(2) How is landscape patchiness important for species?
(3) Are there any threshold effects of patchiness on species survival?

Individual-based landscape modeling has provided a new context for the study of interactions among landscape patchiness, resource redistribution, and animal foraging (Houston et al. 1988; Hyman 1990; Turner et al. 1991; Johnson et al., in press; Milne et al., in press). A landscape simulation model of winter foraging by large ungulates has been developed by Turner et al. (1991) to study elk and bison responses to spatial distribution of resources and landscape patchiness in the northern range of Yellowstone National Park. In agreement with the model predictions, the degree of landscape patchiness affects both foraging efficiency and movement energy costs. When resources are patchily distributed, ungulates forage more efficiently and have higher fragmentation tolerance as the search area and movement scale get larger. The model also suggests that there might be a threshold effect of patchiness on ungulate survival. A distinct decline in ungulate survival, after 90 days of simulation, was caused by a sharp reduction in the daily consumed biomass during the simulation, and by the highly negative energetic balance resulting from drastic increases in movement energy costs. When changes in landscape patchiness are considered, it is apparent that the proportion of resource pixels (P) declines through winter as ungulates are foraging, which means a decrease of available forage biomass (Figure 5a). However, there is no evidence that the decline of P value directly affects ungulate survival after 90 days of simulation. An interesting result relates to a rapid increase in the number of patches in the landscape that shows up around 80 days of simulation (Figure 5b), which means that before the decline in ungulate survival the landscape speeds up its fragmentation rate. Also, resource patches decrease dramatically in size (from 600 to less than 200 pixels) after 90 days of ungulate foraging, suggesting that survival is strongly affected by the decay rate of large resource patches.
Several conclusions can be drawn from the simulations:

(1) As resources become limited, their spatial distribution becomes more important for species survival.
(2) Ungulate survival declines when the landscape increases its degree of patchiness;
(3) The decay of large resource patches in a landscape may strongly influence winter foraging and ungulate survival;
(4) As resource patches are rapidly fragmented, they may have a critical influence on ungulate survival.

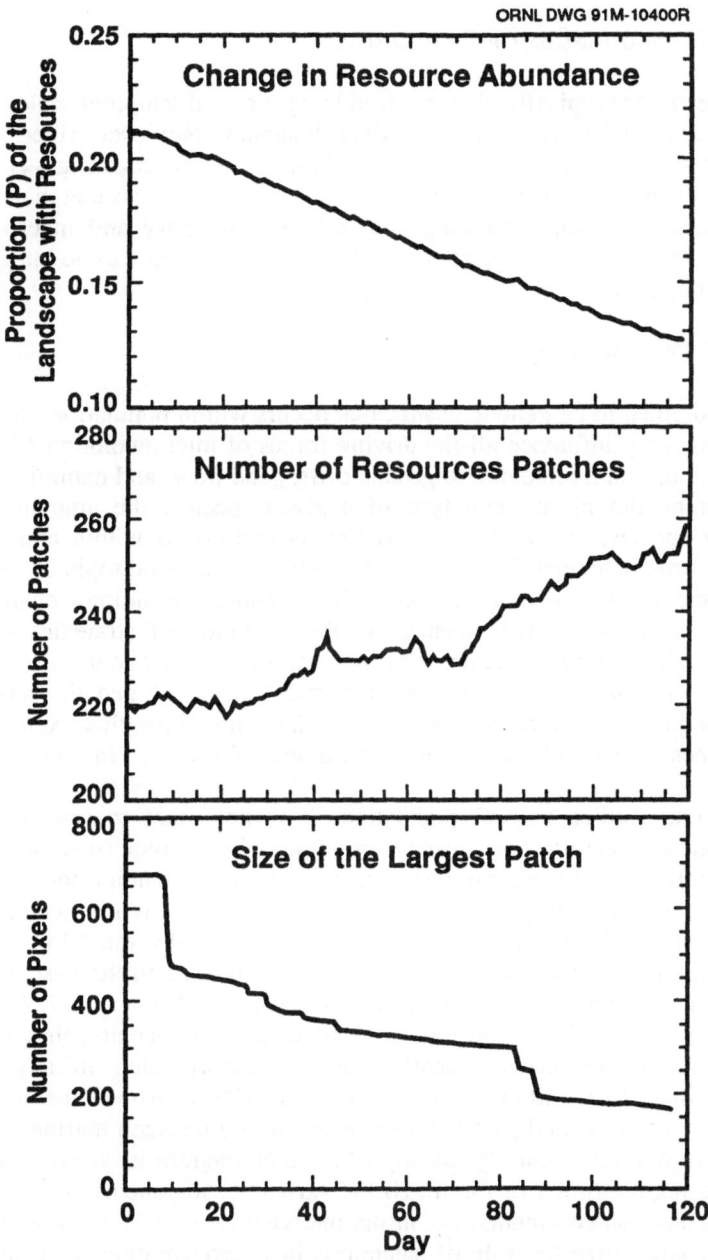

Figure 5. The changes of landscape patchiness by ungulate foraging. (a) Proportions of resource pixels, (b) number of resource patches, and (c) number of pixels for the largest resource patch (from Turner et al. ms.).

Evolutionary Consequences of Patchiness

Evolutionary processes are typically characterized by spatial and temporal scales much larger than those commonly used to describe ecological dynamics. However, since patchiness is apparent at all spatial and temporal scales of resolution, it is worth assessing its consequences at the evolutionary level. In this section, we outline the conceptual basis that makes patchiness a relevant phenomenon for understanding both microevolutionary and macroevolutionary processes. We also show how ecological and evolutionary processes can be integrated within the framework of populations' responses to patchy environments.

Microevolutionary Consequences

Microevolution is defined to be genetic change that occurs within populations and species. In principle, patchiness may influence all the driving forces of microevolution (the mutational input of new variation, and its fate due to genetic drift, gene flow, and natural selection) that collectively determine the mean phenotype of a given species, the amount of variation maintained in that species, and how that variation is partitioned within and among local populations. The word "patchiness" is one way to refer to the general topic of environmental heterogeneity. There is an enormous literature in evolutionary biology dealing with the consequences of heterogeneity, for instance, in the evolution of strategies to cope with heterogeneity (e.g., "bet-hedging," Seger and Brockmann 1987; the evolution of dispersal and diapause strategies, Levin et al. 1984), and the maintenance of genetic variation (e.g., Hedrick 1986). Rather than attempt to review this unwieldy literature, this section outlines the relationship between some of the topics discussed elsewhere in this chapter and microevolution.

Whenever one studies a local population, it is important to remember that that population is the latest successful pass through an immensely complex sieve of past temporal and spatial environments. But the sieve that is relevant to understanding the composition of a given population is a biased sample of the array of available past environments. This biasing results from a number of distinct processes, including habitat selection (Holt 1987), spatial variation in population abundance, asymmetrical dispersal rates, and the basic distributional limits imposed upon a species by its fundamental niche (Holt and Gaines, ms). Although these biases will themselves evolve in response to environmental heterogeneity, they will never be absent, and their combined effects clearly must be incorporated in any analysis of microevolutionary dynamics within a species (Kitchell 1990). Above we used Stommel diagrams to contrast temporal and spatial scales of variability between marine and terrestrial systems. Though we did not explicitly say so, a Stommel diagram is constructed relative to a defined sampling regime (e.g., in Figure 2b the variance components are computed from data from sample points on continents, not in the intervening seas). The suggestion we now explore is that to characterize the role of patchiness in microevolution, one must similarly relativize environmental variation to the array of "sampling units"—individuals—that comprise a given phylogenetic lineage, and that a useful heuristic device for doing so is provided by constructing a "phylogenetic envelope" on a space-time or space-time-environment diagram (see also Holt and Gaines, ms).

Ever since Darwin, it has been conventional to portray biotic diversification in a phylogenetic lineage as a branching tree. The tree is a retrospective view of the historical relationships among organisms in different species. Within a biological species, it is more sensible to display phylogeny (at the level of individuals) as a reticulated web (Maynard Smith 1989). To envisage the relationship between spatial and temporal variability and microevolutionary processes, let us overlay this phylogenetic web onto a space-time diagram. The basic idea is that the phylogenetic web, when placed on a space-time-environment plot, provides the relevant, appropriately biased sample of past environments for analyzing microevolution.

Start with a particular local population (loosely defined as that set of individuals that interbreed to produce the subsequent generation). For simplicity, assume that the organism in question is bisexual with an annual life cycle, and that we census at the beginning of generation t (at the zygote stage). There are several ways one can overlay the phylogenetic web on a space-time plot. The simplest, and the one developed here, is to let the x-axis represent the actual spatial position of individuals in a population and their ancestors; and the y-axis, time. The origin is the spatial position of the population at a given point in time. Each zygote in our sample, of course, had a father and a mother, who themselves were zygotes one year earlier. Their spatiotemporal birth position can also be represented by dots, and their successful production of offspring (one or more) in the current generation is delineated by lines connecting the dots of successive generations (Figure 6).

This graphical representation can be iterated as many generations into the past as is useful; obvious break-off points might be the time of speciation in a peripheral isolate, or invasion from a distant locality. Individuals that leave no descendants can be represented as the terminal ends of such lines. The actual phylogenetic web of successful individual genealogies leading to a particular population is penetrated and surrounded by a penumbra of such termini. Moreover, some ancestors may be members of populations that leave descendants in other contemporaneous populations, and of course some ancestors are more successful at leaving descendants than are others (either due to chance or to selective advantages).

If there is a large number of individuals in the local population, and we are considering evolutionary processes over many generations, this graph will be a very dense, complex skein of entangled threads (which might be more efficiently described by a density function). This skein is surrounded by an envelope, the shape of which describes the slice of the earth relevant to interpreting the genetic composition of one's study population. At a grander level, one could repeat the same process sketched here at various spatial scales, from the home range of single individuals up to the entire ensemble of populations comprising a species range. Figure 7 shows some examples.

The shape of the phylogenetic envelope reflects the interplay of intrinsic biotic factors and external constraints. If the upper boundary of the phylogenetic envelope has a very shallow slope (e.g., Figure 7a), the contemporaneous population is descended from sets of ancestral individuals that are similarly spatially circumscribed. Moreover, if the mean location of these ancestors is the same as their descendants, the population is tied to a particular piece of turf. This may be a reasonable approximation of some oceanic island populations, which are panmictic within-island but receive very few immigrants from outside (Grant and Grant 1989). In this case, microevolution to a first approximation involves within-population microevolutionary processes. Patchiness at small spatial scales within the island can, of course, be very important—for instance, in determining population size (thereby affecting genetic drift) or in setting the relative fitnesses of alternative genotypes or phenotypes—but it seems fair to say that spatial heterogeneity expressed over large spatial scales will not be

294

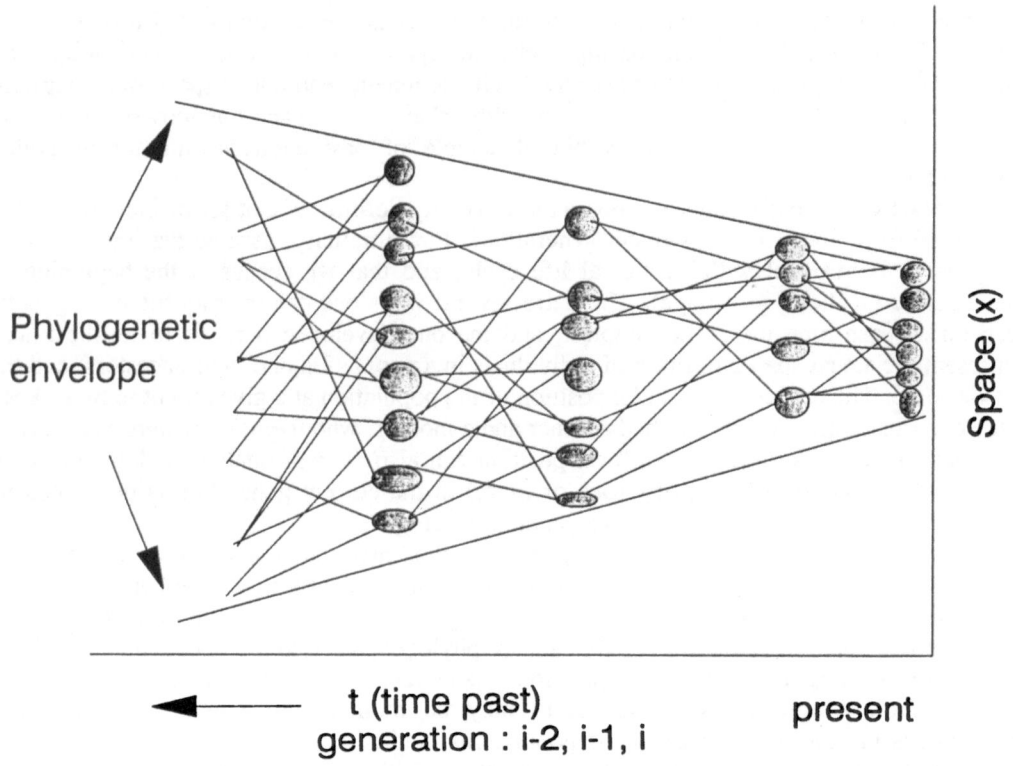

Figure 6. The phylogenetic web in a space-time diagram. In a diploid population with discrete generations, every freshly created zygote has a spatial position, indicated by open circles against the right-hand axis. Each individual descends from two parents, whose own spatial positions at the time of their birth are indicated by open circles one generation past. This graphical depiction of the phylogenetic web that actually progenerated the current population can be iterated as many generations into the past as seems useful. The outer boundary of this web in the space-time plot is called the "phylogenetic envelope."

295

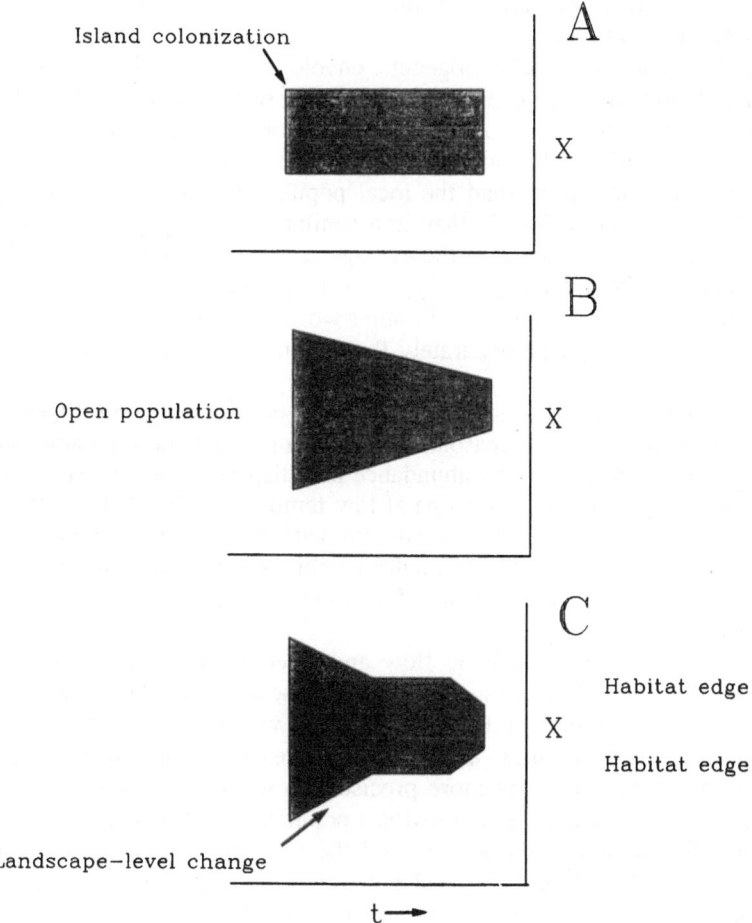

Figure 7. Examples of phylogenetic envelopes. (A) Panmictic population on a homogeneous, oceanic island. Following the initial colonization episode, there is no further dispersal from outside, and the spatial domain of the current population describes that of all ancestral populations, post-colonization. (B) Open population of low vagility on homogeneous portion of a large continent. The ancestors of the current population came from a much larger spatial expanse than contained in that population, and the spatial expanse of that ancestry increases with increasingly distant (in time) ancestors. (C) Open population of low vagility in heterogeneous landscape, which itself changes over a long time scale. As in (B), over a short time scale the spatial extent of the local population's ancestry increases. But this reaches an asymptote at a level set by discrete boundaries between different habitat types (e.g., an island of forest in a sea of prairie). Over still longer time scales, the forest itself moves on the landscape, and was part of a larger contiguous block at some time in the past. Hence, ancestors of the current population may be drawn from a large spatial domain at some sufficiently long temporal scale.

germane to interpreting that population's evolution (except insofar as it influences temporal trends in the local environment).

If the upper boundary of the phylogenetic envelope is steep (Figure 7b), the ancestors of the current population will have been drawn from widely scattered localities. This in effect implies that there is substantial gene flow among populations. This has a number of potentially important implications. First, the population size that is relevant in models of neutral evolution will be somewhat larger than the local population size (Kimura 1983), allowing more variation to be maintained locally than in a similar, isolated population. Second, if the local selective environment differs from the average environment inhabited by the ancestors of the local population, the average phenotypes in the population are likely to be displaced from the local selective optima (or ESSs). In other words, the spatial scale defining the local "population" is too small to gauge accurately the action of natural selection on phenotypic evolution.

Patchiness typically leads to spatial variation in both density and dispersal rates. An alpine plant, for instance, may have reasonable rates of dispersal and high abundances within large patches of alpine tundra, but low abundance and dispersal rates between such patches. This leads to an envelope with a steep slope at low temporal scales that levels off at longer scales (Figure 7c). But at yet longer scales, climatic variation may have led to shifts in entire vegetation zones, so that today's isolated tundra patches are remnants of a much larger past area of tundra; hence the ultimate ancestors of a local population may have been drawn from a much larger area.

The effect of patchiness on gene flow and selection is governed crucially by the temporal stability of patches. Broadly speaking, if a landscape of habitable patches surrounded by an inhospitable matrix is temporally stable (so that within-patch carrying capacities are constant), dispersal tends to be disadvantageous. This leads to a low rate of gene flow and should allow the population to adapt more precisely to local environmental idiosyncrasies. However, if a genetic variant appears in one local population that potentially is advantageous in all, the patchy distribution of the species and the low rate of dispersal between patches slows down the rate at which this allele reaches fixation in the species as a whole. Patchiness thus has opposing effects on local versus species-wide adaptation.

By contrast, if the habitable patches are transient pieces in a shifting mosaic, dispersal is strongly favored. A species occupying transient patches is likely to show frequent local extinctions and recolonizations, which implies that many local populations may ultimately descend from a relatively few ancestral populations. This tends to homogenize the genetic composition of populations (Slatkin 1987). Alleles adapted to particular local environments should have low life expectancies, whereas alleles with species-wide advantages should be able to become fixed relatively quickly.

The effect of spatial variation in the selective environment on adaptive evolution is difficult to gauge in more detail without developing explicit models that specify the nature of the variation and the constraints acting on the character(s) in question. Spatial variation in selection within a population can promote the maintenance of genetic variation if different genotypes are favored in different environments. However, this is easiest to achieve if selection is soft rather than hard (i.e., selections occurs at a life stage different from that at which density-dependence occurs, Jaenike and Holt 1991), and if organisms cannot vary their phenotype plastically to match different environments. Spatial variation in selection between populations can lead to different alleles' being favored in different populations, and a polymorphism will be maintained if dispersal rates are not too great and if spatial variation in local population density is not large (as with competing species, Levin 1974). Such polymorphisms are vulnerable to invasion by phenotypically plastic strategies that exhibit the "best" phenotype in each environment.

In an ecological community, different local populations are likely to exhibit different phylogenetic envelopes. Those populations descending from a long line of resident, ancestral populations are more likely to exhibit finely honed local adaptation than populations recently derived from a melange of spatially separated ancestral populations.

Ideally, one should add other dimensions to these figures in order to characterize the abiotic and biotic factors that influence population growth rates and fitness in this lineage. For the reasons sketched above, organisms within a species x spatial units apart are more likely to inhabit more similar environments than two randomly placed sample points, and this will be expressed to different degrees in different species.

The phylogenetic envelope describes the appropriately biased sample of the environment needed for an analysis of microevolution relevant to a defined local population or species. One could in principle carry out the spectral analysis sketched in Figure 1, for a weighted space-time sample, where the weights are simply the actual occurrence of a species in given environments leading to species-specific Stommel diagrams. Because of range shifts, habitat selection, etc., variance peaks in the general physical spectrum might not correspond to peaks in the spectrum as experienced by a particular phylogenetic lineage. Species assemblages are highly variable in space and time, and biotic components of the selective regime are likely to be similarly variable for any particular species. But different species are likely to experience quite different spatiotemporal patterns of variability in biotic components of selection.

Just as species richness is likely to show peaks on a Stommel diagram, so should intraspecific variability (e.g., as measured by mean heterozygosity, or the additive genetic variance for a quantitative character). If an entire species goes through a bottleneck on a long-time scale (e.g., due to glaciation), variance should be depleted shortly after the bottleneck and cumulate up to the next bottleneck, so there will be a variance peak at this time-scale. It would be particularly interesting to compare Stommel diagrams for species richness and for intraspecific genetic variation.

Macroevolutionary Consequences

So far, patchiness has been shown to affect ecological processes as well as microevolutionary change in both marine and terrestrial systems. In this section we are concerned with patchiness as it affects the origin and subsequent evolutionary changes in the biological properties of an existing higher taxon. In particular, we will propose that differences in physical and biotic scales of variability among systems (terrestrial and marine in particular) not only promote ecological differences or the prevalence of different adaptive strategies among them, but are capable of giving rise to different macroevolutionary dynamics. This is an inquiry into the mechanisms responsible for the transmission of change from ecological to microevolutionary up to macroevolutionary scales.

For ecological phenomena, patchiness expresses itself as affecting persistence and distribution of local species populations (e.g., Kareiva 1990, Hassell 1991), whereas for macroevolutionary phenomena, heterogeneity, represented by major environmental changes and disturbance events, affects extinction and diversification rates of taxa, giving rise to patterns of taxonomic diversity over the Phanerozoic (Benton 1988, Knoll 1991, Raup and

Sepkoski 1984, Valentine and Jablonski 1983b). It is apparent that the effect of patchiness at ecological, spatial, and temporal scales differs from that at macroevolutionary scales, but they are intimately related. In the following paragraphs we elaborate on such a relationship in the context of organism responses to patchy environments.

The link between ecological and evolutionary dynamics can be exemplified by taking, as a starting point, a simple single-species metapopulation model proposed by Levins (1969) to study the extinction of species by way of the dynamics of the extinction and founding of its local populations:

$$dp/dt = mp (1 - p) - ep \qquad (1)$$

where (p) represents the proportion of patches occupied, (m) is the colonization rate of empty patches and (e) is the extinction rate of occupied patches (for further elaboration of this model see Hanski 1982, 1991; Gotelli 1991). From this model it is readily apparent that taxa with different colonization or extinction rates will be characterized by different metapopulation dynamics, giving rise to contrasting patterns of local, regional, and global extinction and diversification. Under this scenario, a large-scale irreversible shift in landscape heterogeneity could increase the extinction or colonization rate of a particular group of species, thus affecting both their local population persistence and the extinction probability of their lineage over evolutionary time. The above example stresses that adaptations to cope with heterogeneity in ecological timeframes are also relevant for understanding patterns arising at the macroevolutionary level. For example, ecological requirements related to habitat or resource use are known to be correlated with speciation and extinction rates within a lineage (Rensch 1959, Eldredge 1979). In particular, generalist species capable of using different resources in alternative environments (such as a lineage might encounter through time) or specialist species whose resource patches are abundant, widespread, and persist through time, are subjected to less directional selection, and their lineages to low speciation and extinction rates, as exemplified for African large mammals (Vrba 1980, 1987). Within a species' geographic range, spatial heterogeneity in the environment provides the template for the kind of population structure envisioned by Levins (1969); a necessary condition for the effective action of processes responsible for the origination of higher taxonomic units as envisioned by the shifting-balance theory of evolution (e.g., Wright 1982).

Dispersal and life history traits are known to be important in affecting ecological and macroevolutionary dynamics in both terrestrial and marine environments. Dispersal capabilities, hence propensity to isolation of local populations, affect speciation rate. Dispersal is an axis along which it is possible to gain insights into the differences between terrestrial and marine ecosystems. In particular, the capability of long distance dispersal in most marine organisms, usually related to hydrodynamic phenomena, contrasts with the limited dispersal of most terrestrial species. Long-distance dispersal in marine systems prevents genetic tracking of local biological changes that might otherwise result in high isolation and subsequent speciation. In fact, terrestrial mammal speciation rates are seven or more times higher than those of Late Tertiary bivalves (Stanley 1973). Additionally, within marine systems, variation among species with regard to modes of dispersal and larval characteristics greatly influence speciation/extinction dynamics (for a review, see Jablonski and Lutz 1983). For example, it is known that weak dispersal in marine gastropods that lack planktotrophic larvae results in narrow geographic ranges, and consequently in higher rates of extinction as well as of speciation. In contrast, species with planktotrophic larvae generally have low extinction rates and low speciation rates (Valentine and Jablonski 1983a,b). Thus dispersability and related life history traits are strongly correlated with patterns arising at macroevolutionary level and also affect ecological time processes related to local and regional persistence of populations in patchy environments.

At large spatial and temporal scales, distinct patterns of environmental heterogeneity will promote selective extinction and diversification of particular taxa with different ecological and life history traits. A good example comes from pelagic ecosystems, where macroevolutionary trends have been driven by changes in the circulation patterns throughout the world's oceans, affecting the vertical and horizontal heterogeneity of the environment due to changes in temperature gradients over large spatial scales. Those changes have regulated species dynamics by fostering episodes of proliferation and extinction of species with different life history strategies and trophic relations within pelagic ecosystems (see Lipps 1986 for a review). It would be of great interest to compare macroevolutionary patterns of extinction and diversification of taxa for ecosystems that differ in terms of spatial and temporal variability in the environment (e.g., deep-sea benthic vs. terrestrial vs. pelagic marine) as shown for Stommel diagrams (Figure 2). Unfortunately, the vagaries of the fossil record for some systems, and the difficulties in correlating marine and non-marine stratigraphic sequences, make these comparisons difficult. However, their relevance to understanding current patterns of taxonomic diversity among ecosystems (e.g., Ray and Grassle 1991) makes these comparisons worth pursuing.

Finally, it is worth mentioning that patches are not necessarily restricted to landscape-level heterogeneity but can also be discerned at larger spatial scales. Depending on the scale of resolution, terrestrial and marine systems could be thought of as two large patches with different patterns of variability giving rise to different macroevolutionary dynamics (Boucot 1983, Benton 1988). The same is true for continental biomes, and major oceanographic systems (e.g., Westrop 1991). At larger spatial scales, the dynamics of fragmentation and connection that patches such as seas and continents underwent during the Phanerozoic have been an important determinant of clade diversity through time (e.g., Flessa and Sepkoski 1978). Further, dispersal and isolation within landmasses and seas, through the creation of corridors and barriers, respectively, have dramatically affected the composition of actual biotas and the subsequent history of ecological interactions (e.g., The great American biotic interchange, see Stheli and Webb 1985, Vermeij 1991).

The Stommel diagram teaches us that patchiness affects biological phenomena across a continuum of spatial and temporal scales. Environmental variability affects ecological dynamics at the population, community, and landscape levels, as well as micro and macroevolutionary change. We believe that more effort must be directed to questions emphasizing the connection between processes operating at different time and spatial scales, with special concern directed to the relationship between ecology and macroevolution.

PERSPECTIVES AND CONCLUSIONS

Spatial heterogeneity and temporal variability have far-reaching ecological and evolutionary consequences. In particular, different patterns of spatiotemporal variability in terrestrial as compared to marine ecosystems may not only affect the prevalence of distinct life history strategies among the organisms that inhabit them, but may also afford different macroevolutionary patterns. The same is true when we consider systems within marine or terrestrial environments, such as terrestrial biomes or pelagic vs. benthic marine systems. One way to assess the spatial and temporal variability that characterize a system is through Stommel diagrams, which depict the relevant spatial and temporal scales across which the variance in a particular physical or ecological variable is distributed. Although they are constrained in terms of precision at large temporal and spatial scales, their utility in describing the spatiotemporal variability of systems, at those scales that are logistically feasible, has not been fully appreciated.

Comparative studies regarding regional and local persistence of populations in patchy environments that focus on organisms with different life histories, ecological requirements, and body size are badly needed for a better understanding of the extinction processes in both ecological and evolutionary time. A comparative metapopulation biology is required, which would provide a tool to reveal new patterns and deeper insights into the processes of extinction and microevolutionary change, and how they relate to patterns of environmental heterogeneity.

A closer interaction between theoretical and empirical approaches is crucial in addressing patchiness phenomena (e.g., Moloney et al., in press). As recently pointed out by Kareiva (1990), to say that spatial heterogeneity has important consequences for population dynamics and species interactions may appear so obvious to the field ecologist as to be trite. Indeed, a vast number of theoretical studies (reviewed above and elsewhere in this volume) underscoring this point have accumulated over recent years. Yet this large body of theoretical work has not been met with a commensurate amount of experimental work; predictions of the consequences of spatially heterogeneous environments to the population dynamics of plants and animals have largely remained untested. Kareiva (1990) bemoaned this point, citing only one marine and three terrestrial field studies employing experimental manipulations of habitat subdivision. (Indeed, it may be of interest that the single marine study was the only one to fail to detect an effect of habitat subdivision.) This is clearly an area that would benefit from further work, be it in terrestrial or marine ecosystems, although overcoming the formidable logistical constraints of performing experimental manipulations in the pelagic realm remains particularly challenging.

ACKNOWLEDGMENTS

We thank Simon Levin, Tom Powell, and John Steele for providing us with the highly stimulating intellectual environment during the workshop and summer school that resulted in this book. We appreciate their encouragement and discussion in the development of these ideas. We also thank Robyn Burnham, Doug Deutschman, Graciela García-Moliner, Colleen Hatfield, Loren Haury, Fabian Jaksic, Alan Johnson, Bruce Milne, and Sergio Navarrete for providing helpful comment and criticisms on different versions of the chapter. Our special thanks to Tom Powell and John Steele for helpful suggestions that greatly improved the final version of the manuscript. R.D. Holt thanks the University of Kansas and the National Science Foundation for support. Funding for Y. Wu was provided by the University of Wyoming-National Park Service Research Center and by the Ecological Research Division, Office of Health and Environmental Research, U.S. Department of Energy, under contract No. DE-AC05-84OR21400 with Martin Marietta Energy Systems, Inc.

301

REFERENCES

Allen, T.H.F., and T.B. Starr. 1982. *Hierarchy: Perspectives for Ecological Complexity.* University of Chicago Press, Chicago.

Barry, J.P., and P.K. Dayton. 1991. Physical heterogeneity and the organization of marine communities. In: J. Kolasa and S.T.A. Pickett (eds.). *Ecological Heterogeneity.* Springer-Verlag, Heidelberg, pp. 271-320.

Benton, M.J. 1988. Mass extinctions in the fossil record of reptiles: Paraphyly, patchiness, and periodicity. In: G.P. Larwood (ed.). *Extinction and Survival in the Fossil Record.* Claredon Press, Oxford, pp. 269-294.

Boucot, A.J. 1983. Does evolution take place in an ecological vacuum? II. *J. of Paleont.* 57:1-30.

Brandt, S.B., and V.A. Wadley. 1981. Thermal fronts as ecotones and zoogeographic barriers in marine and freshwater systems. *Proc. Ecol. Soc. Aust.* 11:13-26.

Brinton, E. 1962. The distribution of Pacific euphausiids. *Bull. Scripps Inst. Oceanogr.* 8:51-270.

Burrough, P.A. 1986. *Principles of Geographical Information Systems for Land Resources Assessment.* Oxford Science Publications. Monographs on Soil and Resources Survey No 12. Claderon Press, Oxford.

Caswell, H., and J.E. Cohen. 1991. Communities in patchy environments: A model of disturbance, competition, and heterogeneity. In: J. Kolasa and S.T.A. Pickett (eds.). *Ecological Heterogeneity.* Springer-Verlag, Heidelberg, pp. 97-122.

Clark, J.S. 1989. Ecological disturbance as a renewable process: Theory and application to fire history. *Oikos* 56:17-30.

Cohen, D. 1966. Optimizing reproduction in a randomly varying environment. *J. Theor. Biol.* 16:267-282.

Cohen, D., and S.A. Levin. 1985. The interaction between dispersal and dormancy strategies in varying and heterogeneous environments. In: E. Teramoto and M. Yamaguti (eds.). *Mathematical Topics in Population Biology, Morphogenesis and Neurosciences. Proc. Kyoto 1985.* Springer-Verlag, Heidelberg, pp. 110-122.

_____. 1991. Dispersal in patchy environments: The effects of temporal and spatial structure. *Theor. Pop. Biol.* 39:63-99.

Costanza, R., F.H. Sklar, and M.L. White. 1990. Modeling coastal landscape dynamics. *Bioscience* 40(2):91-107.

Davis, M.B. 1987. Invasion of forest communities during the Holocene: Beech and hemlock in the Great Lakes Region. In: A.J. Gray, M.J. Crawley and P.J. Edwards (eds.). *Colonization, Succession and Stability.* Blackwell Scientific Publications, Oxford, pp. 373-393.

Davis, M.B., K.D. Woods, S.L. Webb, and R.P. Futuyma. 1986. Dispersal versus climate: Expansion of Fagus and Tsuga into the Upper Great Lakes region. *Vegetatio* 67:93-103.

Dayton, P.K., and M.J. Tegner. 1984. The importance of scale in community ecology: A kelp forest example with terrestrial analogs. In: P. Price, C. Slobodchikoff, and W. Grand (eds.). *A New Ecology: Novel Approaches to Interactive Systems.* Wiley, New York.

Delcourt, H.R., P.A. Delcourt, and T. Webb. 1983. Dynamical plant ecology: The spectrum of vegetational change in space and time. *Quater. Sci. Rev.* 1:153-175.

Downing, J. 1991. Biological heterogeneity in aquatic ecosystems. In: J. Kolasa, and S.T.A. Pickett (eds.). *Ecological Heterogeneity.* Springer-Verlag, Heidelberg, pp. 160-180.

Eldredge, N. 1979. Alternative approaches to evolutionary theory. *Bull. Carnegie Mus. Natur. Hist.* 13:7-19.

Ferson, S. 1988. Are competition communities stable assemblages? Ph.D. Thesis, Department of Ecology and Evolution, State University of New York at Stony Brook.

Flessa, K.W., and J.J. Sepkoski, Jr. 1978. On the relationship between Phanerozoic diversity and changes in habitable area. *Paleobiology* 3:359-366.

Forman, R.T.T., and M. Godron. 1986. *Landscape Ecology.* Wiley, New York.

Fuentes, E.R., and F.M. Jaksic. 1988. The hump-backed species diversity curve: Why has it not been found among land animals. *Oikos* 53:139-143.

Gadgil, M. 1971. Dispersal: Population consequences and evolution. *Ecology* 52:253-260.

Gotelli, N.J. 1991. Metapopulation models: The rescue effect, the propagule rain, and the core-satellite hypothesis. *Am. Nat.* 138:768-776.

Grant, B.R., and P.R. Grant. 1989. *Evolutionary Dynamics of a Natural Population: The Large Cactus Finch of the Galapagos*. University of Chicago Press, Chicago.

Green, D.G. 1989. Simulated effects of fire, dispersal and spatial pattern on competition within forest mosaics. *Vegetatio* 82:139-153

Hanski, I. 1982. Dynamics of regional distribution: The core and satellite species hypothesis. *Oikos* 38:210-221.

_____. 1991. Single-species metapopulation dynamics: Concepts, models and observations. *Biol. J. Linn. Soc.* 42:17-38.

Hardt, R.A., and R.T.T. Forman. 1989. Boundary form effects on woody colonization of reclaimed surface mines. *Ecology* 70:1252-1260.

Harris, G.P. 1986. *Phytoplankton Ecology: Structure, Function and Fluctuation*. Chapman and Hill, London.

Harrison, S. In press. Local extinction in a metapopulation context: An empirical evaluation. *Ann. Zool. Fennica*.

Hassell, M.P., H.N. Comins, and R.M. May. 1991. Spatial structure and chaos in insect population dynamics. *Nature* 353:255-258.

Hastings, A. 1990. Spatial heterogeneity and ecological models. *Ecology* 71:426-428.

Haury, L.R., J.A. McGowan, and P.H. Wiebe. 1978. Patterns and processes in the time-space scale of plankton distribution. In: J.H Steele (ed.). *Spatial Patterns in Plankton Communities*. Plenum Press, New York.

Hedrick, P.W. 1986. Genetic polymorphisism in heterogeneous environments: A decade later. *Ann. Rev. Ecol. Syst.* 17:535-566.

Holt, R.D. 1987. Population dynamics and evolutionary processes: The manifold role of habitat selection. *Ev Ecol.* 1:331-347.

Holt, R.D., and M.S. Gaines. Ms. The analysis of adaptation in heterogeneous landscapes, with implications for the evolution of fundamental niches.

Houston, M.D., D. De Angelis, and W. Post. 1988. New computer models unify ecological theory. *Bioscience* 38:682-691.

Huffaker, C.B. 1958. Experimental studies on predation: Dispersion factors and predator-prey oscillations. *Hilgardia* 27:343-383.

Hyman, J.B. 1990. A landscape approach to the study of herbivory. Ph.D. Dissertation, University of Tennessee, Knoxville.

Jablonski, D., and R.A. Lutz. 1983. Larval ecology of marine benthic invertebrates: Paleobiological implications. *Biol. Rev.* 58:21-89.

Jaenike, J., and R.D. Holt. 1991. Genetic variation for habitat preference: Evidence and explanations. *Am. Nat.* 137:67-90.

Johnson, A.R., J.A. Wiens, B.T. Milne, and T.O. Crist. In press. Animal movements and population dynamics in heterogeneous landscapes. *Landscape Ecology*.

Kareiva, P. 1987. Habitat fragmentation and the stability of predator-prey interactions. *Nature* 326:388-340.

_____. 1990. Population dynamics in spatially complex environments: Theory and data. *Phil. Trans. R. Soc Lond. B* 330:175-190.

Kareiva, P., and M. Anderson. 1988. Spatial aspects of species interactions. In: A. Hastings (ed.). *Community Ecology*. Springer-Verlag, New York, pp. 35-50.

Kimura, M. 1983. *The Neutral Theory of Molecular Evolution*. Cambridge University Press, Cambridge.

Kitchell, J.A. 1990. The reciprocal interaction of organism and effective environment: Learning more about "and". In: R.M. Ross and W. D. Allmon (eds.). *Causes of Evolution: A Paleontological Perspective*. University of Chicago Press, Chicago, pp. 151-172.

Knoll, A.H. 1991. Environmental context of evolutionary change: An example from the end of the Proterozoic Eon. In: *New Perspectives on Evolution*. Wiley-Liss, pp. 77-85.

Kolasa, J., and S.T. Pickett (eds.). 1991. *Ecological Heterogeneity*. Springer-Verlag, Heidelberg.

Le Fèvre, J. 1986. Aspects of the biology of frontal systems. *Adv. Mar. Biol.* 23:163-299.

Levin, S.A. 1974. Dispersion and population interactions. *Am. Nat.* 108:207-228.

_____. 1976. Population dynamics models in heterogeneous environments. *Ann. Rev. Ecol. Syst.* 7:287-310.

_____. 1978. Population models and community structure in heterogeneous environments. In: S.A. Levin (ed.). *Mathematical Association of America, Study in Mathematical Biology II: Populations and Communities*. Math. Assoc. Am., Washington, D.C.

_____. 1985. Ecological and evolutionary aspects of dispersal. In: E. Teramoto and M. Yamaguti (eds.). *Mathematical Topics in Population Biology, Morphogenesis and Neurosciences. Kyoto 1985*. Springer-Verlag, Heidelberg.

Levin, S.A., D. Cohen, and A. Hastings. 1984. Dispersal strategies in patchy environments. *Theor. Pop. Biol.* 26:165-191.

Levins, R. 1969. Some demographic and genetic consequences of environmental heterogeneity for biological control. *Bull. Entomol. Soc. of America* 15:237-240.

Lipps, J.H. 1986. Extinction dynamics in pelagic ecosystems. In: D.K. Elliot (ed.). *Dynamics of Extinction.* John Wiley & Sons, New York, pp. 87-104

McGowan, J.A. 1972. The nature of oceanic ecosystems. In: C.B. Miller (ed.). *The Biology of the Oceanic Pacific*. Oregon State University Press, Corvallis, pp. 9-28.

Margalef, R. 1963. On certain unifying principles in ecology. *Am. Nat.* 97:357-374.

Maynard Smith, J. 1989. Trees, bundles, or nets?. *Trends Ecol. Evol.* 4:302-304.

Menge, B.A., J. Lubchenco, and L.R. Ashkenas. 1985. Diversity, heterogeneity and consumer pressure in a rocky intertidal community. *Oecologia* 65:394-405.

Milne, B.T. 1992. Spatial aggregation and neutral models in fractal landscapes. *Am. Nat.* 139:32-57.

Milne, B.T., M.G. Turner, J.A. Wiens, and A.R. Johnson. In press. Interactions between the fractal geometry of landscapes and allometric herbivory. *Theor. Pop. Biol.*

Moloney, K.A., S.A. Levin, N.R. Chiariello, and L. Buttel. In press. Pattern and scale in a serpentine grassland. *Theor. Pop. Biol.*

Naiman, R.J., and H. D'Camps (eds.). 1990. The ecology and management of aquatic-terrestrial ecotones. *Man and Biosphere Series, Vol. 4*. The Parthenon Publishing Group, Paris.

Navarrete S.A., and J.C. Castilla. 1990. Barnacle walls as mediators of intertidal mussel recruitment: Effects of patch size on the utilization of space. *Mar. Ecol. Prog. Ser.* 68:113-119.

O'Neill. R.V., D.L. De Angelis, J.B. Waide, and T.F.H. Allen. 1986. *A Hierarchical Concept of Ecosystems*. Princeton University Press, Princeton, NJ.

O'Neill, R.V., A.R. Johnson, and A.W. King. 1989. A hierarchical framework for the analysis of scale. *Lands. Ecol.* 3:193-205.

Pagel, M.D., R.M. May, and A.R. Collie. 1991. Ecological aspects of the geographical distribution and diversity of mammalian species. *Am. Nat.* 137:791-815

Paine, R.T., and S.A. Levin. 1981. Intertidal landscapes: Disturbance and the dynamics of pattern. *Ecol. Monogr.* 51:145-178.

Pickett, S.T.A., and P.S. White (eds.). 1985. *The Ecology of Natural Disturbance and Patch Dynamics*. Academic Press, New York.

Pimm, S.L., H.L. Jones, and J. Diamond. 1988. On the risk of extinction. *Am. Nat.* 132:757-785.

Powell, T.M. 1989. Physical and biological scales of variability in lakes, estuaries and the coastal ocean. In: J. Roughgarden, R. M. May, and S.A. Levin (eds.). *Perspectives in Ecological Theory*. Princeton University Press, Princeton.

Raup, D.M., and J.J. Sepkoski Jr. 1984. Periodicity of extinctions in the geological past. *Proc. natn. Acad. Sci., U.S.A.* 81:901-805.

Ray, G.C. 1991. Coastal-zone biodiversity patterns. *Bioscience* 41:490-498.

Ray, G.C., and J.F. Grassle. 1991. Marine biological diversity. *Bioscience* 41:453-457.

Reeve, J. 1990. Stability, variability and persistence in host-parasitoid systems. *Ecology* 71:422-426.

Rensch, B. 1959. *Evolution Above the Species Level*. Methuen, London.

Roff, D.A. 1975. Population stability and the evolution of dispersal in a heterogeneous environment. *Oecologia* 19:217-237.

Root, T. 1988. Energy constraints on avian distributions and abundances. *Ecology* 69:330-339.

Schopf, T.J.M. 1972. Varieties of paleobiologic experience. In: T.J.M. Schopf (ed.). *Models in Paleobiology*. Freeman, Cooper and Co., San Francisco.

Seger, J., and H.J. Brockmann. 1987. What is bet-hedging? *Oxford Surv. Evol. Biol.* 4:182-211.

Shorrocks, B., and I. R. Swingland (eds.). 1990. *Living in a Patchy Environment*. Oxford University Press.

Slatkin, M. 1987. Gene flow and the geographic structure of natural populations. *Science* 236:787-792.

Stamps, J.A., M. Buechner, and V.V. Krishnan. 1987. The effects of edge permeability and habitat geometry on emigration from patches of habitat. *Am. Nat.* 129:533-552.

Stanley, S.M. 1973. Effects of competition on rates of evolution, with special reference to bivalve mollusks and mammals. *Syst. Zool.* 22:486-506.

Stearns S.C., and R.E. Crandall. 1981. Bet-hedging and persistence as adaptations of colonizers. In: *Evolution Today. Proceedings of 2nd International Congress on Systematics and Evolution 1980.* Vancouver, B.C. pp. 371-384.

Steele, J.H. 1978. Some comments on plankton patches. In: J.H. Steele (ed.). *Spatial Pattern in Plankton Communities. NATO Conference Series, Series IV, Marine Science, Vol. 3.* Plenum Press, New York and London, pp. 1-20.

_____. 1989. The ocean 'landscape'. *Lands. Ecol.* 3:185-192.

_____. 1991a. Can ecological theory cross the land-sea boundary?. *J. Theor. Biol.* 153:425-436.

_____. 1991b. Marine functional diversity. *Bioscience* 41(7):470-474.

Stehli, F.G., and S.D. Webb (eds.). 1985. *The Great American Biotic Interchange.* Plenum Publishing Co., New York.

Stommel, H. 1963. Varieties of oceanographic experience. *Science* 139:572-576.

Strathmann, R.R. 1990. Why life histories evolve differently in the sea. *Amer. Zool.* 30:197-207.

Taylor, A. 1990. Metapopulations, dispersal and predatory-prey dynamics: An overview. *Ecology* 71:429-436.

Taylor, R.J., and P.J. Pekins. 1991. Territory boundary avoidance as a stabilizing factor in wolf-deer interactions. *Theor. Pop. Biol.* 38:115-128.

Turchin, P.B. 1986. Modeling the effect of host patch size on mexican beetle emigration. *Ecology.* 67:124-132.

Turner, M.G., R.H. Gardner, V.H. Dale, and R.V. O'Neill. 1989. Predicting the spread of disturbance across heterogeneous landscapes. *Oikos* 55:121-129.

Turner, M.G., and W.H. Romme. 1991. Landscape dynamics in crown fire ecosystems. In: R.D. Laven and P.N. Omi (eds.). *Pattern and Process in Crown Fire Ecosystems.* Princeton University Press, Princeton, NJ.

Turner, M.G., Y. Wu, W.H. Romme, and L.L. Wallace. Ms. Simulation of winter foraging by large ungulates in a heterogeneous landscape.

Valentine, D.W., and D. Jablonski. 1983a. Larval adaptations and patterns of Brachiopod diversity in space and time. *Evolution* 37:1052- 1061.

_____. 1983b. Speciation in the shallow sea: General patterns and biogeographic controls. In: R. W. Simms, J.H. Price, and P.E.S. Whalley (eds.). *Evolution, Time and Space: The Emergence of the Biosphere.* Academic Press, London and New York, pp. 201-226.

Vermeij, G. 1991. When biotas meet: Understanding biotic interchange. *Science* 253:1099-1104.

Vrba, E.S. 1980. Evolution, species and fossils: How does life evolve? *S. Afr. J. Sci.* 76:61-84.

_____. 1987. Ecology in relation to speciation rates: Some case histories of Miocene-rcent mammal clades. *Evol. Ecol.* 1:283-300.

Westrop, S.R. 1991. Intercontinental variation in mass extinction patterns: Influence of biogeographic structure. *Paleobiology* 17:363-368.

Wiens, J.A. 1989. Spatial scaling in ecology. *Funct. Ecol.* 3:385-397.

Wiens, J.A., C.S. Crawford,and J.R. Gosz. 1985. Boundary dynamics: A conceptual framework for studying landscape ecosystems. *Oikos* 45:421-427.

Wright, S. 1982. Character change, speciation, and the higher taxa. *Evolution* 36:427-443.

Wu, Y. 1991. Fire history and potential fire behavior in a Rocky Mountain foothill landscape. Ph. D. Dissertation, University of Wyoming, Laramie, Wyoming.

Yarie, J. 1981. Forest fire cycles and life tables: A case study from interior Alaska. *Can. J. For. Res.* 11:554-562.

CONTRIBUTORS

MARK R. ABBOTT
College of Oceanography
Oregon State University
Ocean Admininstration Building 104
Corvallis, OR 97331-5503

STEVE BOLLENS
Biology Department
Woods Hole Oceanographic Institution
Woods Hole, MA 02543

GAY A. BRADSHAW
USDA Forest Service
PNW Research Station
Corvallis, OR 97331

CHERYL J. BRIGGS
Department of Biological Sciences
University of California
Santa Barbara, CA 93109

DAVID BURROWS
Environmental Systems Research Institute, Inc.
380 New York Street
Redlands, CA 92373

STEVEN R. CARPENTER
Department of Biology
University of Notre Dame
Notre Dame, IN 46556

HAL CASWELL
Biology Department
Woods Hole Oceanographic Institution
Woods Hole, MA 02543

W. MICHAEL CHILDRESS
Center for Biosystems Modelling
Department of Industrial Engineering
Texas A & M University
College Station, TX 77843

JAMES S. CLARK
University of Georgia
Department of Botany
2502 Plant Sciences
Athens, GA 30602

JOEL E. COHEN
Laboratory of Populations
Rockefeller University
New York, NY 10021-6399

KENDRA L. DALY
Graduate Program in Ecology
University of Tennessee
Knoxville, TN 37996

GEOFF DAIRIKI
School of Oceanography, WB-10
University of Washington
Seattle, WA 98195

FRANK W. DAVIS
Department of Geography
National Center for Geographic Information and
    Analysis
University of California
Santa Barbara, CA 93106

PAUL K. DAYTON
Scripps Institution of Oceanography, A-001
University of California at San Diego
La Jolla, CA 92093

KENNETH L. DENMAN
Institute of Ocean Sciences
Fisheries and Oceans
P.O. Box 6000, 9860 West Saanich Road
Sidney, British Columbia, Canada V8L 4B2

DOUGLAS H. DEUTSCHMAN
Section of Ecology and Systematics
Cornell University
Ithaca, NY 14853

ODO DIEKMANN
Centre for Mathematics and Computer Science
P.O. Box 4079, 1009 AB
Amsterdam, The Netherlands

RICHARD DURRETT
Mathematics Department
White Hall
Cornell University
Ithaca, NY 14853

RON J. ETTER
Department of Biology
University of Massachusetts at Boston
Boston, MA 02125

MARIE-JOSEE FORTIN
Department of Ecology and Evolution
SUNY at Stony Brook
Stony Brook, NY 11794-5245

MICHAEL S. GAINES
Department of Systematics and Ecology
The University of Kansas
Lawrence, KS 66045

GRACIELA GARCÍA-MOLINER
Graduate School of Oceanography
University of Rhode Island
Narragansett, RI 02882

CHARLES H. GREENE
Section of Ecology and Systematics
Corson Hall
Cornell University
Ithaca, NY 14853-2701

DANIEL GRÜNBAUM
Department of Zoology
University of Washington
Seattle, WA 98195

WILLIAM S. C. GURNEY
Department of Statistics and Modelling Science
University of Strathclyde
Glasgow G1 1XJ, Scotland, UK

EILEEN E. HOFMANN
Center for Coastal Physical Oceanography
Crittenton Hall
Old Dominion University
Norfolk, VA 23529

ROBERT D. HOLT
Museum of Natural History and
Department of Systematics and Ecology
The University of Kansas
Lawrence, KS 66045

CLAUDIA M. JACOBI
Departamento de Zoologia, IB
Universidade de Sao Paulo
C. P. 11.461
05499 Sao Paulo-SP-Brazil

MIMI A. R. KOEHL
Department of Integrative Biology
University of California, Berkeley
Berkeley, CA 94720

SIDNEY LEIBOVICH
Sibley School of Mechanical & Aerospace
  Engineering
Upson Hall
Cornell University
Ithaca, NY 14853-7501

SIMON A. LEVIN
Department of Ecology and Evolutionary Biology
Eno Hall
Princeton University
Princeton, NJ 08544-1003

and:
Center for Environmental Research
Corson Hall
Cornell University
Ithaca, NY 14853

BAI-LIAN LI
Department of Rangeland Ecology and Management
Texas A & M University
College Station, TX 77843-2126

AUGUSTÍN LOBO
Centro de Estudios Avanzados de Blanes (C.S.I.C.)
Camino de Sta. Barrbara s/n
17300 Blanes (Girona), Spain

PABLO A. MARQUET
Department of Biology
University of New Mexico
Albuquerque, NM 87131

DORAN M. MASON
Chesapeake Biological Laboratory
University of Maryland System
Solomons, MD 20688-0038

KIRK A. MOLONEY
Department of Botany
353 Bessey Hall
Iowa State University
Ames, Iowa 50011-1020

WILLIAM W. MURDOCH
Department of Biological Sciences
University of California
Santa Barbara, CA 93109

307

ROGER M. NISBET
Department of Biological Sciences
University of California
Santa Barbara, CA 93109

MERCEDES PASCUAL
Biology Department
Woods Hole Oceanographic Institution
Woods Hole, MA 02543

JESUS PINEDA
Scripps Institution of Oceanography
University of California, San Diego
La Jolla, CA 92093-0208

THOMAS M. POWELL
Division of Environmental Sciences
University of California, Davis
Davis, CA 95616

ROBERT E. RICKLEFS
Department of Biology
University of Pennsylvania
Philadelphia, PA 19104

NATHAN H. SCHUMAKER
College of Forest Resources
University of Washington
Seattle, WA 98195

JOHN H. STEELE
Woods Hole Oceanographic Institution
Woods Hole, MA 02543

ALLAN STEWART-OATEN
Department of Biological Sciences
University of California
Santa Barbara, CA 93109

HAROLD M. VAN ES
Department of Soil, Crop and Atmospheric
Sciences
Cornell University
Ithaca, NY 14853

DAVID WALLIN
Forest Sciences Laboratory
Oregon State University
3200 Jefferson Way
Corvallis, OR 97331

JIANGUO WU
Department of Botany
School of Arts and Sciences
Miami University
Oxford, OH 45056

YEGANG WU
Oak Ridge National Laboratory
MS 6038
P. O. Box 2008
Oak Ridge, TN 37831

Springer-Verlag
and the Environment

We at Springer-Verlag firmly believe that an international science publisher has a special obligation to the environment, and our corporate policies consistently reflect this conviction.

We also expect our business partners – paper mills, printers, packaging manufacturers, etc. – to commit themselves to using environmentally friendly materials and production processes.

The paper in this book is made from low- or no-chlorine pulp and is acid free, in conformance with international standards for paper permanency.

Lecture Notes in Biomathematics

For information about Vols. 1–54
please contact your bookseller or Springer-Verlag

General Remarks

Lecture Notes are printed by photo-offset from the master-copy delivered in camera-ready form by the authors of monographs, resp. editors of proceedings volumes. For this purpose Springer-Verlag provides technical instructions for the preparation of manuscripts. Volume editors are requested to distribute these to all contributing authors of proceedings volumes. Some homogeneity in the presentation of the contributions in a multi-author volume is desirable.

Careful preparation of manuscripts will help keep production time short and ensure a satisfactory appearance of the finished book. The actual production of a Lecture Notes volume normally takes approximately 8 weeks.

Authors of monographs receive 50 free copies of their book. Editors of proceedings volumes similarly receive 50 copies of the book and are responsible for redistributing these to authors etc. at their discretion. No reprints of individual contributions can be supplied. No royalty is paid on Lecture Notes volumes.

Authors and editors are entitled to purchase further copies of their book for their personal use and other Springer mathematics books at a discount of 33.3 % directly from Springer-Verlag.

Commitment to publish is made by letter of intent rather than by signing a formal contract. Springer-Verlag secures the copyright for each volume.

Addresses:

Professor Simon A. Levin, Princeton University
Department of Ecology and
Evolutionary Biology
Eno Hall, Princeton
New Jersey 08544-1003, USA

Springer-Verlag, Mathematics Editorial
Tiergartenstr. 17
W-6900 Heidelberg
Federal Republic of Germany
Tel.: *49 (6221) 487-410